LABORATORY MANUAL

Human Anatomy
and
Physiology

Fifth Edition

John W. Hole, Jr.

WCB Wm. C. Brown Publishers

Cover image by Dick Oden

The credits section for this book begins on page 433, and is
considered an extension of the copyright page.

ISBN 0-697-05781-X

Some of the laboratory experiments included in this text may be
hazardous if materials are handled improperly or if procedures are
conducted incorrectly. Safety precautions are necessary when you are
working with chemicals, glass test tubes, hot water baths, sharp
instruments, and the like, or for any procedures that generally require
caution. Your school may have set regulations regarding safety
procedures that your instructor will explain to you. Should you have
any problems with materials or procedures, please ask your instructor
for help.

Printed in the United States of America by Wm. C. Brown Publishers,
2460 Kerper Boulevard, Dubuque, IA 52001

10 9 8 7 6 5 4

Contents

Preface

Laboratory Manual for Human Anatomy and Physiology was prepared to be used with the textbook *Human Anatomy and Physiology,* 5th edition, by John W. Hole, Jr. As in the case of the textbook, the laboratory manual is designed for students with minimal backgrounds in the physical and biological sciences who are pursuing careers in allied health fields.

The laboratory manual contains sixty-two laboratory exercises and sixty-one reports, which are integrated closely with the chapters of the textbook. The exercises are planned to illustrate and review anatomical and physiological facts and principles presented in the textbook and to help students investigate some of these ideas in greater detail.

Often the laboratory exercises are short or are divided into several separate procedures. This allows an instructor to select those exercises or parts of exercises that will best meet the needs of a particular program. Also, exercises requiring a minimal amount of laboratory equipment have been included.

The laboratory exercises include a variety of special features that are designed to stimulate interest in the subject matter, to involve students in the learning process, and to guide them through the planned activities.

These special features include the following:

Introduction

The introduction briefly describes the subject of the exercise or the ideas that will be investigated.

Purpose of the Exercise

The purpose provides a statement concerning the intent of the exercise—that is, what will be accomplished.

Learning Objectives

The learning objectives list in general terms what a student should be able to do after completing the exercise.

Materials Needed

This section lists the laboratory materials that are required to complete the exercise and to perform the demonstrations and optional activities.

Procedure

The procedure provides a set of detailed instructions for accomplishing the planned laboratory activities. Usually these instructions are presented in outline form so that a student can proceed through the exercise in stepwise fashion. Frequently, the student is referred to particular sections of the textbook for necessary background information or for review of subject matter presented previously.

The procedures include a wide variety of laboratory activities and, from time to time, direct the student to complete various tasks in the laboratory reports.

Laboratory Reports

A laboratory report to be completed by the student immediately follows each exercise. These reports include various types of review activities, spaces for sketches of microscopic objects, charts for recording observations and experimental results, and questions dealing with the analysis of such data.

It is hoped that as a result of these activities, students will develop a better understanding of the structural and functional characteristics of their bodies and will increase their skills in gathering information by observation and experimentation. Some of the exercises also include demonstrations, optional activities, and useful illustrations.

Demonstrations

Demonstrations appear in separate boxes. They describe specimens, specialized laboratory equipment, or other materials of interest that an instructor may want to display to enrich the student's laboratory experience.

Optional Activities

Optional activities also appear in separate boxes. They are planned to encourage students to extend their laboratory experiences. Some of these activities are open-ended in that they suggest that the student plan an investigation or experiment and carry it out after receiving approval from the laboratory instructor.

Illustrations

Diagrams from the textbook often are used as aids for reviewing subject matter. Other illustrations provide visual instructions for performing steps in procedures or are used to identify parts of instruments or specimens. Micrographs are included to help students identify microscopic structures or to evaluate student understanding of tissues.

In some exercises, the figures include line drawings that are suitable for students to color with colored pencils. This activity may motivate students to observe the illustrations more carefully and help them to locate the special features represented in the figures.

Students can check their work by referring to the corresponding full-color illustrations in the textbook.

Acknowledgments

I would like to express my sincere gratitude to all users of the laboratory manual who sent suggestions for its improvement. I am especially grateful for the contributions of Edward P. Meyertholen, Ball State University, and Kristi Sather-Smith, Hinds Community College. These professors kindly reviewed the manual and examined the manuscript of the new edition. Their thoughtful comments and valuable suggestions were greatly appreciated.

To the student

The exercises in this laboratory manual will provide you with opportunities to observe various anatomical parts and to investigate certain physiological phenomena. Such experiences should help you relate specimens, models, microscope slides, and your own body to what you have learned in the lecture and read about in the textbook.

The following list of suggestions may help to make your laboratory activities more effective and profitable.

1. Prepare yourself before attending the laboratory session by reading the assigned exercise and reviewing the related sections of the textbook. It is important to have some understanding of what will be done in the laboratory before you come to class.

2. Bring your laboratory manual and textbook to each laboratory session. These books are closely integrated and will help you complete most of the exercises.

3. Be on time. During the first few minutes of the laboratory meeting, the instructor often will provide verbal instructions. Make special note of any changes in materials to be used or procedures to be followed. Also listen carefully for information concerning special techniques to be used and precautions to be taken.

4. Keep your work area clean and your materials neatly arranged so that you can locate needed items quickly. This will enable you to proceed efficiently and will reduce the chances of making mistakes.

5. Never eat, drink, or smoke in the laboratory. Consider all materials you use as potentially hazardous.

6. Report to the laboratory instructor immediately any injuries to yourself or damage to equipment.

7. Pay particular attention to the purpose of the exercise, which states what you are to accomplish in general terms, and to the learning objectives, which list what you should be able to do as a result of the laboratory experience. Then, before you leave the class, review the objectives and make sure that you can meet them.

8. Follow the directions in the procedure precisely and proceed only when you understand them clearly. Do not improvise procedures unless you have the approval of the laboratory instructor. Ask questions if you do not understand exactly what you are supposed to do and why you are doing it.

9. Handle all laboratory materials with care. These materials often are fragile and expensive to replace. Whenever you have

questions about the proper treatment of equipment, ask the instructor.

10. Treat all living specimens humanely and try to minimize any discomfort they might experience.

11. Label any tubes or bottles containing chemicals or solutions so that you can identify them easily and correctly.

12. Although at times it will be necessary for you to work with a laboratory partner, try to remain independent when you are making observations, drawing conclusions, and completing the activities in the laboratory reports.

13. Record your observations immediately after making them. In most cases, such data can be entered in spaces provided in the laboratory reports.

14. Read the instructions for each section of the laboratory report before you begin to complete it. Think about the questions before you answer them. Your responses should be based on logical reasoning and phrased in clear and concise language.

15. At the end of each laboratory period, clean your work area and the instruments you have used. Return all materials to their proper places and dispose of wastes as directed by the laboratory instructor. Wash your hands thoroughly before leaving the laboratory.

Body Organization and Terminology

The major features of the human body include certain cavities, a set of membranes associated with these cavities, and a group of organ systems composed of related organs. In order to communicate effectively with each other about the body, investigators have devised names to describe these body features. They also have developed terms to represent the relative positions of body parts, imaginary planes passing through these parts, and body regions.

Purpose of the Exercise

To review the organizational pattern of the human body, to review its organ systems and the organs included in each system, and to become acquainted with the terms used to describe the relative position of body parts, body sections, and body regions.

Learning Objectives

After completing this exercise, you should be able to:

1. Locate and name the major body cavities and identify the membranes associated with each cavity.
2. Name the organ systems of the human organism.
3. List the organs included within each system, and locate the organs in a dissectible manikin.
4. Describe the general functions of each system.
5. Define the terms used to describe the relative positions of body parts.
6. Define the terms used to identify body sections, and identify the plane along which a particular specimen is cut.
7. Define the terms used to identify body regions.

Materials Needed:
  dissectible manikin
  variety of specimens or models sectioned along
    various planes

Procedure A—
Body Cavities and Membranes

1. Review the sections entitled "Body Cavities" and "Thoracic and Abdominopelvic Membranes" in chapter 1 of the textbook.
2. As a review activity, label figures 1.1, 1.2, and 1.3.
3. Locate the following features on the reference plates on pages 32–38 of the textbook and on the dissectible manikin:

dorsal cavity

    cranial cavity

    vertebral cavity

ventral cavity

thoracic cavity

abdominopelvic cavity

    pelvic cavity

diaphragm

mediastinum

oral cavity

nasal cavity

orbital cavity

middle ear cavity

parietal pleura

visceral pleura

pleural cavity

visceral pericardium

parietal pericardium

parietal peritoneum

Figure 1.1 Label the major body cavities.

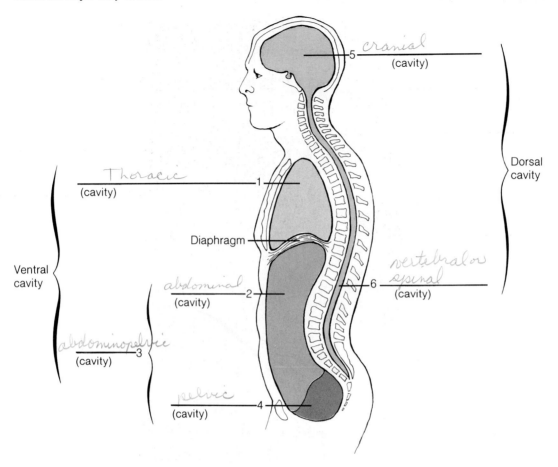

cranial
(cavity) — 5

Thoracic
(cavity) — 1

Diaphragm

Dorsal
cavity

Ventral
cavity

abdominal
(cavity) — 2

vertebral or
spinal
(cavity) — 6

abdominopelvic
(cavity) — 3

pelvic
(cavity) — 4

visceral peritoneum

peritoneal cavity

4. Complete Parts A and B of Laboratory Report 1 on page 9.

Procedure B—Organ Systems

1. Review the section entitled "Organ Systems" in chapter 1 of the textbook.
2. Use the reference plates on pages 32–38 of the textbook and the dissectible manikin to locate the following organs:

integumentary system

   skin

   accessory organs

skeletal system

   bones

   ligaments

   cartilages

muscular system

   skeletal muscles

nervous system

   brain

   spinal cord

   nerves

endocrine system

   pituitary gland

   thyroid gland

   parathyroid glands

   adrenal glands

   pancreas

   ovaries

   testes

   pineal gland

   thymus gland

Figure 1.2   Label the smaller body cavities within the head.

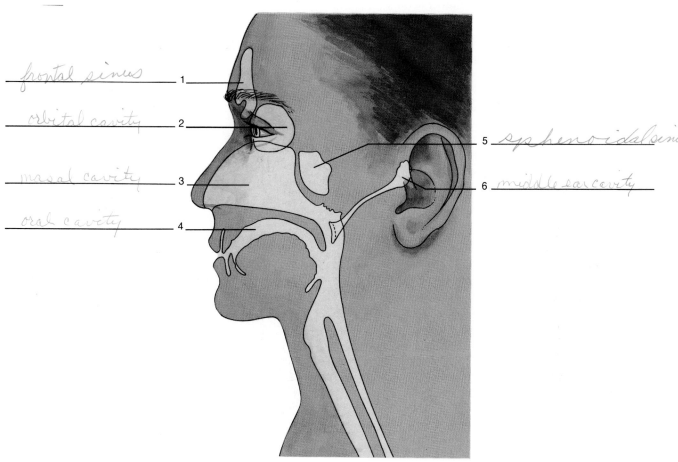

frontal sinus _____ 1

orbital cavity _____ 2

nasal cavity _____ 3

oral cavity _____ 4

5 ___ sphenoidal sinus

6 ___ middle ear cavity

digestive system	bronchi	male reproductive system
mouth	lungs	scrotum
tongue	circulatory system	testes
teeth	heart	epididymides
salivary glands	arteries	vasa deferentia
pharynx	veins	seminal vesicles
esophagus	capillaries	prostate gland
stomach	lymphatic system	bulbourethral glands
liver	lymphatic vessels	penis
gallbladder	lymph nodes	urethra
pancreas	thymus gland	female reproductive system
small intestine	spleen	ovaries
large intestine	urinary system	fallopian tubes
respiratory system	kidneys	uterus
nasal cavity	ureters	vagina
pharynx	urinary bladder	clitoris
larynx	urethra	vulva
trachea		

Figure 1.3 Label the thoracic membranes and cavities in (*a*) and the abdominopelvic membranes and cavity in (*b*).

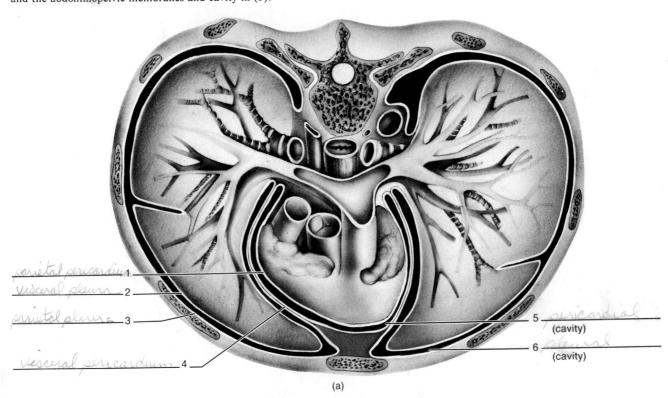

parietal pericardium 1
visceral pleura 2
parietal pleura 3

5 pericardial
(cavity)
6 pleural
(cavity)

visceral pericardium 4

(a)

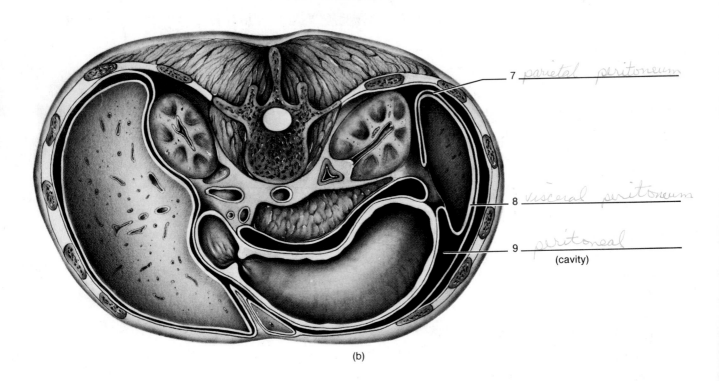

7 parietal peritoneum

8 visceral peritoneum

9 peritoneal
(cavity)

(b)

Figure 1.4 Label (*a*) the regions and (*b*) the quadrants of the abdominal area.

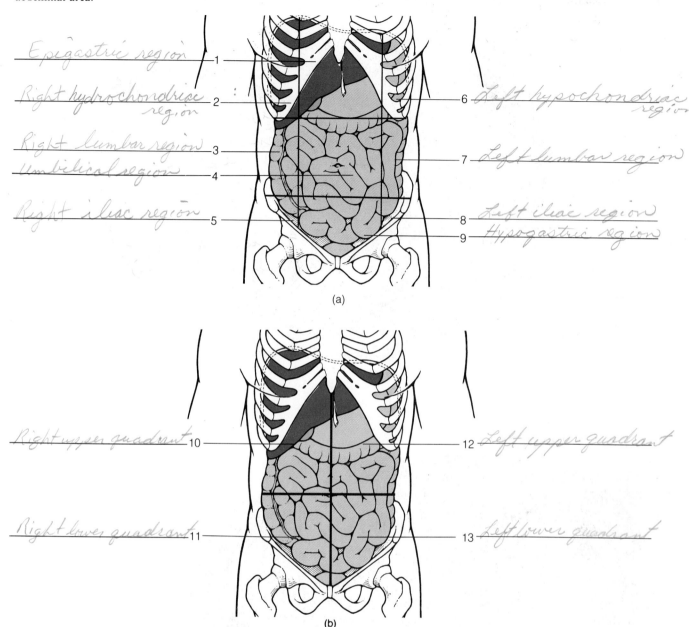

Epigastric region — 1

Right hydrochondriac region — 2

Right lumbar region — 3

Umbilical region — 4

Right iliac region — 5

6 — Left hypochondriac region

7 — Left lumbar region

8 — Left iliac region

9 — Hypogastric region

(a)

Right upper quadrant — 10

Right lower quadrant — 11

12 — Left upper quadrant

13 — Left lower quadrant

(b)

3. Complete Parts C and D of the laboratory report.

Procedure C—Relative Positions, Sections, and Regions

1. Review the section entitled "Anatomical Terminology" in chapter 1 of the textbook.
2. As a review activity, label figures 1.4, 1.5, and 1.6.
3. Examine the sectioned specimens on the demonstration table, and identify the plane along which each is cut.
4. Complete Parts E, F, G, and H of the laboratory report.

Figure 1.5 Label the planes represented in this illustration.

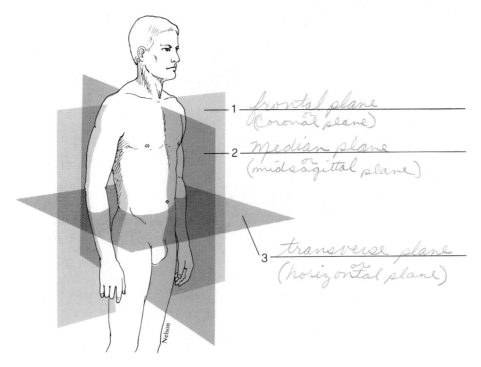

1 _frontal plane_
(coronal plane)

2 _median plane_
(midsagittal plane)

3 _transverse plane_
(horizontal plane)

Figure 1.6 Label these diagrams with terms used to describe body regions: (*a*) anterior regions; (*b*) posterior regions.

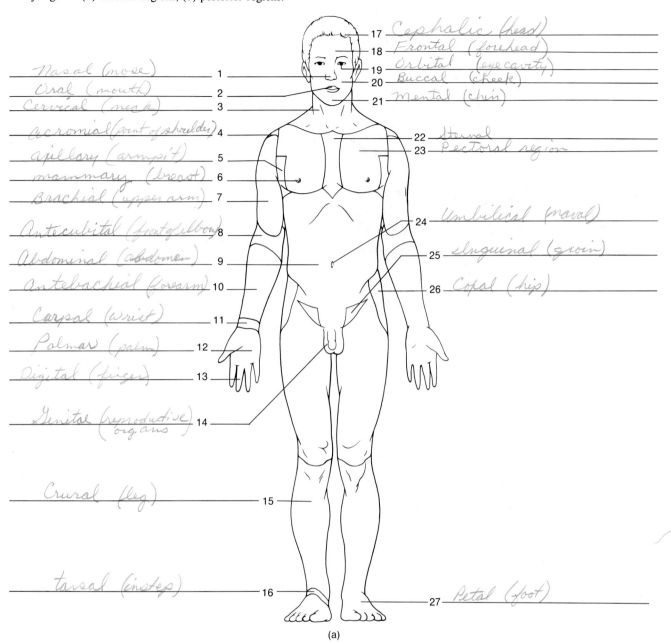

17 Cephalic (head)
18 Frontal (forehead)
19 Orbital (eye cavity)
20 Buccal (cheek)
21 Mental (chin)

1 Nasal (nose)
2 Oral (mouth)
3 Cervical (neck)
4 acromial (point of shoulder)
5 axillary (armpit)
6 mammary (breast)
7 Brachial (upper arm)
8 Antecubital (front of elbow)
9 Abdominal (abdomen)
10 antebrachial (forearm)
11 Carpal (wrist)
12 Palmar (palm)
13 Digital (finger)
14 Genital (reproductive organs)

22 Sternal
23 Pectoral region
24 Umbilical (naval)
25 inguinal (groin)
26 Coxal (hip)

15 Crural (leg)

16 tarsal (instep)

27 Petal (foot)

(a)

Figure 1.6—*Continued*

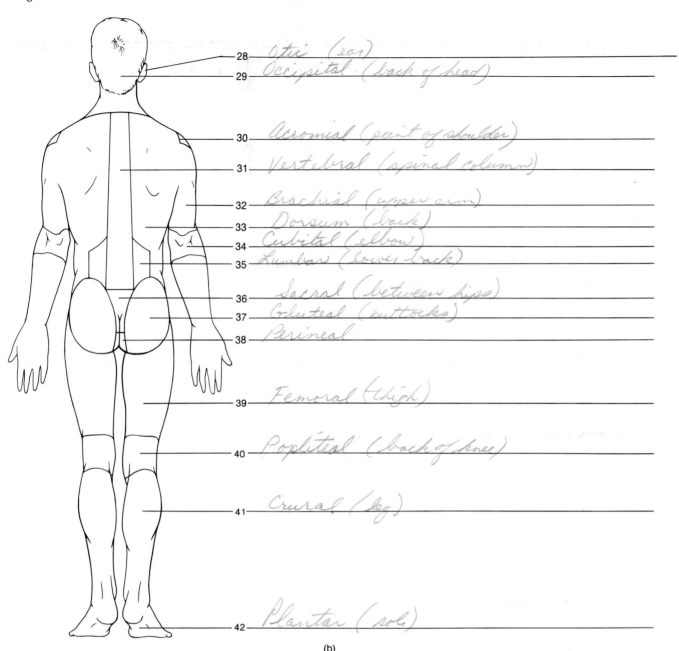

28 — Otic (ear)

29 — Occipital (back of head)

30 — Acromial (point of shoulder)

31 — Vertebral (spinal column)

32 — Brachial (upper arm)

33 — Dorsum (back)

34 — Cubital (elbow)

35 — Lumbar (lower back)

36 — Sacral (between hips)

37 — Gluteal (buttocks)

38 — Perineal

39 — Femoral (thigh)

40 — Popliteal (back of knee)

41 — Crural (leg)

42 — Plantar (sole)

(b)

Name _____

Date _____

Section _____

Body Organization and Terminology

Part A

Match the body cavities in column A with the organs contained in the cavities in column B. Place the letter of your choice in the space provided.

Column A		Column B
A. abdominal cavity	_A_ 1.	liver
	_H_ 2.	lungs
B. cranial cavity	_A_ 3.	spleen
C. middle ear cavity	_A_ 4.	stomach
	_B_ 5.	brain
D. oral cavity	_D_ 6.	teeth
E. orbital cavity	_A_ 7.	gallbladder
F. pelvic cavity	_H_ 8.	trachea
G. vertebral canal	_F_ 9.	urinary bladder
H. thoracic cavity	_E_ 10.	eyes
	_G_ 11.	spinal cord
	_F_ 12.	rectum
	_C_ 13.	ear bones
	_H_ 14.	heart
	_H_ 15.	mediastinum

Part B

Complete the following statements:

1. The membrane on the surface of the lung is called the _Visceral pleura_ .

2. The membrane on the surface of the heart is called the _visceral pericardium_ .

3. The membrane that lines the wall of the abdominopelvic cavity is called the _parietal peritoneum_ .

4. The membrane on the surface of the stomach (in the abdominal cavity) is called the
Visceral peritoneum .

5. The thin watery fluid that occurs between the pleural membranes is called
serous fluid .

6. Epicardium is another name for _visceral pericardium_ .

7. The region between the lungs that separates the thoracic cavity into two compartments is called the
mediastinum .

8. The thin muscular structure that separates the thoracic and abdominopelvic cavities is called the
diaphragm .

Part C

Match the organ systems in column A with the general functions in column B. Place the letter of your choice in the space provided.

Column A		Column B
A. circulatory system	_C_ 1.	secretes hormones
B. digestive system	_D_ 2.	provides outer covering
C. endocrine system	_I_ 3.	produces new organism
D. integumentary system	_G_ 4.	stimulates muscles to contract
E. lymphatic system	_D_ 5.	provides framework for soft tissues
F. muscular system	_H_ 6.	exchanges gases between air and blood
G. nervous system	_J_ 7.	produces blood cells
H. respiratory system	_A_ 8.	transports excess fluid from tissues to blood
I. reproductive system	_F_ 9.	maintains posture
J. skeletal system	_K_ 10.	transports liquid wastes to the outside
K. urinary system	_F_ 11.	produces most body heat
	_K_ 12.	removes liquid and wastes from blood
	_G_ 13.	interprets information from sensory units
	_B_ 14.	converts food molecules into forms that are absorbed
	_H_ 15.	transports gases between lungs and body cells

Part D

Match the organ systems in column A with the organs in column B. Place the letter of your choice in the space provided.

Column A		Column B
A. circulatory system	_C_ 1.	adrenal glands
B. digestive system	_A_ 2.	arteries
C. endocrine system	_G_ 3.	brain
D. integumentary system	_B_ 4.	gallbladder
E. lymphatic system	_A_ 5.	heart
F. muscular system	_L_ 6.	kidneys
G. nervous system	_J_ 7.	larynx
H. reproductive system (female)	_K_ 8.	ligaments
I. reproductive system (male)	_B_ 9.	liver
J. respiratory system	_J_ 10.	lungs
K. skeletal system	_H_ 11.	ovaries
L. urinary system	_C_ 12.	pancreas
	_C_ 13.	parathyroid glands
	_I_ 14.	prostate gland
	_B_ 15.	salivary glands
	_D_ 16.	skin
	_E_ 17.	spleen
	_G_ 18.	spinal cord
	_I_ 19.	testes
	_C_ 20.	thymus gland
	_J_ 21.	trachea
	_H_ 22.	uterus
	_L_ 23.	ureters
	_I_ 24.	urethra
	_H_ 25.	vagina

Part E

Indicate if each of the following sentences makes correct or incorrect usage of the word in boldface type. If the sentence is incorrect, supply a term that will make it correct in the space provided.

1. The mouth is **superior** to the nose. _____ *inferior* _____

2. The stomach is **inferior** to the diaphragm. _____

3. The trachea is **anterior** to the spinal cord. _____

4. The larynx is **posterior** to the esophagus. _____ *anterior* _____

5. The heart is **medial** to the lungs. _____

6. The kidneys are **lateral** to the adrenal glands. _____

7. The hand is **proximal** to the elbow. _____ *distal* _____

8. The knee is **proximal** to the ankle. _____

9. Blood in **deep** blood vessels gives color to the skin. _____ *peripheral* _____

10. A **peripheral** nerve passes from the spinal cord into the leg. _____ *deep* _____

11. The spleen and gallbladder are **ipsilateral.** _____ *contralateral* _____

12. The dermis is the **superficial** layer of the skin. _____

Part F

Name each of the planes represented in figure 1.7(a) and the sections represented in figure 1.7(b).

1. _____ *transverse plane (horizontal plane)* _____
2. _____ *Frontal plane (coronal plane)* _____
3. _____ *Median plane (midsagittal plane)* _____
4. _____ *cross section* _____
5. _____ *oblique section* _____
6. _____ *longitudinal section* _____

Figure 1.7 Name (*a*) the planes and (*b*) the sections represented in these diagrams.

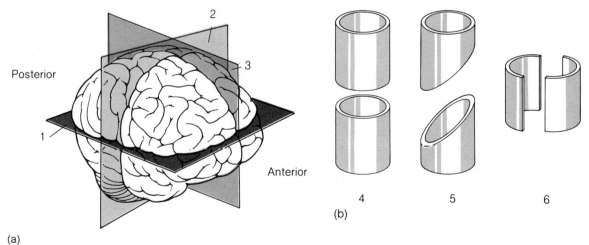

(a)

(b)

Part G

Match the body regions in column A with the locations in column B. Place the letter of your choice in the space provided.

Column A		Column B
A. antebrachium	_F_ 1.	abdomen
B. antecubital	_I_ 2.	ribs
C. axillary	_N_ 3.	reproductive organs
D. brachial	_C_ 4.	armpit
E. buccal	_K_ 5.	elbow
F. celiac	_M_ 6.	forehead
G. cephalic	_O_ 7.	buttocks
H. cervical	_A_ 8.	forearm
I. costal	_L_ 9.	back
J. crural	_H_ 10.	neck
K. cubital	_D_ 11.	upper arm
L. dorsum	_E_ 12.	mouth or inner cheek
M. frontal	_J_ 13.	leg
N. genital	_G_ 14.	head
O. gluteal	_B_ 15.	space in front of elbow

Part H

Match the body regions in column A with the locations in column B. Place the letter of your choice in the space provided.

Column A		Column B
A. inguinal	_J_ 1.	pelvis
B. lumbar	_C_ 2.	breasts
C. mammary	_F_ 3.	ear
D. mental	_K_ 4.	between anus and reproductive organs
E. occipital	_L_ 5.	sole
F. otic	_N_ 6.	middle of thorax
G. palmar	_H_ 7.	chest
H. pectoral	_O_ 8.	navel
I. pedal	_D_ 9.	chin
J. pelvic	_M_ 10.	behind knee
K. perineal	_I_ 11.	foot
L. plantar	_E_ 12.	back of head
M. popliteal	_A_ 13.	abdominal wall near thigh
N. sternal	_B_ 14.	loin
O. umbilical	_G_ 15.	palm

Care and Use of the Compound Microscope

Because the human eye is unable to perceive objects less than 0.1 mm in diameter, a microscope is an essential tool for the study of small structures such as cells. The microscope usually used for this purpose is the *compound microscope*. It is called compound because it utilizes an eyepiece-lens system that magnifies or compounds the magnifying power of an objective-lens system. Such an instrument can magnify images of small objects up to about 1,000 times.

Purpose of the Exercise

To become familiar with the major parts of a compound microscope and their functions, and to make use of the compound microscope to observe small objects.

Learning Objectives

After completing this exercise, you should be able to:

1. Locate and identify the major parts of a compound microscope.
2. Describe the functions of these parts.
3. Calculate the total magnification produced by various combinations of eyepiece and objective lenses.
4. Prepare a simple microscope slide.
5. Make proper use of the microscope to observe small objects.

Materials Needed:
  compound microscope
  lens paper
  microscope slides
  coverslips
  transparent plastic mm rule
  medicine dropper
  dissecting needle (needle probe)

For Demonstrations:

  micrometer scale
  binocular microscope

Procedure

1. Observe a compound microscope, and study figure 2.1 to learn the names of its major parts. Note that the lens system of a compound microscope includes three parts—the condenser, objective lens, and eyepiece.

   Light enters this system from a *substage lamp* (or *mirror*) and usually is concentrated and focused by a *condenser* onto a microscope slide or specimen placed on the *stage*. The condenser, which contains a set of lenses, usually is kept in its highest position possible.

   The *iris diaphragm,* which is located between the light source and the condenser, can be used to increase or decrease the intensity of the light entering the condenser. Locate the lever that operates the iris diaphragm beneath the stage, and move it back and forth. Note how this movement causes the size of the opening in the diaphragm to change. (Some microscopes have a revolving plate beneath the stage with different-sized holes to admit varying amounts of light instead of an iris diaphragm.) Which way do you move the diaphragm to increase the light intensity? _____ Which way to decrease it?

   After light passes through a specimen mounted on a microscope slide, it enters an *objective lens*. This lens projects the light upward into the *body tube,* where it produces a magnified image (real image) of the object being viewed.

Figure 2.1 Major parts of a compound microscope.

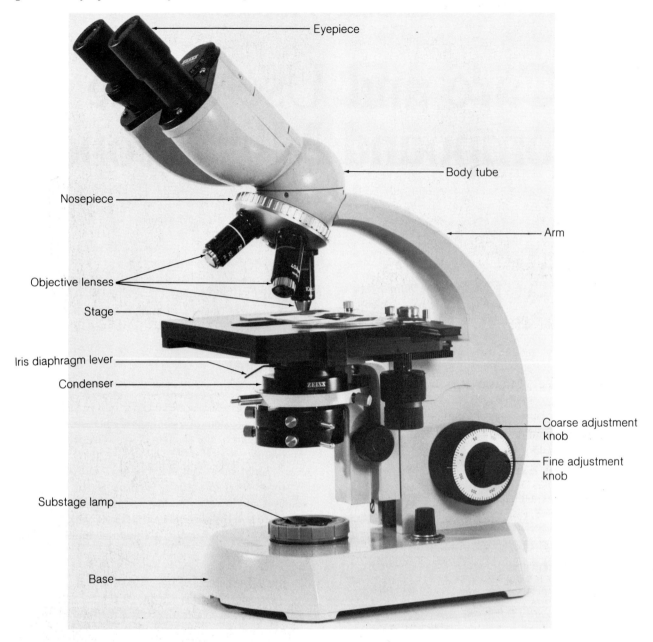

The *eyepiece lens* (ocular) then magnifies this real image to produce another image (virtual image), which is seen by the eye. Typically, the eyepiece lens magnifies the real image ten times (10×). Look for the number in the metal of the eyepiece that indicates its power. What is the eyepiece power of your microscope? _____ (See figure 2.2.)

The objective lenses are mounted in a revolving *nosepiece* so that different magnifications can be achieved by rotating any one of several lenses into position above the specimen. Commonly, this set of lenses includes a scanning lens (4×), a low-power

lens (10×), and a high-dry–power lens (about 45×). Sometimes an oil immersion objective lens (about 100×) is present. Look for the number marked in the metal of each objective that indicates its power. What are the objective lens powers of your microscope? _____

To calculate the *total magnification* achieved when using a particular objective lens, multiply the power of the eyepiece by the power of the objective lens used. Thus, the 10× eyepiece and the 45× objective produce a total magnification of 10 × 45 or 450×.

2. Complete Part A of Laboratory Report 2 on page 19.

Figure 2.2 The powers of this 10× eyepiece (*a*) and this 40× objective lens (*b*) are marked in the metal.

(a)

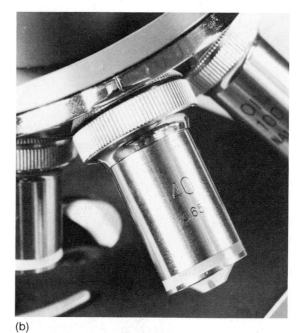

(b)

3. Familiarize yourself with the following list of rules for care of the microscope:
   a. Handle the microscope with great care. It is an expensive and delicate instrument. To move it or carry it, hold it by its *arm* with one hand and support its *base* with the other hand.
   b. Keep the microscope in its *case* or under its *dustcover* when it is not being used.
   c. To clean the lenses, rub them gently with *lens paper* only.
   d. If the microscope has a substage lamp, be sure the electric cord does not hang off the laboratory table where someone might trip over it.
   e. Never remove parts of the microscope or try to disassemble the eyepiece or objective lenses.
   f. If your microscope is not functioning properly, report the problem to your laboratory instructor immediately.
4. Turn on the substage lamp and look through the eyepiece. You will see a lighted circular area called the *field of view.*

   You can measure the diameter of this field of view by focusing the lenses on the millimeter scale of a transparent plastic rule. To do this:
   a. Place the rule on the microscope stage in the spring clamp or under the slide clips. (*Note:* If your microscope is equipped with a mechanical stage, it may be necessary to use a short section cut from a transparent plastic rule. The section should be several millimeters long and can be mounted on a microscope slide for viewing.)

Figure 2.3 When you lower the objective lens, you can prevent it from touching the specimen by watching from the side.

   b. Center the millimeter scale in the beam of light coming up through the condenser and rotate the low-power objective lens into position.
   c. While you watch from the side to prevent the lens from touching anything, lower the objective until it is as close to the rule as possible, using the *coarse adjustment knob* and then using the *fine adjustment knob* (figure 2.3).
   d. Look into the eyepiece, and use the coarse adjustment knob to raise the objective lens until the lines of the millimeter scale come into sharp focus.

Figure 2.4 The divisions of a micrometer scale in an eyepiece can be calibrated against the known divisions of a micrometer slide.

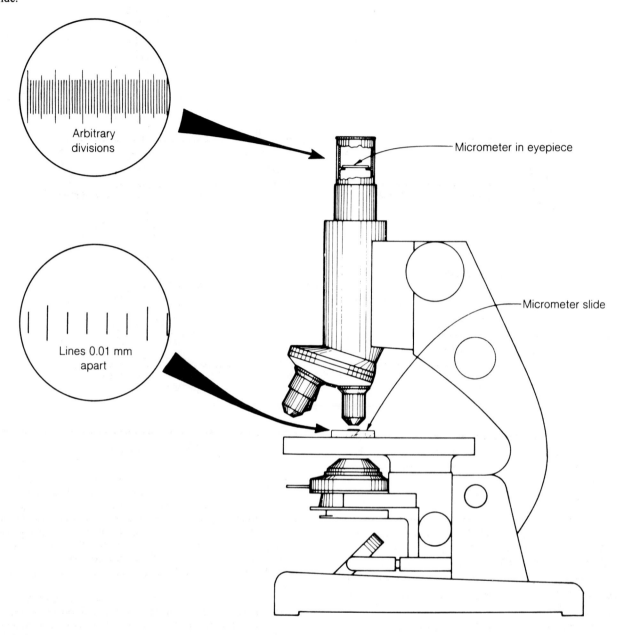

Arbitrary
divisions

Micrometer in eyepiece

Micrometer slide

Lines 0.01 mm
apart

e. Adjust the light intensity by moving the lever of the iris diaphragm so that the field of view is brightly illuminated but comfortable to your eye. At the same time, take care not to overilluminate the field, because in very bright light transparent objects tend to disappear.

f. Position the millimeter rule so that its scale crosses the greatest diameter of the field of view. Also, move the rule so that one of the millimeter marks is against the edge of the field of view.

g. Estimate the distance across the field of view in millimeters.

5. Complete Part B of the laboratory report.

6. Most microscopes are designed to be *parfocal*. This means that when a specimen is in focus with the low-power objective, it will be in focus (or nearly so) when a higher power objective is rotated into position.

Rotate the high-dry objective into position while you watch from the side, and then look at the millimeter scale of the transparent plastic rule. If

Figure 2.5 Steps in the preparation of a wet mount.

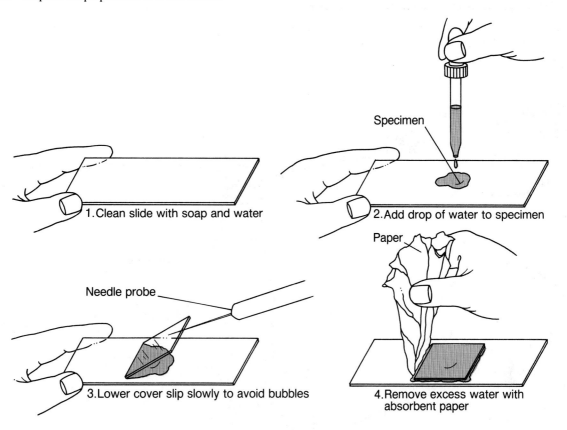

Specimen

1. Clean slide with soap and water

2. Add drop of water to specimen

Needle probe

Paper

3. Lower cover slip slowly to avoid bubbles

4. Remove excess water with absorbent paper

you need to move the high-dry objective to sharpen the focus, use the *fine adjustment knob* only. What might happen if you use the coarse adjustment knob?

_____

_____

Adjust the iris diaphragm so that the field of view is properly illuminated. Once again, adjust the millimeter rule so that the scale crosses the field of view through its greatest diameter, and position the rule so that a millimeter mark is against one edge of the field.

Try to estimate the distance across the field of view in millimeters.

7. Locate the numeral "4" (or "9") on the plastic rule, and focus on it, using the low-power objective. Note how the number appears in the field of view. Move the plastic rule to the right, and note which way the image moves. Slide the rule away from you, and again note how the image moves.

8. Complete Parts C and D of the laboratory report.

Demonstration

A compound microscope is sometimes equipped with a micrometer scale mounted in the eyepiece. Such a scale is subdivided into fifty to one hundred equal divisions (figure 2.4). These arbitrary divisions can be calibrated against the known divisions of a micrometer slide placed on the microscope stage. Once the values of the divisions are known, the length and width of a microscopic object can be measured by superimposing the scale over the magnified image of the object.

Observe the micrometer scale in the eyepiece of the demonstration microscope. Focus the low-power objective on the millimeter scale of a micrometer slide (or a plastic rule), and estimate the distance between the divisions on the micrometer scale in the eyepiece. What is the distance between the finest divisions of the scale in micrometers? _____

9. Prepare several temporary *wet mounts,* using any small, transparent objects of interest, and examine the specimens, using the low-power objective and then a high-dry–power objective to observe their details.

To prepare a wet mount:

a. Carefully clean a microscope slide with soap and water, and dry it with a paper towel.

b. Place a tiny, thin piece of the specimen you want to observe in the center of the slide, and use a medicine dropper to put a drop of water over it. If the specimen is solid, you might want to tease some of it apart with dissecting needles. In any case, the specimen must be thin enough so that light can pass through it. Why is it necessary for the specimen to be so thin?

c. Cover the specimen with a coverslip. Try to avoid trapping bubbles of air beneath the coverslip by lowering it slowly into the drop of water.

d. Remove any excess water from the edge of the coverslip with absorbent paper (figure 2.5).

e. Place the slide under the slide clips on the microscope stage, and position the slide so that the specimen is centered in the light beam passing up through the condenser.

f. Focus on the specimen using the low-power objective, then examine it with high-dry power.

10. If an oil immersion objective lens is available, use it to examine the specimen. To use the oil immersion lens:

a. Center the object you want to study in the high-dry field of view.

b. Rotate the high-dry objective away from the microscope slide, place a small drop of immersion oil on the coverslip, and swing the oil immersion lens into position. To achieve sharp focus, use the *fine adjustment knob* only.

c. You will need to open the iris diaphragm more fully for proper illumination. More light is needed because the oil immersion objective covers a very small lighted area of the microscope slide.

d. Because the oil immersion lens must be very close to the coverslip to achieve sharp focus, care must be taken to avoid breaking the coverslip or damaging the objective lens. For this reason, never lower the objective when you are looking into the eyepiece. Instead, always raise the objective to achieve focus, or prevent the objective from touching the coverslip by watching the microscope slide and coverslip from the side if the objective lens needs to be lowered.

11. When you have finished working with the microscope, remove the microscope slide from the stage, wipe any oil from the objective lens with lens

Figure 2.6 A binocular stereoscopic microscope.

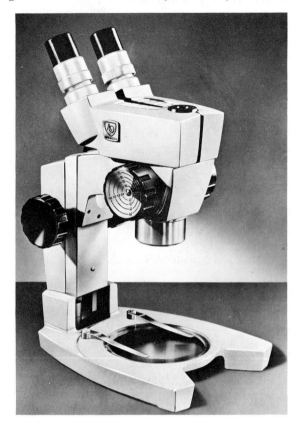

paper, and swing the low-power objective into position. Wrap the electric cord around the base of the microscope and replace the dustcover.

12. Complete Part E of the laboratory report.

Demonstration

A binocular stereoscopic microscope (figure 2.6) is useful for observing the details of relatively large, opaque specimens. Although this type of microscope achieves less magnification than a compound microscope, it has the advantage of producing a three-dimensional image rather than the flat, two-dimensional image of the compound microscope. In addition, the image produced by the stereoscopic microscope is positioned in the same manner as the specimen, rather than being reversed and inverted as it is by the compound microscope.

Observe the binocular microscope. Note that the eyepieces can be pushed apart or together to fit the distance between your eyes. Focus the microscope on the end of your finger. Which way does the image move when you move your finger to the right?

When you move it away? _____
If the instrument has more than one objective, change the magnification to higher power. Use the instrument to examine various small, opaque objects available in the laboratory.

laboratory report 2

Name _____

Date _____

Section _____

Care and Use of the Compound Microscope

Part A

Complete the following:

1. What total magnification will be achieved if the 10× eyepiece and the 10× objective lens are used? _____

2. If the 10× eyepiece and the 100× objective are used? _____

Part B

Complete the following:

1. Sketch the millimeter scale as it appears under low-power magnification.

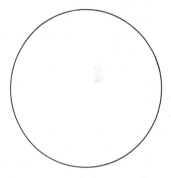

2. What is the diameter of the low-power field of view in millimeters? _____

3. Microscopic objects often are measured in *micrometers*. A micrometer equals 1/1000 of a millimeter and is symbolized by the Greek letter mu (μ). What is the diameter of the low-power field of view in micrometers? _____

4. If a circular object or specimen extends halfway across the low-power field, what is its diameter in millimeters? _____

5. What is its diameter in micrometers? _____

Part C

Complete the following:

1. Sketch the millimeter scale as it appears using the high-dry–power objective.

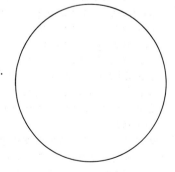

2. What do you estimate the diameter of this field of view to be in millimeters? _____

3. How does the diameter of the low-power field of view compare to that of the high-power field?

_____

4. Why is it more difficult to measure the diameter of the high-dry field of view than the low-power field? _____

_____

5. What change occurred in the light intensity of the field of view when you exchanged the low-power objective for the high-power objective? _____

6. Sketch the numeral "4" (or "9") as it appears through the lenses of the compound microscope.

7. What has the lens system done to the image of the numeral? (That is, is it right side up, upside down, or what?) _____

8. When you moved the rule to the right, which way did the image move? _____

9. When you moved the rule away, which way did the image move? _____

Part D

Match the names of the microscope parts in column A with the descriptions in column B. Place the letter of your choice in the space provided.

Column A		Column B
A. arm	___ 1.	increases or decreases light intensity
B. adjustment knob	___ 2.	platform that supports microscope slide
C. condenser	___ 3.	concentrates light onto specimen
D. eyepiece lens	___ 4.	causes objective lens to move upward or downward
E. field of view	___ 5.	produces magnified image of specimen in body tube
F. iris diaphragm	___ 6.	holds microscope slide in position
G. nosepiece	___ 7.	magnifies real image of specimen
H. objective lens	___ 8.	serves as a handle for carrying microscope
I. slide clip	___ 9.	part to which objective lenses are attached
J. stage	___ 10.	circular area seen through the eyepiece

Part E

Prepare sketches of the objects you observed using the microscope. For each sketch, include the name of the object, the magnification you used to observe it, and its estimated dimensions in millimeters and micrometers.

Laboratory Exercise 3

Cellular Structure

Cells are the "building blocks" from which all larger parts of the human body are formed. They account for the shape, organization, and construction of the body and are responsible for carrying on its life processes.

Purpose of the Exercise

To review the structure and functions of major cellular components and to observe examples of human cells.

Learning Objectives

After completing this exercise, you should be able to:

1. Name and locate the major components of a cell on a model or diagram, and describe the general functions of these components.
2. Prepare a wet mount of cheek-lining cells, stain the cells, and identify the major components of a cheek-lining cell.
3. Locate cells on prepared slides of human tissues, and identify their major components.
4. Identify major cellular components in a transmission electron micrograph.

Materials Needed:
    animal cell model
    clean microscope slides
    coverslips
    flat toothpicks
    medicine dropper
    methylene blue stain (dilute)
    prepared microscope slides of human tissues
    compound microscope

For Optional Activity:

    single-edged razor blade
    plant materials such as leaves, soft stems, fruits, and vegetables

Procedure

1. Review the section entitled "A Composite Cell" in chapter 3 of the textbook.
2. Observe the animal cell model, and identify its major parts.
3. As a review activity, label figures 3.1 and 3.2.
4. Complete Part A of Laboratory Report 3 on page 24.
5. Prepare a wet mount of cheek-lining cells. To do this:
   a. Gently scrape the inner lining of your cheek with the broad end of a flat toothpick.
   b. Stir the toothpick in a drop of water on a clean microscope slide and dispose of the toothpick in the wastebasket.
   c. Cover the drop with a coverslip.
   d. Observe the cheek-lining cells by using the microscope. To report what you observe, sketch a single cheek-lining cell in the space provided in Part B of the laboratory report.
6. Prepare a second wet mount of cheek-lining cells, but this time add a drop of dilute methylene blue stain to the cells. Cover the liquid with a coverslip, and observe the cells with the microscope. Add to your sketch any additional structures you observe in the stained cells.
7. Answer the questions in Part B of the laboratory report.
8. Observe each of the prepared slides of human tissues, using the microscope. To report what you observe, sketch a single cell of each type in the space provided in Part C of the laboratory report.
9. Answer the questions in Part C of the laboratory report.
10. Study figure 3.3. These electron micrographs represent extremely thin slices of cells. Note that each micrograph contains a section of a nucleus and some cytoplasm. Compare the organelles shown in these micrographs with the organelles included in the electron micrographs of the textbook chapter 3.
11. Complete Parts D and E of the laboratory report.

Figure 3.1 Label the parts of this composite cell.

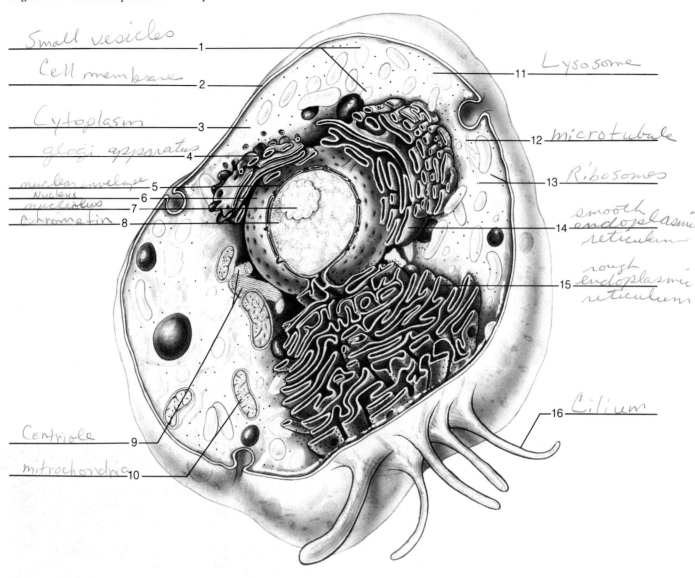

Small vesicles ___ 1

Cell membrane ___ 2

Cytoplasm ___ 3

glogi apparatus ___ 4

nuclear envelope ___ 5

Nucleus ___ 6

nucleolus ___ 7

chromatin ___ 8

Centriole ___ 9

mitrochondria ___ 10

11 ___ Lysosome

12 ___ microtubule

13 ___ Ribosomes

14 ___ smooth endoplasmic reticulum

15 ___ rough endoplasmic reticulum

16 ___ Cilium

Optional Activity

Investigate the microscopic structure of various plant materials. To do this, prepare very tiny, thin slices of plant specimens using a single-edged razor blade. Keep the slices in a container of water until you are ready to observe them. To observe a specimen, place it in a drop of water on a clean microscope slide, and cover it with a coverslip. Use the microscope and view the specimen using low- and high-power magnifications. Add a drop of dilute methylene blue stain and note if any additional structures become visible. How are the microscope structures of the plant specimens similar to the human tissues you observed? _____

How are they different? _____

Figure 3.2 Label the parts of the cell membrane.

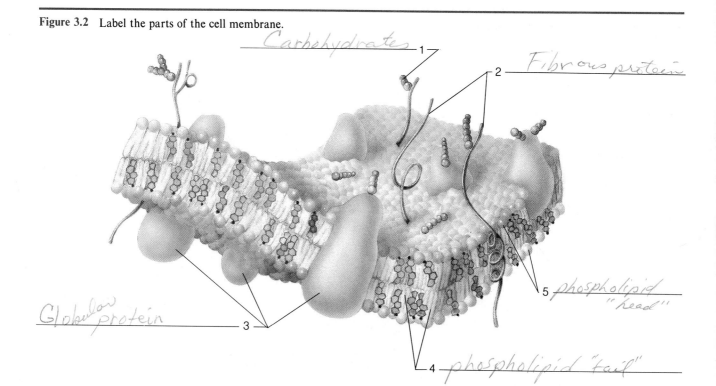

Carbohydrates 1

Fibrous protein 2

Globular protein 3

5 *phospholipid "head"*

4 *phospholipid "tail"*

Figure 3.3 Transmission electron micrographs of cellular components. Which of the structures indicated by arrows can you identify?

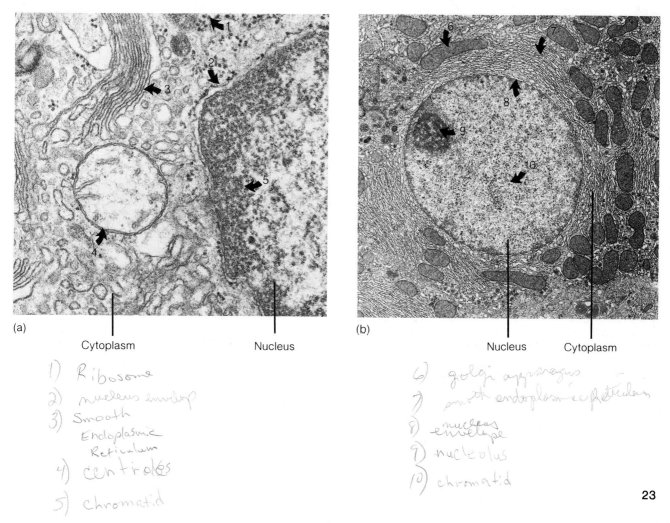

(a)

Cytoplasm   Nucleus

(b)

Nucleus   Cytoplasm

1) Ribosome
2) nucleus envelop
3) Smooth Endoplasmic Reticulum
4) centrioles
5) chromatid

6) golgi apparagus
7) smooth endoplasmic reticulum
8) nucleus envelope
9) nucleolus
10) chromatid

laboratory report 3

Name _____

Date _____

Section _____

Cellular Structure

Part A

Match the cellular components in column A with the descriptions in column B. Place the letter of your choice in the space provided.

Column A

A. cell membrane
B. centrosome
C. chromatin
D. cilium
E. cytoplasm
F. endoplasmic reticulum
G. Golgi apparatus
H. inclusion
I. lysosome
J. microfilament
K. microtubule
L. mitochondrion
M. nuclear envelope
N. nucleolus
O. nucleus
P. peroxisome
Q. ribosome
R. vesicle

Column B

_C_ 1. loosely coiled fibers within nucleus
_H_ 2. lifeless cellular product
_L_ 3. organelle whose inner membrane forms partitions
_B_ 4. nonmembranous organelle located near nucleus
_Q_ 5. small RNA-containing particles attached to endoplasmic reticulum
_R_ 6. membranous sac formed by the pinching off of cell membrane
_N_ 7. dense body of RNA within the nucleus
_K_ 8. thin, hollow threadlike parts
_F_ 9. network of sacs, canals, and vesicles
_A_ 10. outermost limit of living material of cell
_E_ 11. occupies space between cell membrane and nucleus
_G_ 12. flattened membranous sacs near nucleus
_D_ 13. hairlike part associated with basal body
_J_ 14. tiny rod that helps cell to shorten
_I_ 15. membranous sac that contains digestive enzymes
_P_ 16. contains enzymes that decompose hydrogen peroxide
_M_ 17. double-layered membrane with "pores"
_O_ 18. spherical organelle that contains chromosomes

Part B

Complete the following:

1. Sketch a single cheek-lining cell. Label the cellular components you recognize.

2. What effect did the methylene blue stain have on the cheek-lining cells? _____
_____

3. What advantage may be gained by staining cells? _____
_____

Part C

Complete the following:

1. Sketch a single cell of each kind you observed in the prepared slides of human tissues. Name the tissue and label the cellular components you recognize.

2. What do the various kinds of cells in these tissues have in common? _____
_____

3. What are the main differences you observed among these cells? _____
_____

Part D

Identify the structures indicated by the arrows in figures 3.3(a) and 3.3(b).

1. _____   6. _____

2. _____   7. _____

3. _____   8. _____

4. _____   9. _____

5. _____   10. _____

Part E

Answer the following questions after observing the transmission electron micrographs in figure 3.3.

1. What cellular structures were visible in the transmission electron micrographs that were not apparent in the cells you observed using the microscope? _____

   _____

2. Before they can be observed by using a transmission electron microscope, cells are sliced into very thin sections. What disadvantage does this procedure present in the study of cellular parts? _____

   _____

Movements through Cell Membranes

A cell membrane functions as a gateway through which chemical substances and small particles may enter or leave a cell. These substances move through the membrane by physical processes, such as diffusion, osmosis, and filtration, or by physiological processes, such as active transport, phagocytosis, or pinocytosis.

Purpose of the Exercise

To demonstrate some of the physical processes by which substances move through cell membranes.

Learning Objectives

After completing this exercise, you should be able to:

1. Define diffusion and identify examples of diffusion.
2. Define osmosis and identify examples of osmosis.
3. Distinguish between hypertonic, hypotonic, and isotonic solutions.
4. Define filtration and identify examples of filtration.

Materials Needed:
For Procedure A—Diffusion:

   petri dish
   methylene blue stain (or food coloring)
   medicine dropper

For Procedure B—Osmosis:

   thistle tube
   molasses (or Karo syrup)
   selectively permeable membrane (dialysis tubing)
   ring stand and clamp
   beaker
   string or rubber band

For Procedure C—Hypertonic, Hypotonic, and Isotonic Solutions:

   test tubes
   test tube rack

   10 ml graduated cylinder
   medicine dropper
   uncoagulated animal blood
   distilled water
   0.85% NaCl (aqueous solution)
   2.0% NaCl (aqueous solution)
   clean microscope slides
   coverslips
   microscope

For Procedure D—Filtration:

   glass funnel
   filter paper
   ring stand and ring
   beaker
   powdered charcoal
   1% glucose (aqueous solution)
   1% starch (aqueous solution)
   test tubes
   water bath (boiling water)
   Benedict's solution
   IKI solution

Procedure A—Diffusion

1. Review the section entitled "Diffusion" in chapter 3 of the textbook.
2. To demonstrate diffusion:
   a. Place a petri dish half-filled with water on a piece of white paper and wait until the water surface is still.
   b. Use a medicine dropper to gently place a drop of methylene blue stain on the surface (figure 4.1).
   c. Observe the stain for a few minutes.
3. Complete Part A of Laboratory Report 4 on page 30.

Optional Activity

Repeat the demonstration of diffusion, using one petri dish filled with ice-cold water and a second dish filled with very hot water. At the same moment, add a drop of stain to each dish and watch the stain as before. What difference do you note in the rate of diffusion in the two dishes? How do you explain this difference? _____

_____

_____

_____

Procedure B—Osmosis

1. Review the section entitled "Osmosis" in chapter 3 of the textbook.
2. To demonstrate osmosis:
   a. Plug the tube end of a thistle tube with your finger.
   b. Fill the bulb with molasses.
   c. Cover the bulb opening with a piece of moist selectively permeable membrane.
   d. Tie the membrane tightly in place with string or fasten it with a rubber band.
   e. Immerse the bulb end of the tube in a beaker of water.
   f. Support the upright portion of the tube with a clamp on a ring stand.
   g. Mark the level of the molasses in the tube.
   h. From time to time observe its level (figure 4.2).
3. Complete Part B of the laboratory report.

Optional Activity

Repeat the demonstration of osmosis, but this time use two thistle tubes. Fill the bulbs with different concentrations of molasses—25% molasses in water for one and 50% molasses in water for the other. Set the tubes up as before and observe the rates at which the levels of molasses change. What difference did you note between the rates of change in the two tubes? How do you explain this difference? _____

_____

_____

_____

Procedure C—Hypertonic, Hypotonic, and Isotonic Solutions

1. Review the section entitled "Osmosis" in chapter 3 of the textbook.
2. To demonstrate hypertonic, hypotonic, and isotonic solutions:
   a. Place three test tubes in a rack and mark them as tube 1, tube 2, and tube 3. (*Note:* One set of tubes can be used to supply test samples for the entire class.)

Figure 4.1 To demonstrate diffusion, place a drop of stain on the surface of some water.

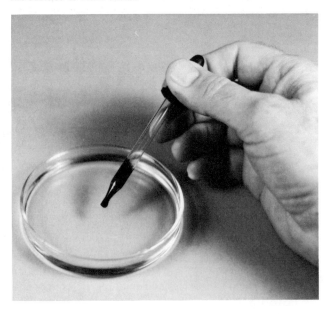

   b. Add 5 ml of distilled water to tube 1; add 5 ml of 0.85% NaCl to tube 2; and add 5 ml of 2.0% NaCl to tube 3.
   c. Place three drops of fresh uncoagulated animal blood into each of the tubes, and gently mix the blood with the solutions.
   d. Using three separate medicine droppers, remove a drop from each tube and place the drops on three separate microscope slides marked 1, 2, and 3.
   e. Cover the drops with coverslips and observe the blood cells, using the high-dry power of the microscope.
3. Complete Part C of the laboratory report.

Procedure D—Filtration

1. Review the section entitled "Filtration" in chapter 3 of the textbook.
2. To demonstrate filtration:
   a. Place a glass funnel in the ring of a ring stand over an empty beaker. Fold a piece of filter paper in half and then in half again. Open one thickness of the filter paper to form a cone, and place it in the funnel.
   b. Prepare a mixture of powdered charcoal, water, glucose solution, and starch solution in a beaker and pour some of the mixture into the funnel until it reaches the top of the filter paper cone (figure 4.3).
   c. Test some of the filtrate in the beaker for the presence of glucose. To do this, place five drops of filtrate in a clean test tube and add 5 ml of Benedict's solution. Place the test tube in a water bath of boiling water for two minutes and

Figure 4.2 (*a*) Fill the bulb of the thistle tube with molasses; (*b*) tie a piece of selectively permeable membrane over the bulb opening; and (*c*) immerse the bulb in a beaker of water.

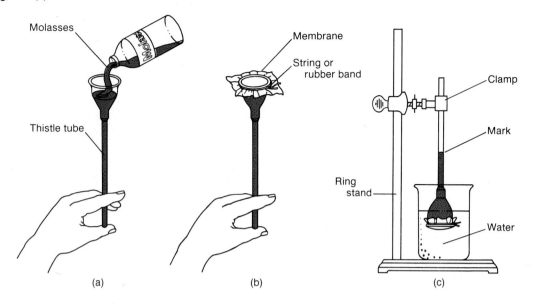

(a)　　　　(b)　　　　(c)

Figure 4.3 Apparatus used to illustrate filtration.

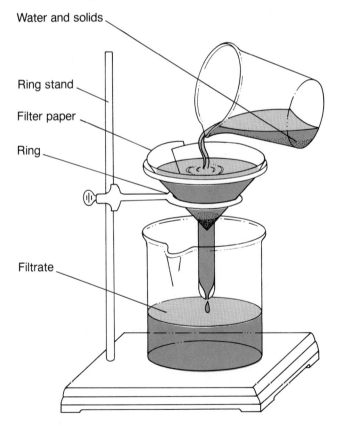

Figure 4.4 Heat the filtrate and Benedict's solution in a boiling water bath for two minutes.

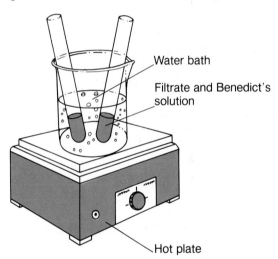

then allow the liquid to cool slowly. If the color of the solution changes to green, yellow, or red, glucose is present (figure 4.4).

　d. Test some of the filtrate in the beaker for the presence of starch. To do this, place 3 ml of filtrate in a test tube and add a few drops of IKI solution. If the color of the solution changes to blue-black, starch is present.

3. Complete Part D of the laboratory report.

laboratory report 4

Name _____

Date _____

Section _____

Movements through Cell Membranes

Part A

Complete the following:

1. Describe what happened after you placed a drop of stain on the water surface. _____
_____

2. Briefly define *diffusion.* _____
_____

3. Indicate which of the following provides an example of diffusion by answering *yes* or *no.*
   a. A perfume bottle is opened and soon the odor can be sensed in all parts of the room. _____
   b. A sugar cube is dropped into a cup of hot water, and, without being stirred, all of the liquid becomes sweet tasting. _____
   c. Water molecules move from a faucet through a garden hose when the faucet is turned on. _____
   d. A person blows air molecules into a balloon by exhaling forcefully. _____
   e. A crystal of blue copper sulfate is placed in a test tube of water. The next day the solid is gone, but the water is colored. _____

Part B

Complete the following:

1. What happened to the level of the molasses in the thistle tube during the laboratory period? _____
_____

2. How do you explain this change? _____
_____

3. How does this experiment demonstrate the development of *osmotic pressure?* _____
_____
_____

4. Briefly define *osmosis.* _____
_____

5. Indicate which of the following involves osmosis by answering *yes* or *no.*
   a. A fresh potato is peeled, weighed, and soaked in a strong salt solution. The next day it is discovered that the potato has lost weight. _____
   b. Garden grass wilts after being exposed to dry chemical fertilizer. _____
   c. Air molecules escape from a punctured tire as a result of high pressure inside. _____
   d. Plant seeds soaked in water swell and become several times as large as before. _____
   e. When the bulb of a thistle tube filled with water is sealed by a selectively permeable membrane and submerged in a beaker of molasses, the water level in the tube falls. _____

Part C

Complete the following:

1. In the spaces below, sketch a few blood cells from each of the test tubes.

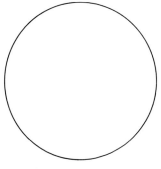 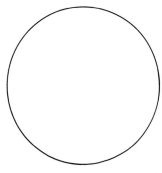 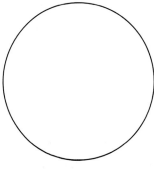

Tube 1
(distilled water)

Tube 2
(0.85% NaCl)

Tube 3
(2.0% NaCl)

2. Based on your results, which tube contained a solution that was hypertonic to the blood cells? _____ Give the reason for your answer. _____
_____

3. Which tube contained a solution that was hypotonic to the blood cells? _____ Give the reason for your answer. _____
_____

4. Which tube contained a solution that was isotonic to the blood cells? _____ Give the reason for your answer. _____
_____

Part D

Complete the following:

1. Which of the substances in the mixture you prepared passed through the filter paper into the filtrate? _____
_____

2. What evidence do you have for your answer to question 1? _____
_____
_____

3. What force was responsible for the movement of substances through the filter paper? _____
_____

4. What substances did not pass through the filter paper? _____
_____

5. What factor prevented these substances from passing through? _____
_____
_____

6. Briefly define *filtration.* _____
_____
_____

7. Indicate which of the following involves filtration by answering *yes* or *no*.

a. Urea molecules leave a patient's blood and enter the dialyzing fluid of an artificial kidney where the urea concentration is lower. _____

b. Oxygen molecules move into a cell and carbon dioxide molecules leave a cell because of differences in the concentrations of these substances on either side of the cell membrane. _____

c. Blood pressure forces water molecules from the blood outward through the thin wall of a blood capillary. _____

d. Urine is forced from the urinary bladder through the tubular urethra by muscular contractions. _____

e. Air molecules enter the lungs through the airways when air pressure is greater outside these organs than inside. _____

Laboratory Exercise 5

Life Cycle of a Cell

The life cycle of a cell consists of the series of changes it undergoes from the time it is formed until it reproduces. Typically, a newly formed cell grows to a certain size and then splits in half to form two daughter cells. This reproductive process involves two major steps: (1) division of the cell's nuclear parts, which is called *mitosis,* and (2) division of the cell's cytoplasm.

Purpose of the Exercise

To review the stages in the life cycle of a cell and to observe cells in various stages of their life cycles.

Learning Objectives

After completing this exercise, you should be able to:

1. Describe the life cycle of a cell.
2. Identify the stage in the life cycle of a particular cell.
3. Arrange into a correct sequence a set of models or drawings of cells in various stages of their life cycles.

Materials Needed:
models of animal mitosis
microscope slides of whitefish blastula
compound microscope

For Demonstration:

microscope slide of human leukocytes in mitosis
oil immersion lens

Procedure

1. Review the section entitled "Life Cycle of a Cell" in chapter 3 of the textbook.
2. As a review activity, label the various stages of the cell's life cycle represented in figure 5.1.
3. Observe the animal mitosis models, and review the major events in a cell's life cycle represented by each of them. Be sure you can arrange these models in a correct sequence if their positions are changed.
4. Complete Part A of Laboratory Report 5 on page 36.
5. Obtain a slide of whitefish mitosis.
   a. Examine the slide, using the microscope. The tissue on this slide was obtained from a developing embryo (blastula) of a fish, and many of the embryonic cells are undergoing mitosis. Note that the chromosomes of these reproducing cells are stained darkly (figure 5.2).
   b. Search the tissue for cells in various stages of reproduction. Note that there are several sections on the slide. If you cannot locate different mitotic stages in one section, examine the cells of another.
   c. Each time you locate a cell in a different stage, sketch it in an appropriate circle on Laboratory Report 5, Part B.
6. Complete Parts C and D of the laboratory report.

Demonstration

See if you can locate some human chromosomes by examining a prepared slide of human leukocytes, using the oil immersion lens of a microscope. The cells on this slide were cultured in a special medium and were stimulated to undergo mitosis. The mitotic process was arrested in metaphase by exposing the cells to a chemical called colchicine, and the cells were caused to swell osmotically. As a result of this treatment, the chromosomes were spread apart. Sets of paired chromosomes (chromatids) should be visible when they are magnified about 1,000× (figure 5.3).

Figure 5.1 Label the major stages in the life cycle of this cell.

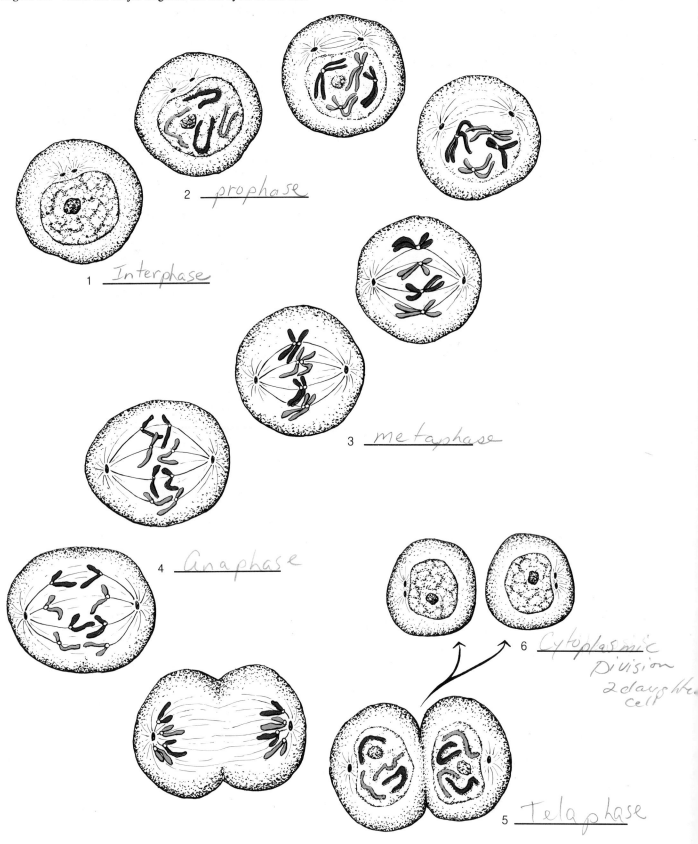

1 Interphase

2 prophase

3 metaphase

4 Anaphase

5 Telaphase

6 Cytoplasmic
Division
2 daughter
cell

Figure 5.2 Cells of a whitefish blastula. (Photograph by Carolina Biological Supply Company.)

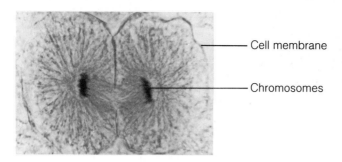

— Cell membrane

— Chromosomes

Figure 5.3 A set of human chromosomes. How many chromosomes are in this set?

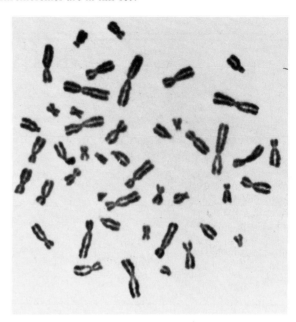

Name _____

Date _____

Section _____

Life Cycle of a Cell

Part A

Complete the chart:

Stage	Major Events Occurring
Mitosis Prophase	
Metaphase	
Anaphase	
Telophase	
Cytoplasmic division	
Interphase	

Part B

Sketch cells in different stages of mitosis to illustrate the whitefish cell's life cycle. Label the stage represented by each. Also label the major cellular parts represented in the sketches.

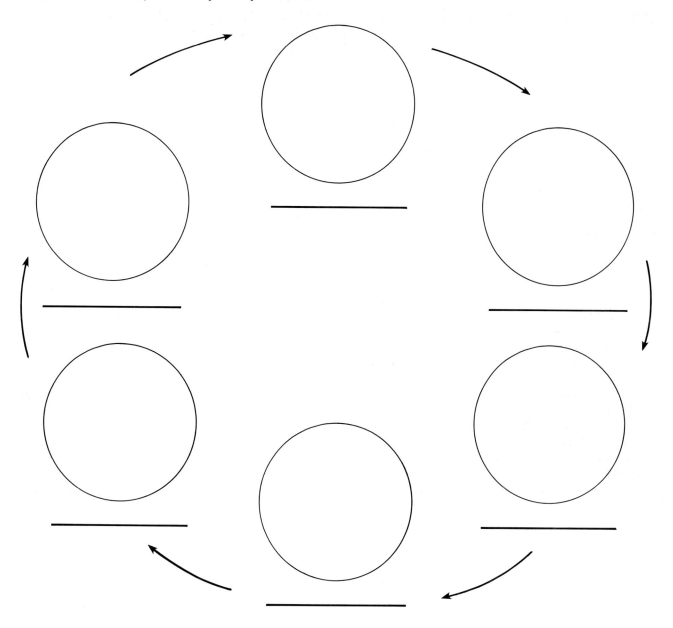

Part C

Complete the following:

1. In what ways are the daughter cells, which result from mitosis, similar? _____
   _____
   _____

2. How are the daughter cells different? _____
   _____
   _____

3. Distinguish between mitosis and cytoplasmic division. _____
   _____
   _____

Figure 5.4 Identify the mitotic stage of the cell in each of these micrographs (a–d).

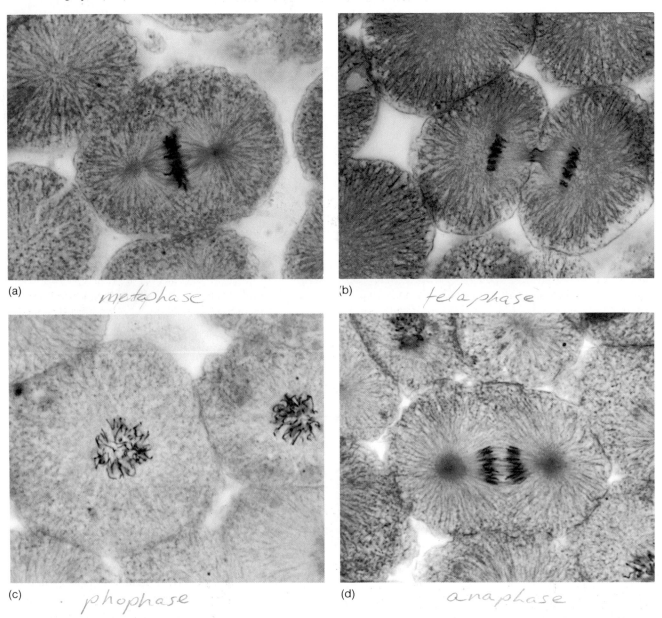

(a) *metaphase*

(b) *telaphase*

(c) *phophase*

(d) *anaphase*

Part D

Identify the mitotic stage represented by each of the micrographs in figure 5.4.

a. ___metaphase___

b. ___telaphase___

c. ___prophase___

d. ___anaphase___

Epithelial Tissues

Atissue is composed of a layer or mass of cells—cells that are similar in size, shape, and function.

Within the human body, there are four major types of tissues: (1) epithelial tissue, which covers body surfaces; (2) connective tissue, which binds and supports parts; (3) muscle tissue, which makes movement possible; and (4) nerve tissue, which conducts impulses from one part of the body to another and helps to control and coordinate body activities.

Purpose of the Exercise

To review the characteristics of epithelial tissue and to observe examples of this tissue.

Learning Objectives

After completing this exercise, you should be able to:

1. Describe the general characteristics of epithelial tissue.
2. List six types of epithelial tissue.
3. Describe the special characteristics of each type of epithelial tissue.
4. Identify examples of epithelial tissue.

Materials Needed:

compound microscope
prepared slides of the following epithelial tissues:
  simple squamous epithelium (lung)
  simple cuboidal epithelium (kidney)
  simple columnar epithelium (small intestine)
  pseudostratified columnar epithelium (trachea)
  stratified squamous epithelium (esophagus)
  transitional epithelium (urinary bladder)

Procedure

1. Review the section entitled "Epithelial Tissues" in chapter 5 of the textbook.
2. Complete Part A of Laboratory Report 6 on page 40.
3. Use the microscope to observe the prepared slides of types of epithelial tissues. As you observe each tissue, look for the special features described in the textbook, such as cell size, shape, and arrangement, that distinguish it from the others.
4. Complete Part B of the laboratory report.
5. Test your ability to recognize each type of epithelial tissue. To do this, have a laboratory partner select one of the prepared slides, cover its label, and focus the microscope on the tissue. Then, see if you can identify the tissue correctly.
6. Complete Part C of the laboratory report.

laboratory report 6

Name _____

Date _____

Section _____

Epithelial Tissues

Part A

Match the tissues in column A with the characteristics in column B. Place the letter of your choice in the space provided.

Column A

A. simple squamous epithelium
B. simple cuboidal epithelium
C. simple columnar epithelium
D. pseudostratified columnar epithelium
E. stratified squamous epithelium
F. transitional epithelium

Column B

E or F

_E__ 1. consists of many layers of cells
_D__ 2. functions to move sex cells
_A__ 3. single layer of flattened cells
_D__ 4. nuclei located at different levels within cells
_A__ 5. forms walls of capillaries
_D__ 6. forms linings of respiratory passages
_E__ 7. younger cells cuboidal, older cells flattened
_F__ 8. forms inner lining of urinary bladder
_D__ 9. cells commonly possess cilia
_B__ 10. has centrally located spherical nuclei
_C__ 11. forms lining of digestive tract
_C__ 12. nuclei located near basement membrane
D or _C__ 13. cells longer than wide
_E__ 14. forms lining of mouth
_B__ 15. forms lining of ducts in glands

Part B

In the space that follows, sketch a few cells of each type of epithelium you observed. For each, identify the tissue, label the major parts of its cells, and indicate the magnification used.

Simple squamous epithelium (__43__X)

Simple cuboidal epithelium (__43__X)

Simple columnar epithelium (__10__X)

Pseudostratified columnar epithelium (__43__X)

Stratified squamous epithelium (__43__X)

Transitional epithelium (__43__X)

Figure 6.1 Identify each of the epithelial tissues shown in these micrographs (*a–f*). Magnifications: *a* (×400); *c* (×500); *d* (×250); *e* (×100).

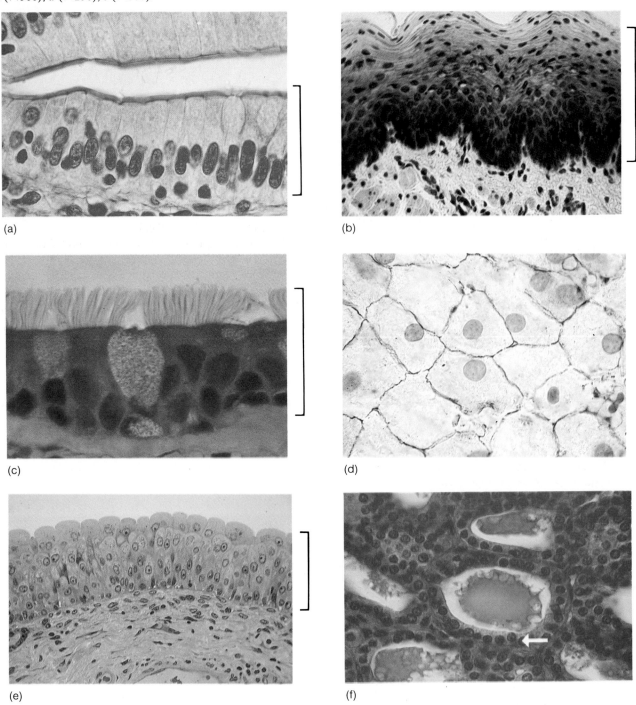

(a)

(b)

(c)

(d)

(e)

(f)

Part C

Identify the epithelial tissues in the micrographs shown in figure 6.1.

a. _Simple columnar_

b. _stratified squamous_

c. _pseudostratified columnar_

d. _simple squamous_

e. _transitional_

f. _simple cuboidal_

Laboratory Exercise 7

Connective Tissues

Connective tissues contain a variety of cell types and occur in all parts of the body. They bind structures together, provide support and protection, fill spaces, store fat, and produce blood cells.

Purpose of the Exercise

To review the characteristics of connective tissue and to observe examples of the major types of this tissue.

Learning Objectives

After completing this exercise, you should be able to:

1. Describe the general characteristics of connective tissue.
2. List the major types of connective tissue.
3. Describe the special characteristics of each of the major types of connective tissue.
4. Identify the major types of connective tissue on microscope slides.

Materials Needed:
   compound microscope
   prepared slides of the following:
      loose connective tissue
      adipose tissue
      fibrous connective tissue
      elastic connective tissue
      reticular connective tissue
      hyaline cartilage
      elastic cartilage
      fibrocartilage
      bone

Procedure

1. Review the section entitled "Connective Tissues" in chapter 5 of the textbook.
2. Complete Part A of Laboratory Report 7 on page 44.
3. Use a microscope to observe the prepared slides of various connective tissues. As you observe each tissue, look for the special features described in the textbook that distinguish it from the others.
4. Complete Part B of the laboratory report.
5. Test your ability to recognize each of these connective tissues by having a laboratory partner select a slide, cover its label, and focus the microscope on this tissue. Then, see if you identify the tissue correctly.
6. Complete Part C of the laboratory report.

Name _____

Date _____

Section _____

Connective Tissues

Part A

Match the tissues in column A with the characteristics in column B. Place the letter of your choice in the space provided.

Column A		Column B
A. adipose tissue	_E_ 1.	forms framework of ear
B. bone	_A_ 2.	functions as heat insulator
C. hyaline cartilage	_H_ 3.	contains jellylike intercellular material
D. elastic connective tissue	_B_ 4.	cells arranged around osteonic canal
E. elastic cartilage	_A_ 5.	commonly called fat
F. fibrocartilage	_H_ 6.	binds skin to underlying organs
G. fibrous connective tissue	_G_ 7.	main tissue of tendons and ligaments
H. loose connective tissue	_A_ 8.	provides stored energy supply
I. reticular connective tissue	_C_ 9.	forms soft part of the nose
	_F_ 10.	forms pads between adjacent sections of backbone
	_B_ 11.	cellular processes that extend outward through canaliculi
	_C_ 12.	forms supporting rings of trachea
	_A_ 13.	cells greatly swollen with nuclei pushed to sides
	_G_ 14.	forms protective layer of eye
	_B_ 15.	matrix contains collagen and mineral salts
	_D_ 16.	occurs in attachments between vertebrae
	_I_ 17.	forms supporting tissue in walls of liver and spleen

Part B

In the space that follows, sketch a few cells from each of the types of connective tissues you observed. For each sketch, identify the tissue, label the major parts of the cells, and indicate the magnification used.

cell mem · nucleus · Fat · cytoplasm *Adipose* ~~Loose connective~~ tissue (43 X)	*Fibroblast · Nucleus · Elastic fiber · Collagenous Fiber · ground substance* (*areolar tissue*) ~~Adipose~~ tissue (43 X) *Loose Connective*
Nucleus *Collagenous Fibers* *nucleolus* *Fibroblast · nucleus* Fibrous connective tissue (43 X)	*Fibroblasts · Elastic Fibers* *Collagenous Fibers* *Aorta* Elastic connective tissue (43 X)
Fibroblast · Reticular Fiber Reticular connective tissue (43 X)	*Lacuna · Intercellular material · Chondrocyte · nucleus* Hyaline cartilage (43 X)

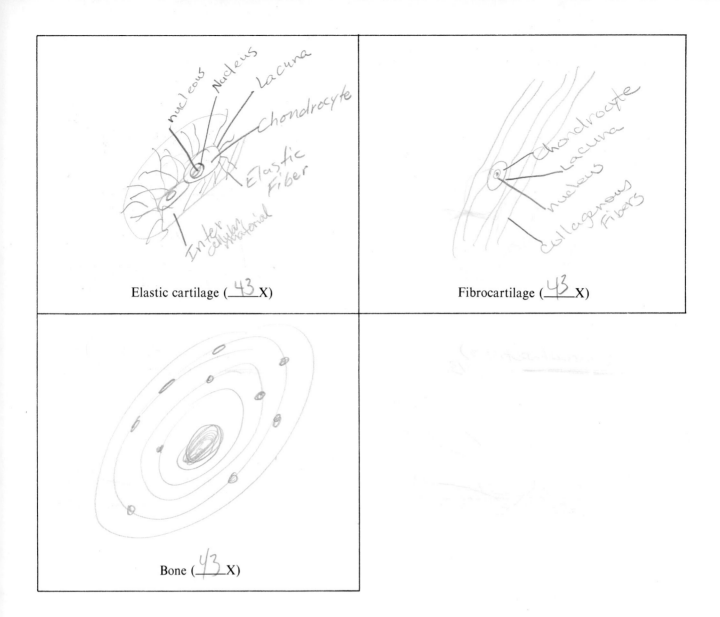

Elastic cartilage (__43__X)

Fibrocartilage (__43__X)

Bone (__43__X)

Part C

Identify the connective tissues in micrographs a–i shown in figure 7.1.

a. ___Hyaline cartilage___

b. ___Loose connective___

c. ___Elastic cartilage___

d. ___Fibrocartilage___

e. ___bone tissue___

f. ___adipose tissue___

g. ___Fibrous connective tissue___

h. ___Riticular connective tissue___

i. ___Elastic connective tissue___

Figure 7.1 Identify each of the connective tissues shown in these micrographs (*a–i*). Magnifications: *a* (×250); *c* (×100); *f* (×250); *h* (×110); *i* (×100).

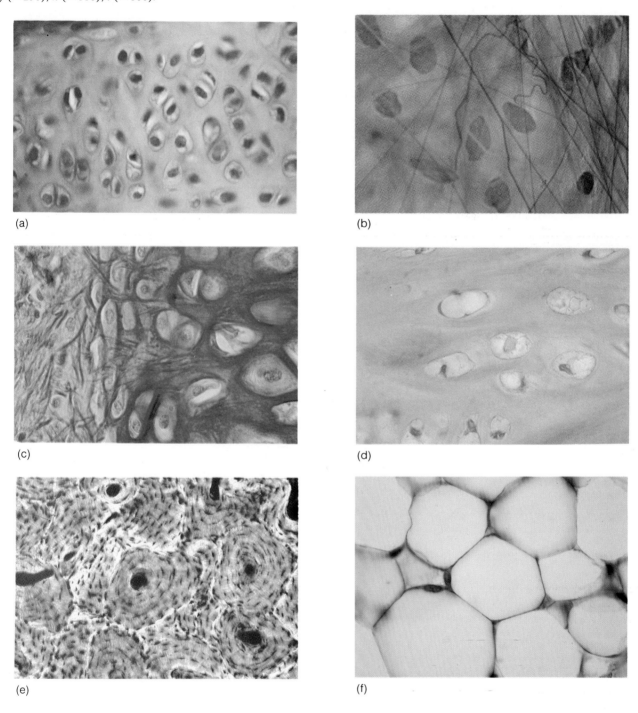

(a)

(b)

(c)

(d)

(e)

(f)

Figure 7.1– *Continued*

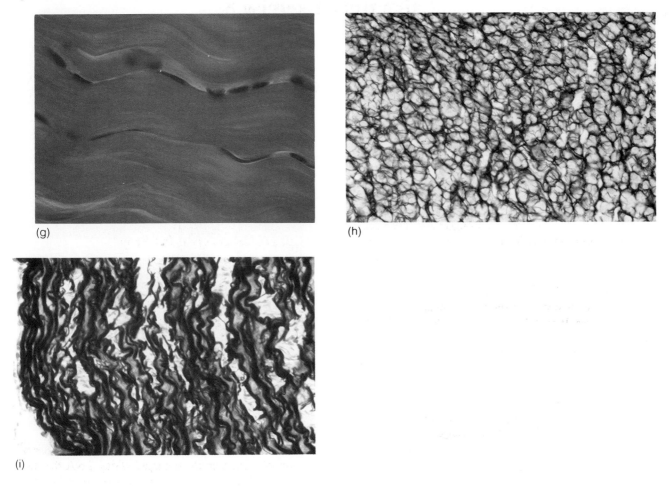

(g)

(h)

(i)

Laboratory Exercise 8

Muscle and Nerve Tissues

Muscle tissues are characterized by the presence of elongated cells or muscle fibers that can contract. As they shorten, these fibers pull at their attached ends and cause body parts to move. The three types of muscle tissues are skeletal muscle, smooth muscle, and cardiac muscle.

Nerve tissue occurs in the brain, spinal cord, and associated nerves. It contains nerve cells, which are the basic units of the nervous system, and neuroglial cells, which perform supportive and protective functions.

Purpose of the Exercise

To review the characteristics of muscle and nerve tissues, and to observe examples of these tissues.

Learning Objectives

After completing this exercise, you should be able to:

1. Describe the general characteristics of muscle tissue.
2. List the three types of muscle tissue.
3. Describe the special characteristics of each type of muscle tissue.
4. Identify examples of muscle tissues.
5. Describe the general characteristics of nerve tissue.
6. Identify nerve tissue.

Materials Needed:
compound microscope
prepared slides of the following:
    skeletal muscle tissue
    smooth muscle tissue
    cardiac muscle tissue
    nerve tissue (spinal cord smear and/or
       cerebellum, Golgi)

Procedure

1. Review the sections entitled "Muscle Tissues" and "Nerve Tissue" in chapter 5 of the textbook.
2. Complete Part A of Laboratory Report 8 on page 50.
3. Using the microscope, observe each of the types of muscle tissue on the prepared slides. Look for the special features of each type, as described in the textbook.
4. Observe the prepared slide of nerve tissue, and identify nerve cells, nerve fibers, and neuroglial cells.
5. Complete Parts B and C of the laboratory report.

laboratory report 8

Name _____

Date _____

Section _____

Muscle and Nerve Tissues

Part A

Complete the following statements:

1. Because skeletal muscles can be controlled by conscious effort, they are said to be _voluntary_.

2. The cross-markings on skeletal muscle fibers are called _striations_.

3. Cells of _skeletal_ muscle tissue have many nuclei.

4. Muscles are sometimes paralyzed as a result of damage to the _nerve_ that are associated with them.

5. _smooth_ muscle tissue is found in the walls of hollow internal organs.

6. Muscle tissue that cannot be controlled by conscious effort is called _invol_.

7. _smooth_ muscle tissue is responsible for the movement of food through the digestive tube.

8. Intercalated disks are found between the cells of _cardiac_ muscle tissue.

9. The fibers of skeletal and _cardiac_ muscle tissue are striated.

10. _cardiac_ muscle tissue makes up the bulk of the heart.

11. Organs that consist mainly of nerve tissue include the _brain spinal cord & nerves_.

12. Nerve cells also are called _neurons_.

13. Nerve cells are sensitive to _changes to surrounding (receptors)_

14. The general functions of nerve cells include _transmitting info from place to place (coordinate)_

15. The general functions of neuroglial cells include _connective tis func. support to nerve to neuron. Bind tissue together. provide clean up,_

50

Part B

In the space that follows, sketch a few cells or fibers of each of the three types of muscle tissue and of nerve tissue as they appear through the microscope. Identify each tissue sketched, label the major parts of the cells or fibers, and indicate the magnification used.

Skeletal muscle tissue (____X)

Smooth muscle tissue (_43_X)

Cardiac muscle tissue (____X)

Nerve tissue (_43_X)

Figure 8.1 Identify the tissues illustrated by these micrographs (*a–d*). Magnifications: *a* (×250); *b* (×450); *c* (×400); *d* (×250).

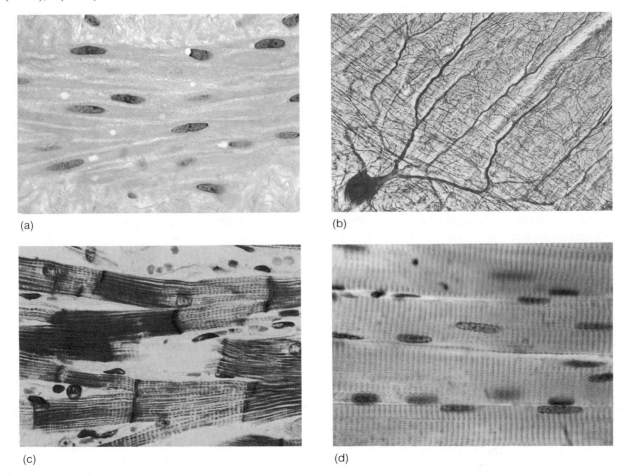

(a)

(b)

(c)

(d)

Part C

Identify each type of tissue illustrated in figure 8.1.

a. _____ smooth muscle

b. _____ nerve (neuron)

c. _____ cardiac (intercalary disc)

d. _____ skeletal (striated)

Integumentary System

The integumentary system includes the skin, hair, nails, sebaceous glands, and sweat glands. These organs provide a protective covering for deeper tissues, aid in regulating body temperature, retard water loss, house sensory receptors, synthesize various chemicals, and excrete small quantities of wastes.

Purpose of the Exercise

To observe the organs and tissues of the integumentary system and to review the functions of these parts.

Learning Objectives

After completing this exercise, you should be able to:

1. Name the organs of the integumentary system.
2. Describe the major functions of these organs.
3. Distinguish between epidermis, dermis, and the subcutaneous layer.
4. Identify the layers of the skin, a hair follicle, an arrector pili muscle, a sebaceous gland, and a sweat gland on a microscope slide or diagram.

Materials Needed:

hand magnifier or dissecting microscope
compound microscope
prepared microscope slide of human scalp or axilla
forceps
microscope slide and coverslip

For Optional Activity:

prepared slide of heavily pigmented human skin

Procedure

1. Review the sections entitled "Skin and Its Tissues" and "Accessory Organs of the Skin" in chapter 6 of the textbook.
2. As a review activity, label figures 9.1, 9.2, and 9.3.
3. Complete Part A of Laboratory Report 9 on page 57.

4. Use the hand magnifier or dissecting microscope and proceed as follows:
   a. Observe the skin, hair, and nails on your hand.
   b. Compare the type and distribution of hairs on the front and back of your lower arm.
   c. Pull out a single hair with forceps, and mount it on a microscope slide under a coverslip.
   d. Observe the root and shaft of the hair, and note the scalelike parts that make up the shaft.
5. Complete Part B of the laboratory report.
6. Use low-power magnification of the compound microscope and proceed as follows:
   a. Observe the prepared slide of human scalp or axilla.
   b. Locate the epidermis, dermis, and subcutaneous layer, a hair follicle, an arrector pili muscle, a sebaceous gland, and a sweat gland.
   c. Focus on the epidermis with high power, and locate the stratum corneum and stratum basale. Note how the shapes of the cells in these two layers differ.
   d. Observe the dense, irregular connective tissue that makes up the bulk of the dermis.
7. Complete Part C of the laboratory report.
8. Using low-power magnification, locate a hair follicle that has been sectioned longitudinally through its bulblike base, a sebaceous gland close to the follicle, and a sweat gland. Observe the detailed structure of these parts with high-power magnification.
9. Complete Parts D and E of the laboratory report.

Optional Activity

Observe the prepared slide of heavily pigmented human skin with low-power magnification. Note that the pigment is most abundant in the epidermis. Focus on this region with the high-power objective. The pigment-producing cells or melanocytes are located among the deeper layers of epidermal cells and the underlying connective tissue cells of the dermis. Differences in skin color are due primarily to the amount of pigment (melanin) produced by these cells. The number of melanocytes in the skin is about the same for members of all racial groups.

Figure 9.1 Label this section of skin.

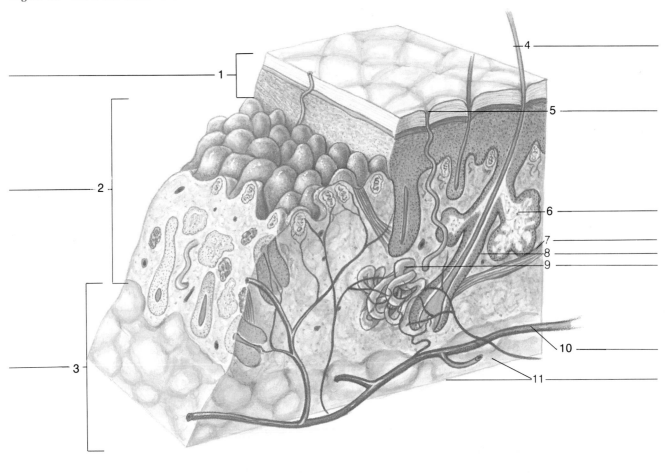

Figure 9.2 Label the epidermal layers in this section of skin.

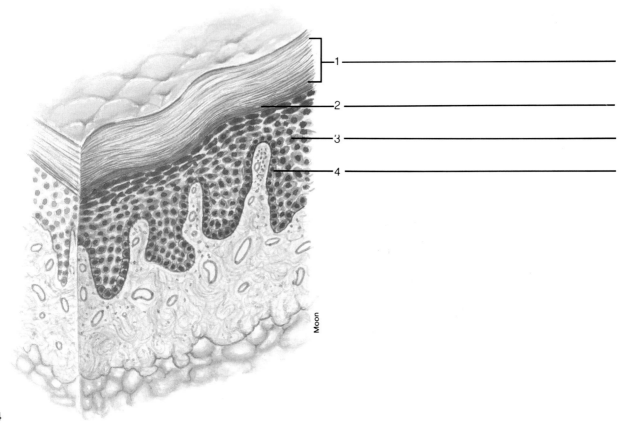

Moon

Figure 9.3   Label the features associated with this hair
follicle.

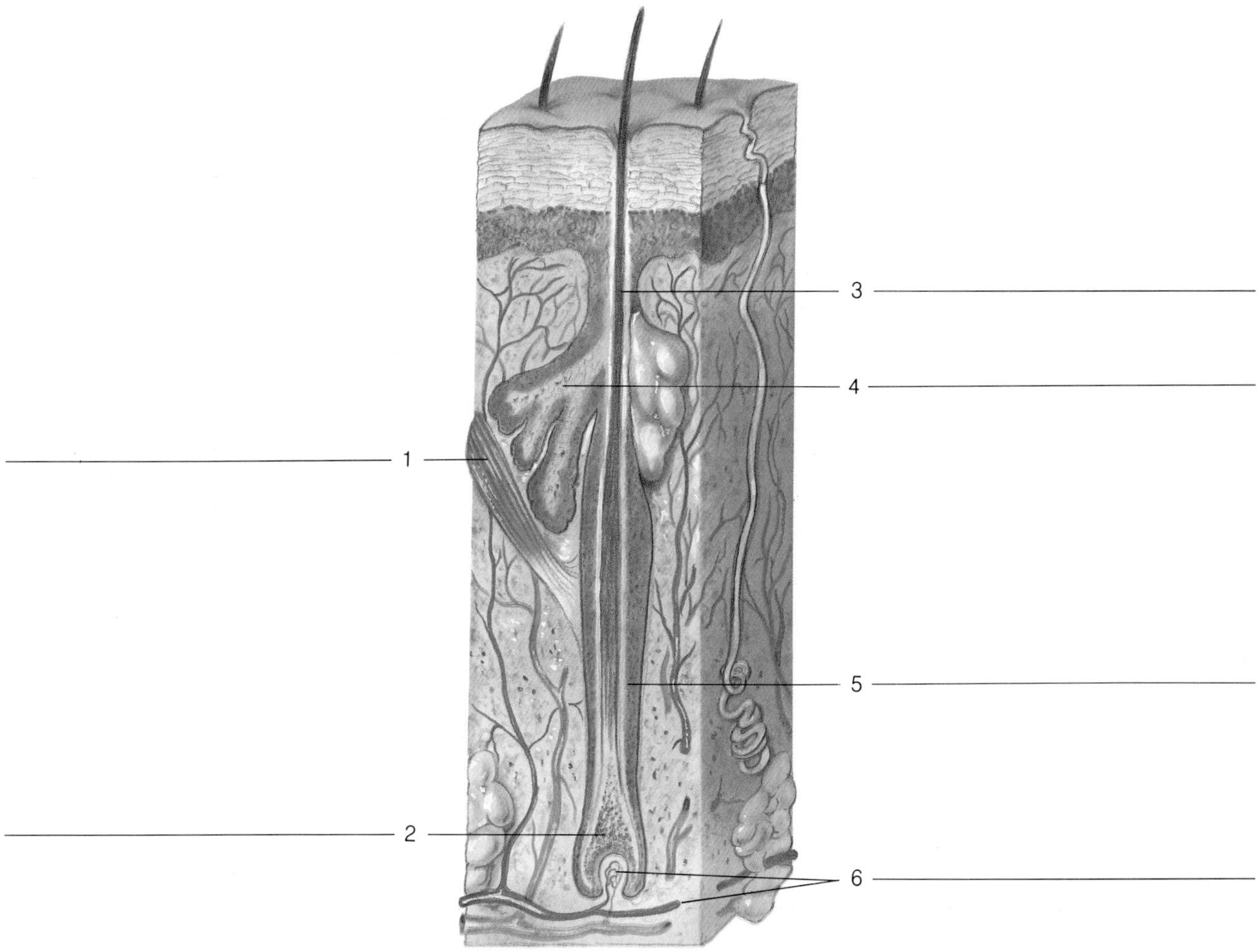

1 _____

2 _____

3 _____

4 _____

5 _____

6 _____

laboratory report 9

Name _____

Date _____

Section _____

Integumentary System

Part A

Match the structures in column A with the descriptions and functions in column B. Place the letter of your choice in the space provided.

Column A	Column B
A. apocrine gland	___ 1. an oily secretion
B. arrector pili muscle	___ 2. outermost layer of epidermis
C. dermis	___ 3. development stimulated by sex hormones
D. eccrine gland	___ 4. epidermal pigment
E. epidermis	___ 5. inner layer of skin
F. hair follicle	___ 6. responds to elevated body temperature
G. keratin	___ 7. pigment-producing cell
H. keratinization	___ 8. outer layer of skin
I. melanin	___ 9. gland usually associated with hair follicle
J. melanocyte	___ 10. hard protein of nails
K. sebum	___ 11. binds skin to underlying organs
L. sebaceous gland	___ 12. reproductive layer of epidermis
M. stratum corneum	___ 13. process of hardening epidermal cells
N. stratum basale	___ 14. tubelike part that forms hair
O. subcutaneous layer	___ 15. causes hair to stand on end

Part B

Complete the following:

1. How does the skin of your palm differ from that on the back of your hand? _____

_____

_____

2. Describe the differences you observed in the type and distribution of hair on the front and back of your lower arm. _____

_____

3. Explain how the scalelike parts of a hair are formed. _____

_____

4. What is the source of pigment in hair? _____

_____

Part C

Complete the following:

1. Distinguish between epidermis, dermis, and subcutaneous layer. _____
   _____
   _____

2. How do the cells of stratum corneum and stratum basale differ? _____
   _____
   _____

3. Explain these differences (from question 2). _____
   _____
   _____

4. What special qualities does the connective tissue of the dermis have? _____
   _____
   _____

Part D

Complete the following:

1. What kind of tissue forms the inner lining of a hair follicle? _____

2. In which layer of skin are follicles usually found? _____

3. In what way are sebaceous glands associated with hair follicles? _____
   _____

4. What kinds of cells make up sebaceous glands? _____
   _____

5. In which layer of skin are sweat glands usually found? _____

Part E

Identify the numbered features in the micrographs of figure 9.4.

1. _____   8. _____

2. _____   9. _____

3. _____   10. _____

4. _____   11. _____

5. _____   12. _____

6. _____   13. _____

7. _____

Figure 9.4 Identify the features indicated in these micrographs of human skin (*a–d*). Magnifications: *b* (×50); *d* (×175).

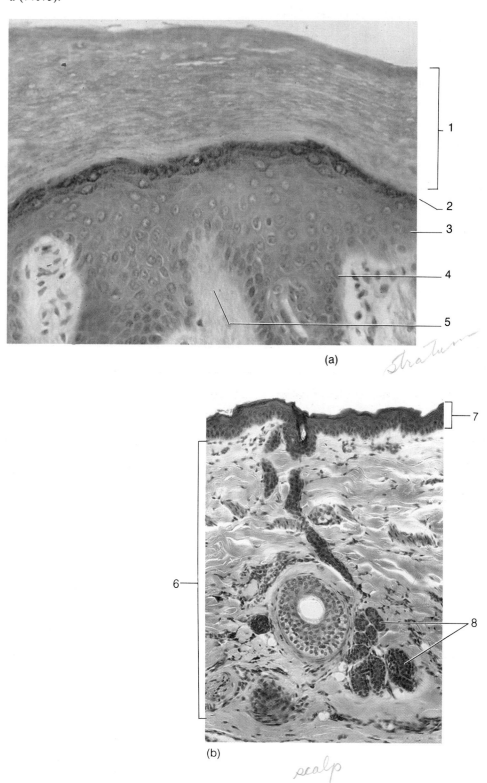

(a)

Stratum

(b)

scalp

Figure 9.4—*Continued*

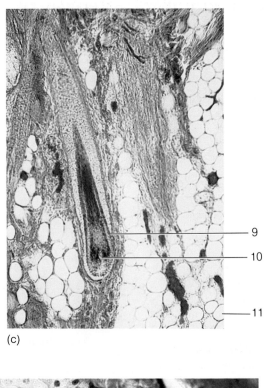

(c)

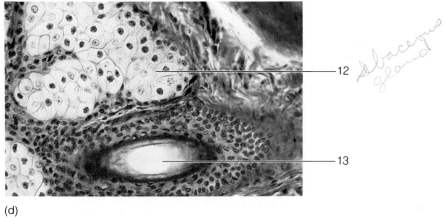

sebaceous gland

(d)

Laboratory Exercise 10

Structure and Classification of Bone

A bone represents an organ of the skeletal system. As such, it is composed of a variety of tissues including bone tissue, cartilage, fibrous connective tissue, blood, and nerve tissue.

Although various bones of the skeleton vary greatly in size and shape, they have much in common structurally and functionally.

Purpose of the Exercise

To review the way bones are classified and to examine the macroscopic structure of a long bone.

Learning Objectives

After completing this exercise, you should be able to:

1. Name four groups of bones based on their shapes, and give an example for each group.
2. Locate and name the macroscopic parts of a long bone.
3. Describe the functions of various parts of a bone.
4. Distinguish between compact and spongy bone.

Materials Needed:
   prepared microscope slide of ground compact bone
   human bone specimens including long, short, flat, and irregular types
   human long bone, sectioned longitudinally
   fresh bones, sectioned longitudinally and transversely
   stereoscopic (dissecting) microscope

For Demonstration:

   fresh chicken bones
   dilute hydrochloric acid

Procedure

1. Review the section entitled "Bone" in chapter 5 of the textbook.
2. Reexamine the microscopic structure of bone tissue by observing a prepared microscope slide of ground compact bone. Use the figures of bone tissue in chapter 5 of the textbook to locate the following features:

   lamella   osteonic canal

   osteocyte   canaliculus

   lacuna   osteon

3. Review the section entitled "Bone Structure" in chapter 7 of the textbook.
4. As a review activity, label figures 10.1 and 10.2.
5. Observe the individual bone specimens and arrange them into groups, according to shape.
6. Complete Part A of Laboratory Report 10 on page 64.
7. Examine the sectioned bones and locate the following:

   epiphysis

   articular cartilage

   diaphysis

   periosteum

   compact bone

   spongy bone

   trabeculae

   medullary cavity

   endosteum

   marrow

61

Figure 10.1 Label the major parts of this long bone.

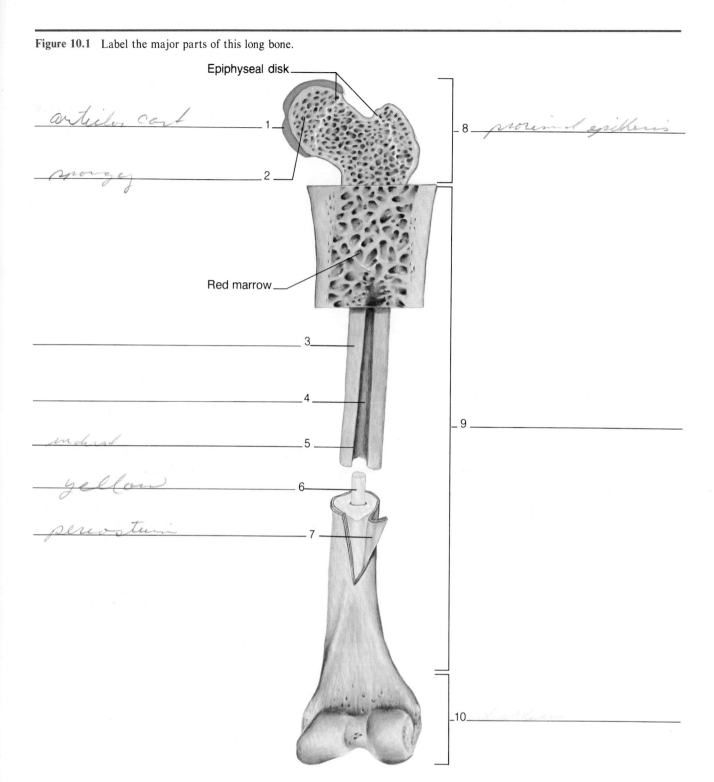

Epiphyseal disk

1 _articular cart_

2 _spongy_

Red marrow

3

4

5 _indist_

6 _yellow_

7 _periosteum_

8 _proximal epiphysis_

9

10

Figure 10.2 Label the features associated with the microscopic structure of bone.

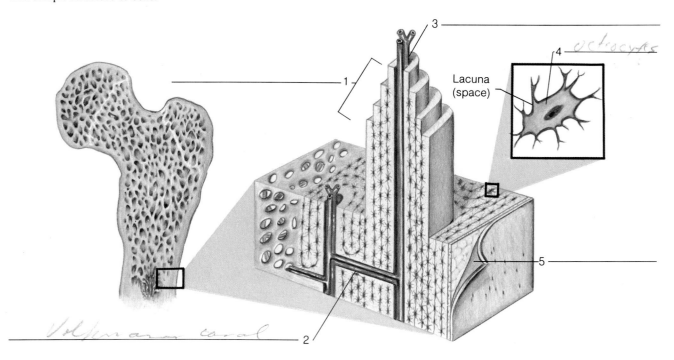

osteocytes

Lacuna (space)

Volkmann canal

8. Use the dissecting microscope to observe the compact bone and spongy bone of the sectioned specimens. Also examine the marrow in the medullary cavity and the spaces within the spongy bone of the fresh specimen.

9. Complete Part B of the laboratory report.

Demonstration

Examine a piece of fresh chicken bone and a chicken bone that has been exposed overnight to dilute hydrochloric acid. This acid treatment removes the inorganic salts from the bone matrix. Rinse the bones in water and note the texture and flexibility of each. Based on your observations, what quality of the fresh bone seems to be due to the inorganic salts that were removed by the acid treatment?

_____

_____

Examine the specimen of chicken bone that has been exposed to high temperature. This treatment removes the protein and other organic substances from the bone matrix. What quality of the fresh bone seems to be due to these organic materials?

_____

_____

Name _____

Date _____

Section _____

Structure and Classification of Bone

Part A

Complete the following statements:

1. A bone that is platelike is classified as a(an) _flat_ bone.

2. The bones of the wrist are examples of _short_ bones.

3. The bone of the upper arm is an example of a(an) _long_ bone.

4. Vertebrae of the backbone are examples of _irregular_ bones.

5. The kneecap is an example of a(an) _sesamoid round_ bone.

6. The bones that form a protective covering around the back and sides of the brain are examples of _flat_ bones.

7. Distinguish between the epiphysis and the diaphysis of a long bone. _____
_____

8. Describe where cartilage is found in a long bone. _____
_____

9. Describe where fibrous connective tissue is found in a long bone. _____
_____

10. Distinguish between the periosteum and the endosteum. _____
_____
_____

Part B

Complete the following:

1. What differences did you note between the macroscopic structure of compact bone and spongy bone? _____
Compact - solid   osteon   stronger
spongy - hollow spaces   peticular plates   lighter in weight

2. How are these differences related to the functions of these types of bone? _____
_____
_____

3. From your observations, how does the marrow in the medullary cavity compare with the marrow in the spaces of the spongy bone? _yellow_
red marrow ends of bones
_____

Organization of the Skeleton

The skeleton can be divided into two major portions: (1) the axial skeleton, which consists of the bony and cartilaginous parts that support and protect the organs of the head, neck, and trunk, and (2) the appendicular skeleton, which consists of the bones of the limbs and those that anchor the limbs to the axial skeleton.

Purpose of the Exercise

To review the organization of the skeleton, the major bones of the skeleton, and the terms used to describe skeletal structures.

Learning Objectives

After completing this exercise, you should be able to:

1. Distinguish between the axial skeleton and the appendicular skeleton.
2. Locate and name the major bones of the human skeleton.
3. Define the terms used to describe skeletal structures and locate examples of such structures on the human skeleton.

Materials Needed:
  articulated human skeleton

For Demonstration:
  X-ray films of skeletal structures

Procedure

1. Review the section entitled "Organization of the Skeleton" in chapter 7 of the textbook. (Note that pronunciations of the names for major skeletal structures are included within the narrative of chapter 7.)

2. As a review activity, label figure 11.1.
3. Examine the human skeleton, and locate the following parts. Palpate as many of the corresponding bones in your own skeleton as possible.

axial skeleton

  skull

    cranial bones

    facial bones

  hyoid bone

  vertebral column

    vertebrae

    intervertebral disks

    sacrum

    coccyx

  thoracic cage

    ribs

    sternum

appendicular skeleton

  pectoral girdle

    scapula

    clavicle

  upper limbs

    humerus

    radius

    ulna

    carpals

    metacarpals

    phalanges

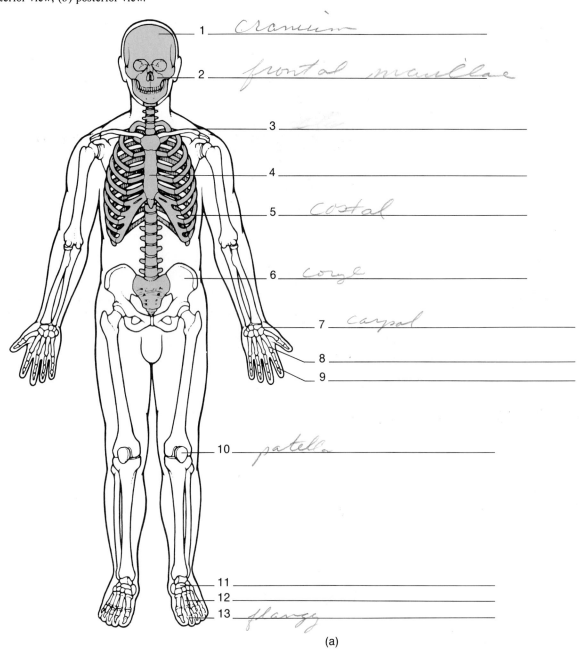

1 _cranium_

2 _frontal maxillae_

3 _____

4 _____

5 _costal_

6 _coxal_

7 _carpal_

8 _____

9 _____

10 _patella_

11 _____

12 _____

13 _flangs_

(a)

Figure 11.1—*Continued*

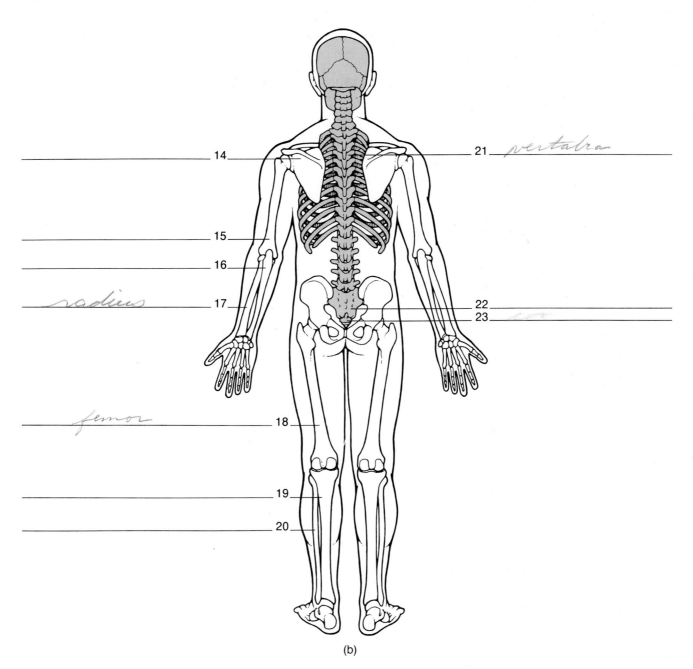

14 _____

15 _____

16 _____

17 _____ *radius*

femor _____ 18 _____

19 _____

20 _____

21 _____ *vertabra*

22 _____

23 _____ *a*

(b)

pelvic girdle

   coxal bone

  lower limbs

   femur

   tibia

   fibula

   patella

   tarsals

   metatarsals

   phalanges

4. Study chart 7.4 in chapter 7 of the textbook. Locate an example of each of the following features in the human skeleton:

condyle

crest

epicondyle

facet

fissure

fontanel

foramen

fossa

fovea

head

linea

meatus

process

ramus

sinus

spine

suture

trochanter

tubercle

tuberosity

5. Complete Parts A, B, and C of Laboratory Report 11 on pages 69 and 70.

Demonstration

Images on X-ray films are produced by allowing X ray from an X-ray tube to pass through a body part and expose photographic film that is positioned on the opposite side of the part. The image that appears on the film after it is developed reveals the presence of parts with different densities. Bone, for example, is very dense tissue and is a good absorber of X ray. Thus, bone generally appears light on the film. Air-filled spaces, on the other hand, absorb almost no X ray, and they appear as dark areas on the film. Liquids and soft tissues absorb intermediate quantities of X ray, so they usually appear in various shades of gray.

Examine the available X-ray films of skeletal structures by holding each film in front of a light source. Identify as many of the bony parts as you can.

laboratory report 11

Name _____

Date _____

Section _____

Organization of the Skeleton

Part A

Complete the following statements:

1. The extra bones that sometimes develop between the flat bones of the skull are called _____ .

2. Sesamoid bones occurring in some tendons function to _crushed or worn away by friction_

3. _cranium_ is another term for braincase.

4. The _hyoid_ bone supports the tongue.

5. The _coccyx stem_ at the distal end of the vertebral column is composed of several fused vertebrae.

6. The ribs are attached to the _sternum_ anteriorly.

7. The thoracic cage is composed of _____ pairs of ribs.

8. The scapulae and clavicles together form the _pectoral girdle_

9. The humerus, radius, and _____ articulate to form the elbow joint.

10. The wrist is composed of eight bones called _____ .

11. The coxal bones are attached posteriorly to the _sacrum_ .

12. The coxal bones, sacrum, and coccyx together form the _pelvis_ .

13. The _patella_ covers the anterior surface of the knee.

14. The bones that articulate with the distal ends of the tibia and fibula are called _tarsal_ .

15. All finger and toe bones are called _phalanges_ .

Part B

Match the terms in column A with the definitions in column B. Place the letter of your choice in the space provided.

Column A		Column B
A. condyle	_C_ 1.	nearly flat surface
B. crest	_F_ 2.	deep depression
C. facet	_A_ 3.	rounded process
D. fontanel	_E_ 4.	opening or passageway
E. foramen	_G_ 5.	line of union
F. fossa	_B_ 6.	narrow, ridgelike projection
G. suture	_B_ 7.	soft region between bones of skull

Part C

Match the terms in column A with the definitions in column B. Place the letter of your choice in the space provided.

Column A			**Column B**
A.	fovea	C 1.	tubelike passageway
B.	head	A 2.	tiny pit or depression
C.	meatus	G 3.	small, knoblike process
D.	sinus	E 4.	thornlike projection
E.	spine	B 5.	enlargement at end of bone
F.	trochanter	D 6.	hollow space within bone
G.	tubercle	F 7.	relatively large process

The Skull

A human skull consists of twenty-two bones that, except for the lower jaw, are firmly interlocked along sutures. Eight of these immovable bones make up the braincase, or cranium, and thirteen more immovable bones form the facial skeleton.

Purpose of the Exercise

To examine the structure of the human skull and to identify the bones and major features of the skull.

Learning Objectives

After completing this exercise, you should be able to:

1. Distinguish between the cranium and the facial skeleton.
2. Locate and name the bones of the skull and their major features.
3. Locate and name the sutures of the cranium.
4. Locate and name the sinuses of the skull.

Materials Needed:
    human skull, articulated
    human skull, disarticulated
    human skull, sagittal section

For Demonstration:

    fetal skull

Procedure

1. Review the section entitled "Skull" in chapter 7 of the textbook.
2. As a review activity, label figures 12.1, 12.2, 12.3, 12.4, and 12.5.
3. Examine the cranial bones of the articulated human skull and the sectioned skull. Also observe the corresponding disarticulated bones. Locate the following bones and features in the laboratory specimens and at the same time palpate as many of these bones and features in your own skull as possible.

frontal bone
    supraorbital foramen (or notch)
    frontal sinuses
parietal bones
    sagittal suture
    coronal suture
occipital bone
    lambdoidal suture
    foramen magnum
    occipital condyles
temporal bones
    squamosal suture
    external auditory meatus
    mandibular fossae
    mastoid process
    styloid process
    carotid canal
    jugular foramen
    zygomatic process
sphenoid bone
    sella turcica
    sphenoidal sinuses
ethmoid bone
    cribriform plates
    perpendicular plate
    superior nasal concha
    middle nasal concha
    ethmoidal sinuses
    crista galli

Figure 12.1 Label the anterior features of the skull.

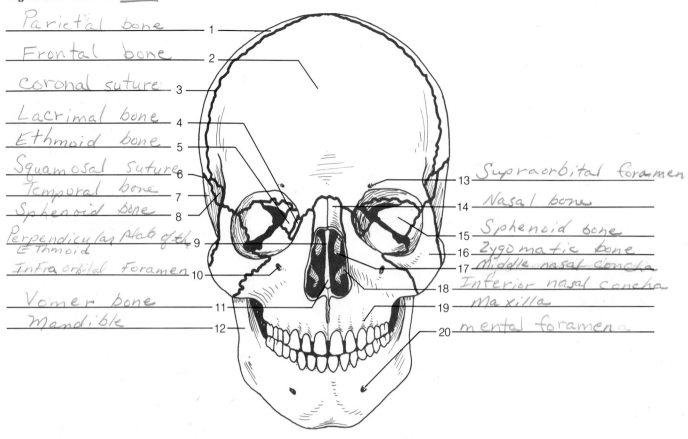

Parietal bone — 1
Frontal bone — 2
Coronal suture — 3
Lacrimal bone — 4
Ethmoid bone — 5
Squamosal suture — 6
Temporal bone — 7
Sphenoid bone — 8
Perpendicular plate of the Ethmoid — 9
Infraorbital foramen — 10
Vomer bone — 11
Mandible — 12

13 — Supraorbital foramen
14 — Nasal bone
15 — Sphenoid bone
16 — Zygomatic bone
17 — Middle nasal concha
18 — Inferior nasal concha
19 — Maxilla
20 — mental foramen

4. Complete Parts A and B of Laboratory Report 12 on page 77.
5. Examine the facial bones of the articulated and sectioned skulls and the corresponding disarticulated bones. Locate the following:

maxillary bones

   maxillary sinuses

   palatine processes

   alveolar process

   alveolar arch

palatine bones

zygomatic bones

   temporal process

   zygomatic arch

lacrimal bones

nasal bones

vomer bone

inferior nasal conchae

mandible

   ramus

   mandibular condyle

   coronoid process

   alveolar arch

   mandibular foramen

   mental foramen

6. Complete Parts C and D of the laboratory report.

Optional Activity

Use colored pencils to differentiate the bones illustrated in figures 12.1–12.5. Select a different color for each individual bone in the series. This activity should help you locate various bones that are shown in different views in the figures. You can check your work by referring to the corresponding figures in the textbook, which are presented in full color.

Figure 12.2 Label the lateral features of the skull.

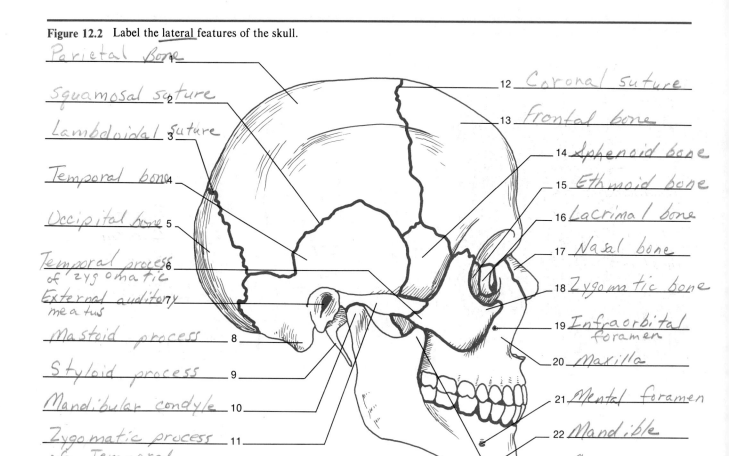

1 Parietal Bone
2 Squamosal suture
3 Lambdoidal suture
4 Temporal bone
5 Occipital bone
6 Temporal process of zygomatic
7 External auditory meatus
8 Mastoid process
9 Styloid process
10 Mandibular condyle
11 Zygomatic process of Temporal

12 Coronal suture
13 Frontal bone
14 Sphenoid bone
15 Ethmoid bone
16 Lacrimal bone
17 Nasal bone
18 Zygomatic bone
19 Infraorbital foramen
20 Maxilla
21 Mental foramen
22 Mandible
23 Coronoid process

7. Study chart 7.8 of chapter 7 in the textbook. Locate the following features of the human skull:

carotid canal

condylar canal

foramen lacerum

foramen magnum

foramen ovale

foramen rotundum

foramen spinosum

greater palatine foramen

hypoglossal canal

incisive foramen

inferior orbital fissure

infraorbital foramen

internal acoustic meatus

jugular foramen

mandibular foramen

mental foramen

optic canal

stylomastoid foramen

superior orbital fissure

supraorbital foramen

8. Complete Parts E and F of the laboratory report.

Demonstration

Examine the fetal skull. Note that the skull is incompletely developed and that the cranial bones are separated by fibrous membranes. These membranous areas are called fontanels or "soft spots." The fontanels close as the cranial bones grow together. The posterior fontanel usually closes within a few months after birth, while the anterior one may not close until the middle or end of the second year. What other features characterize the fetal skull? _____

Figure 12.3 Label the inferior features of the skull.

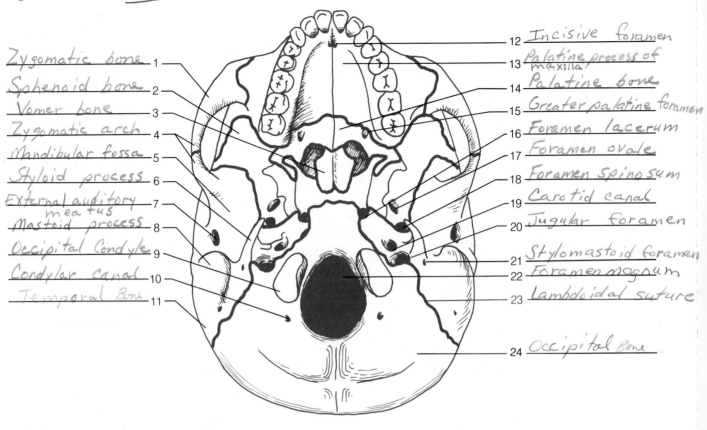

Zygomatic bone —1

Sphenoid bone —2

Vomer bone —3

Zygomatic arch —4

Mandibular fossa —5

Styloid process —6

External auditory
meatus —7

Mastoid process —8

Occipital Condyle —9

Condylar canal —10

Temporal Bone —11

12— Incisive foramen

13— Palatine process of
maxilla

14— Palatine bone

15— Greater palatine foramen

16— Foramen lacerum

17— Foramen ovale

18— Foramen spinosum

19— Carotid canal

20— Jugular foramen

21— Stylomastoid foramen

22— Foramen magnum

23— Lambdoidal suture

24— Occipital Bone

Figure 12.4 Label the features of the floor of the cranial cavity.

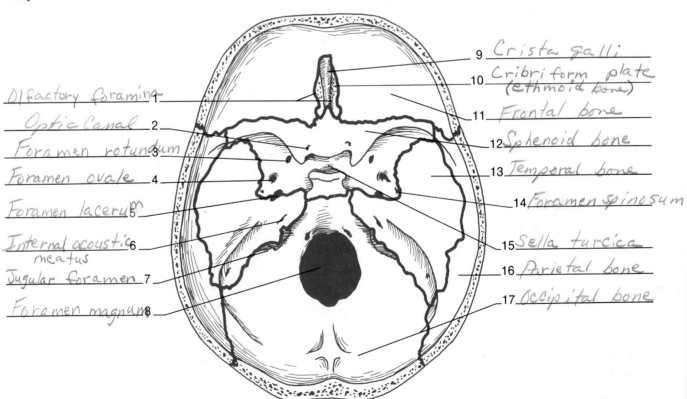

Olfactory foramina —1

Optic Canal —2

Foramen rotundum —3

Foramen ovale —4

Foramen lacerum —5

Internal acoustic
meatus —6

Jugular foramen —7

Foramen magnum —8

9— Crista galli

10— Cribriform plate
(ethmoid bone)

11— Frontal bone

12— Sphenoid bone

13— Temporal bone

14— Foramen spinosum

15— Sella turcica

16— Parietal bone

17— Occipital bone

Figure 12.5 Label the features of the <u>sagittal</u> section of the skull.

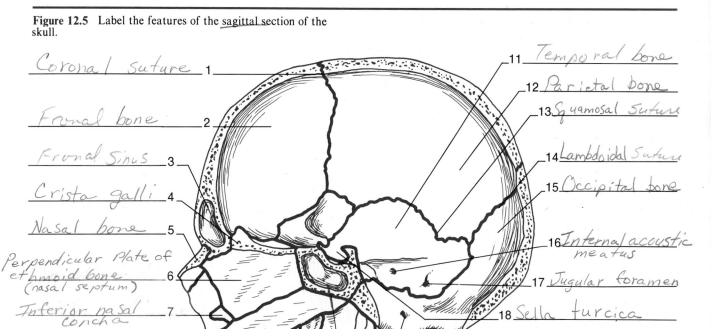

Coronal suture _____ 1

Frontal bone _____ 2

Frontal Sinus _____ 3

Crista galli _____ 4

Nasal bone _____ 5

Perpendicular Plate of ethmoid bone (nasal septum) _____ 6

Inferior nasal concha _____ 7

Maxilla _____ 8

Palatine process of maxilla _____ 9

mandible _____ 10

11 _____ Temporal bone

12 _____ Parietal bone

13 _____ Squamosal suture

14 _____ Lambdoidal suture

15 _____ Occipital bone

16 _____ Internal acoustic meatus

17 _____ Jugular foramen

18 _____ Sella turcica

19 _____ Foramen magnum

20 _____ Hypoglossal canal

21 _____ Mastoid process

22 _____ Styloid process

23 _____ Sphenoidal sinus

24 _____ Palatine bone

25 _____ Vomer bone

26 _____ Alveolar arch

laboratory report 12

Name _____

Date _____

Section _____

The Skull

Part A

Match the bones in column A with the features in column B. Place the letter of your choice in the space provided.

Column A		Column B
A. ethmoid bone	_A_ 1.	perpendicular plate
B. frontal bone	_A_ 2.	cribriform plate
C. occipital bone	_A_ 3.	crista galli
D. parietal bone	_F_ 4.	external auditory meatus
E. sphenoid bone	_C_ 5.	foramen magnum
F. temporal bone	_F_ 6.	mandibular fossa
	_F_ 7.	mastoid process
	_A_ 8.	middle nasal concha
	_C_ 9.	occipital condyle
	_E_ 10.	sella turcica
	_F_ 11.	styloid process
	_B_ 12.	supraorbital foramen

Part B

Complete the following statements:

1. The _____Coronal_____ suture joins the frontal bone to the parietal bone.

2. The parietal bones are fused along the midline by the _____Sagittal_____ suture.

3. The _____Lambdoidal_____ suture joins the parietal bones to the occipital bone.

4. The temporal bones are joined to the parietal bones along the _____Squamosal_____ sutures.

5. Name the three cranial bones that contain sinuses. _____Frontal bone, Ethmoid bone, Sphenoid bone_____

6. Name a facial bone that contains a sinus. _____Maxilla_____

Part C

Match the bones in column A with the characteristics in column B. Place the letter of your choice in the space provided.

Column A		Column B
A. inferior nasal concha	E 1.	forms bridge of nose
B. lacrimal bone	C 2.	only movable bone in the facial skeleton
C. mandible	C 3.	contains coronoid process
D. maxillary bone	H 4.	creates prominence of cheek below and to side of eye
E. nasal bone	D 5.	contains sockets of upper teeth
F. palatine bone	G 6.	forms inferior portion of nasal septum in back
G. vomer bone	H 7.	forms anterior portion of zygomatic arch
H. zygomatic bone	A 8.	scroll-shaped bone
	D 9.	forms anterior roof of mouth
	C 10.	contains mental foramen
	F 11.	forms posterior roof of mouth
	B 12.	scalelike part in medial wall of orbit

Part D

Identify the features of the skull indicated in figures 12.6 and 12.7.

1. Frontal bone
2. Frontal sinus
3. Sphenoid bone
4. Maxilla
5. mandible
6. Coronal suture
7. Parietal bone
8. Occipital bone
9. Internal acoustic meatus
10. Occipital condyle
11. Frontal bone
12. Crista galli
13. Foramen lacerum
14. Jugular foramen
15. Ethmoid bone
16. Sphenoid bone
17. Foramen ovale
18. Foramen spinosum
19. Foramen magnum
20. Occipital bone

Part E

Match the passageways in column A with the structures transmitted in column B. Place the letter of your choice in the space provided.

Column A		Column B
A. foramen magnum	C 1.	maxillary division of trigeminal nerve
B. foramen ovale	A 2.	nerve fibers of spinal cord
C. foramen rotundum	G 3.	optic nerve
D. incisive foramen	F 4.	vagus and accessory nerves
E. internal acoustic meatus	D 5.	nasopalatine nerves
F. jugular foramen	B 6.	mandibular division of trigeminal nerve
G. optic canal	E 7.	vestibular and cochlear nerves

Figure 12.6 Identify the features indicated on this sagittal section of a skull.

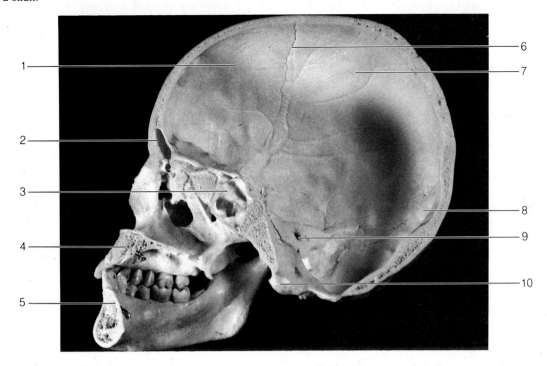

1
2
3
4
5
6
7
8
9
10

Figure 12.7 Identify the features indicated on this skull section.

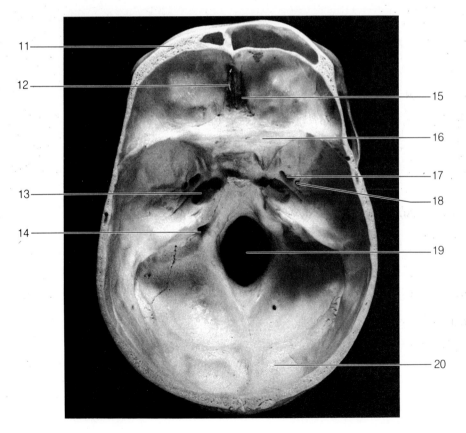

11
12
13
14
15
16
17
18
19
20

Figure 12.8 Identify the features indicated on this skull.

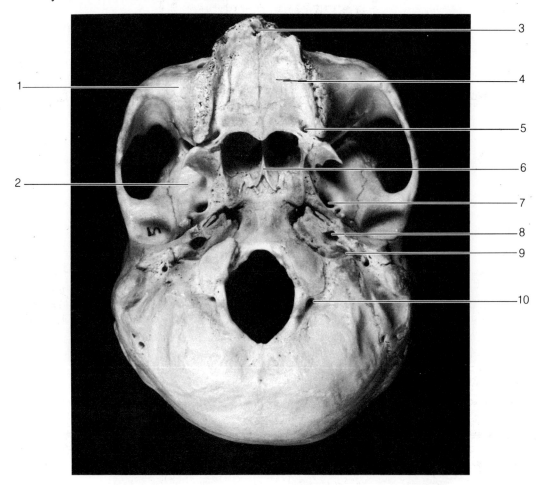

Part F

Identify the features of the skull indicated in figure 12.8.

1. _Maxilla_
2. _Sphenoid bone_
3. _Incisive foramen_
4. _Palatine process of maxilla_
5. _Greater palatine foramen_
6. _Vomer bone_
7. _Foramen ovale_
8. _Carotid Canal_
9. _Jugular foramen_
10. _Condylar canal_

Laboratory Exercise 13

Vertebral Column and Thoracic Cage

The vertebral column extends from the skull to the pelvis and forms the vertical axis of the human skeleton. The column is composed of many vertebrae, which are separated from one another by cartilaginous intervertebral disks and are held together by ligaments.

The thoracic cage surrounds the thoracic and upper abdominal cavities. It includes the ribs, the thoracic vertebrae, the sternum, and the costal cartilages.

Purpose of the Exercise

To examine the vertebral column and the thoracic cage of the human skeleton and to identify the bones and major features of these parts.

Learning Objectives

After completing this exercise, you should be able to:

1. Identify the major features of the vertebral column.
2. Name the parts of a typical vertebra.
3. Distinguish between a cervical, thoracic, and lumbar vertebra.
4. Identify the parts of the thoracic cage.
5. Distinguish between true and false ribs.

Materials Needed:
   human skeleton, articulated
   samples of cervical, thoracic, and lumbar vertebrae

Procedure A—The Vertebral Column

1. Review the section entitled "Vertebral Column" in chapter 7 of the textbook.
2. As a review activity, label figures 13.1, 13.2, 13.3, 13.4, and 13.5.

3. Examine the vertebral column of the human skeleton and locate the following parts and features. At the same time, locate as many of the corresponding parts and features in your own skeleton as possible.

   vertebrae

   intervertebral disks

   sacrum

   coccyx

   cervical curvature

   thoracic curvature

   lumbar curvature

   pelvic curvature

4. Examine the available samples of cervical, thoracic, and lumbar vertebrae and locate the following:

   body

   pedicles

   vertebral foramen

   laminae

   spinous process

   vertebral arch

   transverse process

   superior articulating processes

   inferior articulating processes

   intervertebral foramina

   cervical vertebra

      transverse foramina

      vertebra prominens

Figure 13.1 Label the parts and features of the vertebral column.

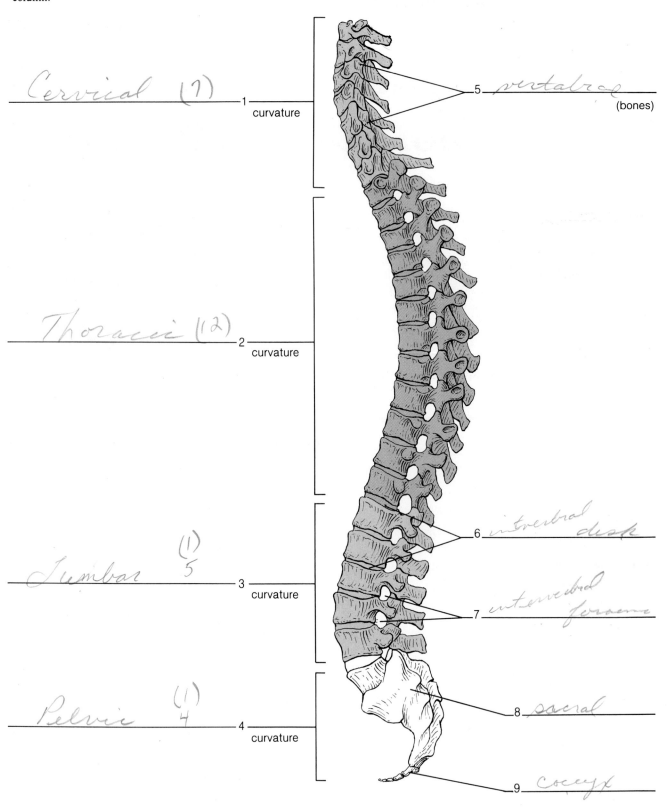

Cervical (7) — 1 curvature

Thoracic (12) — 2 curvature

Lumbar (1) 5 — 3 curvature

Pelvic (1) 4 — 4 curvature

5 vertabrae (bones)

6 intervebral disk

7 intervebral foramen

8 sacral

9 coccyx

Figure 13.2 Label (*a*) the superior features of a typical vertebra and (*b*) the lateral features of adjacent vertebrae.

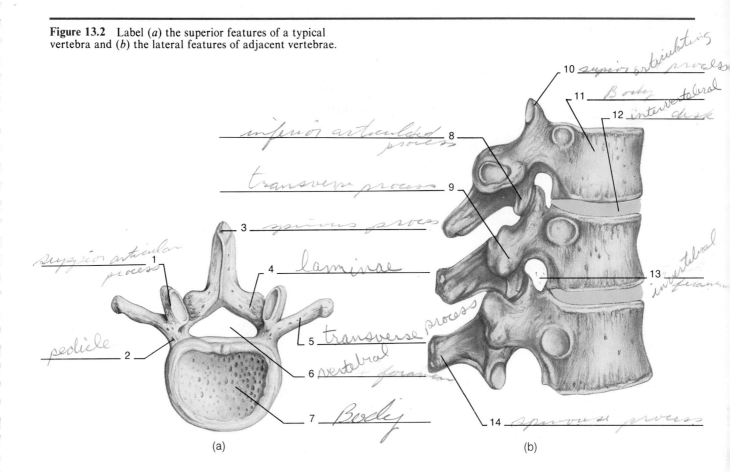

10 _superior articulating process_
11 _Body_
12 _intervertebral disk_

8 _inferior articulated process_
9 _transverse process_
3 _spinous process_
4 _laminae_
5 _transverse process_
6 _vertebral foramen_
7 _Body_

1 _superior articular process_
2 _pedicle_

13 _intervertebral foramen_
14 _spinous process_

(a)   (b)

atlas

axis

   odontoid process

thoracic vertebra

lumbar vertebra

5. Examine the sacrum and coccyx. Locate the following features:

  sacrum

  superior articulating process

  dorsal sacral foramen

  ventral sacral foramen

  sacral promontory

  sacral canal

  median sacral crest

  sacral hiatus

  coccyx

6. Complete Part A of Laboratory Report 13 on page 87.

Procedure B—The Thoracic Cage

1. Review the section entitled "Thoracic Cage" in chapter 7 of the textbook.
2. As a review activity, label figure 13.6.
3. Examine the thoracic cage of the human skeleton, and locate the following parts:

  ribs

    head

    tubercle

    true ribs

    false ribs

    floating ribs

  costal cartilages

  sternum

    manubrium

    body (gladiolus)

    xiphoid process

4. Complete Parts B and C of the laboratory report.

Figure 13.3 Label the superior features of (*a*) the atlas and
(*b*) the axis.

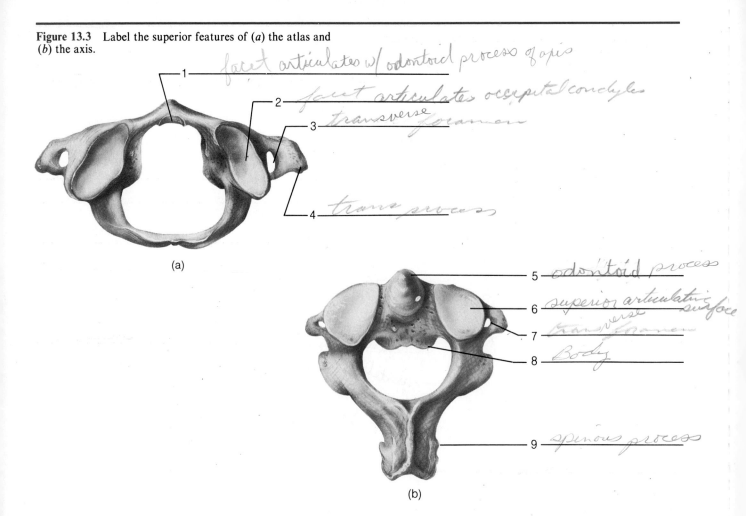

1 — facet articulates w/ odontoid process of axis

2 — facet articulates occipital condyles

3 — transverse foramen

4 — trans process

(a)

5 — odontoid process

6 — superior articulating surface

7 — transverse foramen

8 — Body

9 — spinous process

(b)

Figure 13.4 Label the parts of the (*a*) cervical, (*b*) thoracic, and (*c*) lumbar vertebrae.

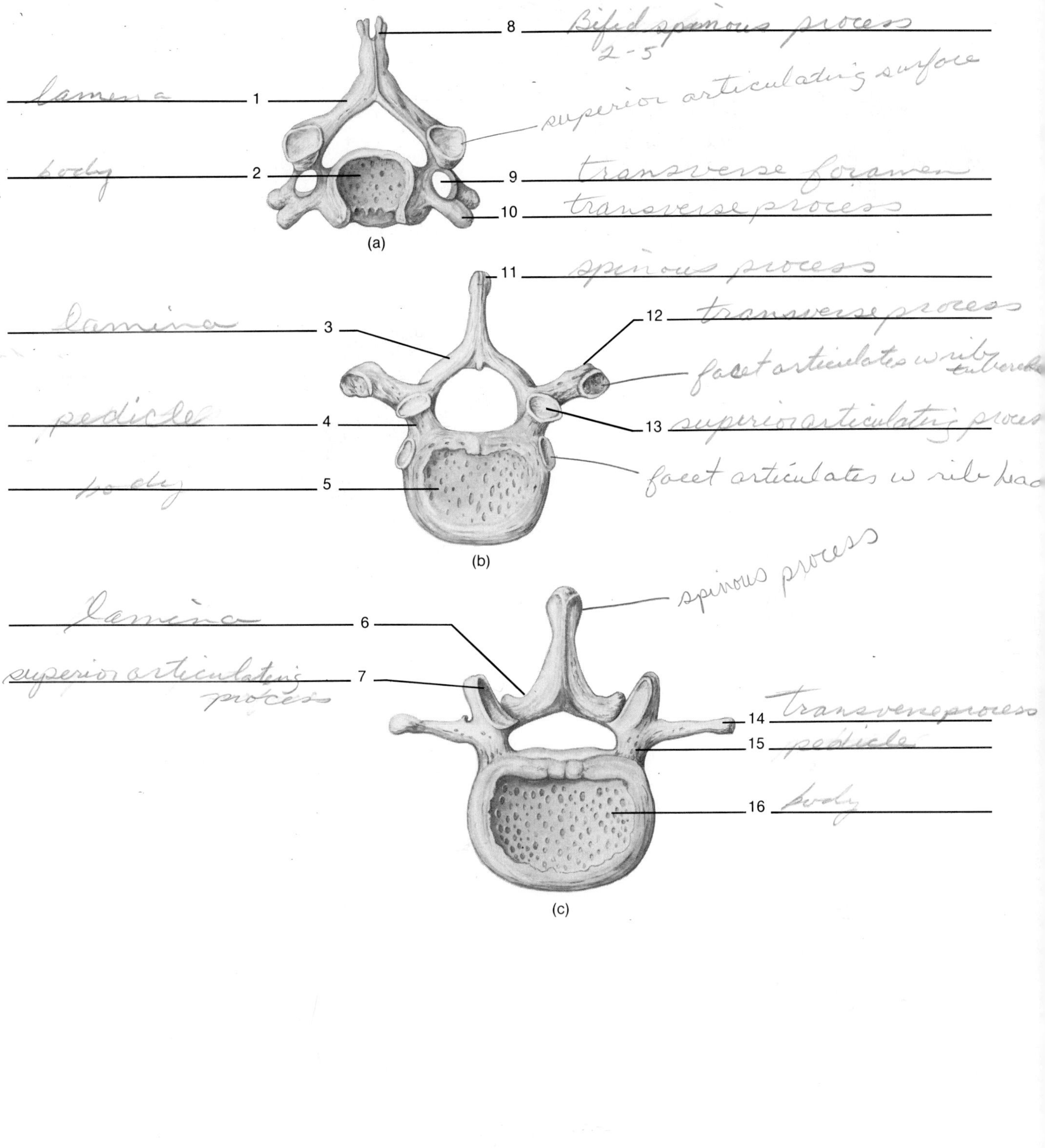

lamina — 1

body — 2

8 — Bifid spinous process 2-5

superior articulating surface

9 — transverse foramen

10 — transverse process

(a)

lamina — 3

pedicle — 4

body — 5

11 — spinous process

12 — transverse process

facet articulates w rib tubercle

13 — superior articulating process

facet articulates w rib head

(b)

lamina — 6

superior articulating process — 7

spinous process

14 — transverse process

15 — pedicle

16 — body

(c)

Figure 13.5 Label the parts of the sacrum: (*a*) anterior view;
(*b*) posterior view.

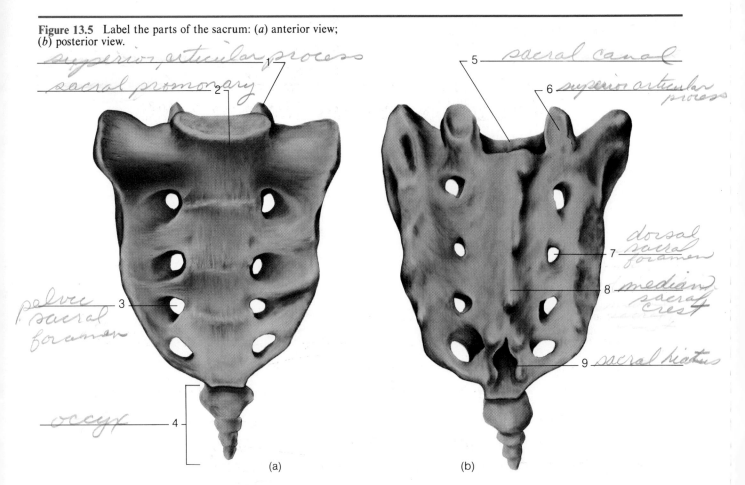

superior articular process
sacral promontory

sacral canal
superior articular process

pelvic sacral foramen

dorsal sacral foramen
median sacral crest

coccyx

sacral hiatus

(a)

(b)

Figure 13.6 Label the parts and features of the thoracic cage.

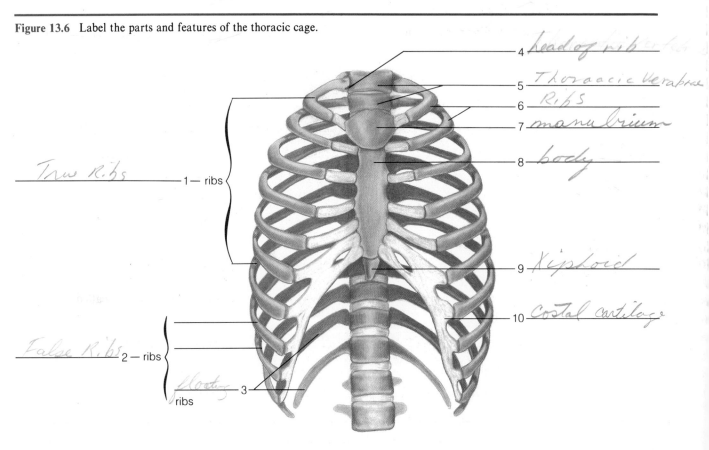

head of rib
Thoracic vertebrae
Ribs
manubrium
body

True Ribs

1 — ribs

xiphoid

Costal cartilage

False Ribs
2 — ribs

floating
ribs
3

laboratory report 13

Name _____

Date _____

Section _____

Vertebral Column and Thoracic Cage

Part A

Complete the following statements:

1. The vertebral column encloses and protects the _spinal cord_.

2. The number of separate bones in the vertebral column of an infant is _33_.

3. The thoracic and pelvic curvatures of the vertebral column are called _primary_ curves.

4. The _Centrum (Body)_ of the vertebrae support the weight of the head and trunk.

5. The _Intervertebral disk_ separate adjacent vertebrae, and they soften the forces created by walking.

6. The pedicles, laminae, and spinous processes of a vertebra form the _vertebral arch_.

7. The intervertebral foramina provide passageways for _spinal cord_.

8. Transverse foramina of cervical vertebrae serve as passageways for _transverse artery_.

9. The first vertebra also is called the _atlas_.

10. When the head is moved from side to side, the first vertebra pivots around the _Axis_ of the second vertebra. _odontoid process_

11. The _Lumbar_ vertebrae have the largest and strongest bodies.

12. The number of vertebrae that fuse to form the sacrum is _5_.

13. The joint between a coxal bone of the pelvis and the sacrum is called the _sacroiliac_ joint.

14. The upper, anterior margin of the sacrum that projects forward is called the _____.

15. An opening called the _sacral hiatus_ exists at the tip of the sacral canal. _sacrum promontery_

Part B

Complete the following statements:

1. Men have twelve pairs of ribs and women have _12_ pairs.

2. The last two pairs of ribs that have no cartilaginous attachments to the sternum are sometimes called _floating_ ribs.

3. The tubercles of the ribs articulate with the _transverse process_ of the thoracic vertebrae.

4. Costal cartilages are composed of _connective_ tissue. _hyaline car_

5. The manubrium articulates with the _clavicals_ on its superior border.

6. List three general functions of the thoracic cage. _protection of internal organs, place pectoral girdle, plays role for breathing (intercostal muscle)_

Figure 13.7 Identify the features indicated in this X-ray film of the neck.

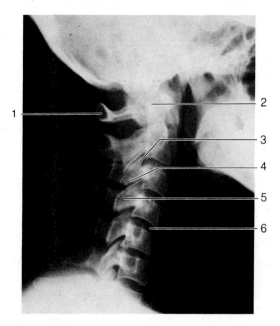

Part C

Identify the features indicated in the X-ray film of the neck in figure 13.7.

1. _Spinous process_
2. _atlas_
3. _superior articulating process_
4. _inferior articulating process_
5. _transverse process_
6. _intervertebral disk_

Pectoral Girdle and Upper Limb

The pectoral girdle consists of two clavicles and two scapulae. These parts function to support the upper limbs and to serve as attachments for various muscles that move these limbs.

Each upper limb includes a humerus, radius, ulna, and several carpals, metacarpals, and phalanges. These bones form the framework of the arm, wrist, palm, and fingers. They also function as levers when the limbs are moved.

Purpose of the Exercise

To examine the bones of the pectoral girdle and upper limb and to identify the major features of these parts.

Learning Objectives

After completing this exercise, you should be able to:

1. Locate and identify the bones of the pectoral girdle and their major features.
2. Locate and identify the bones of the upper limb and their major features.

Materials Needed:
human skeleton, articulated

Procedure A—The Pectoral Girdle

1. Review the section entitled "Pectoral Girdle" in chapter 7 of the textbook.
2. As a review activity, label figures 14.1 and 14.2.
3. Examine the bones of the pectoral girdle and locate the following parts and features. At the same time, locate as many of the corresponding parts and features in your own skeleton as possible.

clavicle

medial end

lateral end

scapula

spine

head

acromion process

coracoid process

glenoid cavity

4. Complete Part A of Laboratory Report 14 on page 95.

Figure 14.1 Label the bones and features of the pectoral girdle.

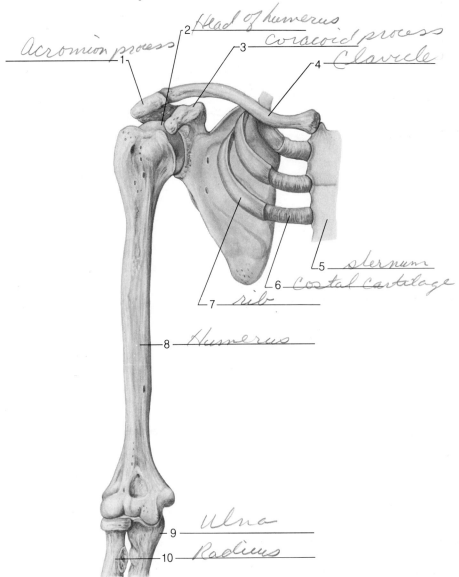

Acromion process 1

2 Head of humerus

3 Coracoid process

4 Clavicle

5 sternum

6 Costal cartilage

7 rib

8 Humerus

9 Ulna

10 Radius

Figure 14.2 Label (*a*) the posterior surface of the scapula and (*b*) the lateral aspect of the scapula.

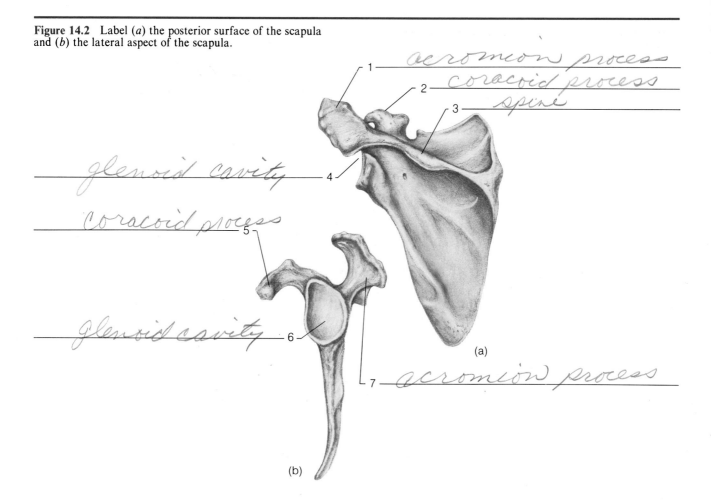

1 _acromion process_
2 _coracoid process_
3 _spine_

4

glenoid cavity

coracoid process 5

glenoid cavity 6

(a)

7 _acromion process_

(b)

Procedure B—The Upper Limb

1. Review the section entitled "Upper Limb" in chapter 7 of the textbook.
2. As a review activity, label figures 14.3, 14.4, and 14.5.
3. Examine the bones of the upper limb and locate the following parts and features:

humerus

    head

    greater tubercle

    lesser tubercle

    intertubercular groove

    surgical neck

    deltoid tuberosity

    capitulum

    trochlea

    medial epicondyle

    lateral epicondyle

    coronoid fossa

    olecranon fossa

radius

    head

    radial tuberosity

    styloid process

    ulnar notch

ulna

    trochlear notch (semilunar notch)

    olecranon process

    styloid process

    head

    radial notch

Figure 14.3 Label (*a*) the posterior surface and (*b*) the anterior surface of the left humerus.

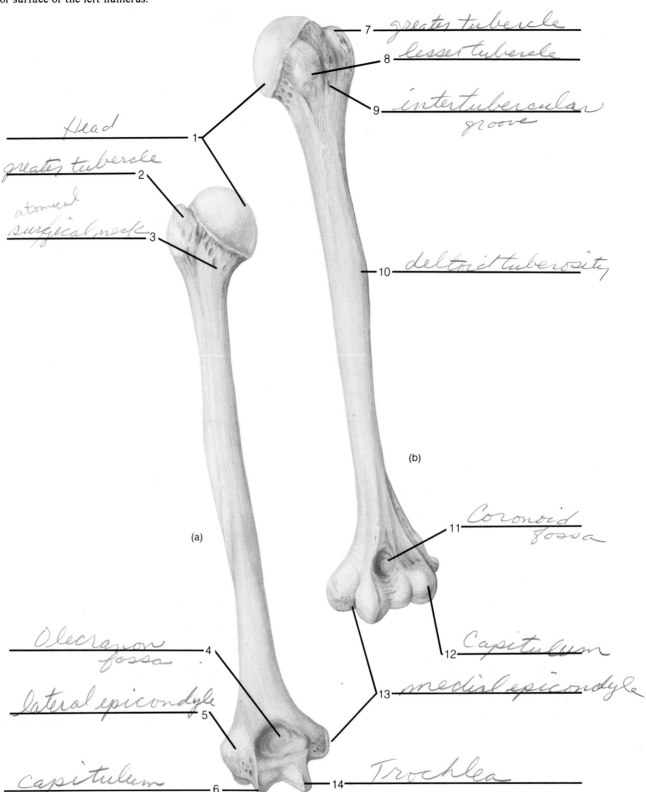

7 _greater tubercle_

8 _lesser tubercle_

9 _intertubercular groove_

Head _____ 1

greater tubercle _____ 2

atomical surgical neck _____ 3

10 _deltoid tuberosity_

(b)

11 _Coronoid fossa_

(a)

Olecranon fossa _____ 4

lateral epicondyle _____ 5

12 _Capitulum_

13 _medial epicondyle_

capitulum _____ 6

14 _Trochlea_

Figure 14.4 Label the major features of the radius and ulna.

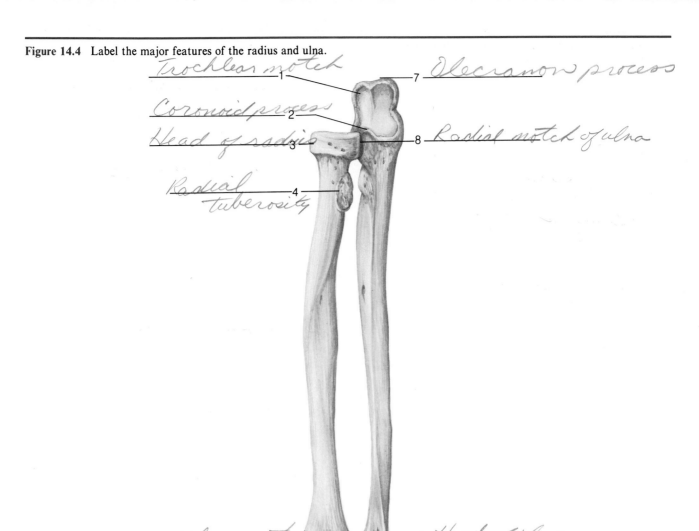

1. Trochlear notch
2. Coronoid process
3. Head of radius
4. Radial tuberosity
5. Ulnar notch of radius
6. Styloid process
7. Olecranon process
8. Radial notch of ulna
9. Head of ulna
10. Styloid process

hand

  carpal bones

    pisiform

    triangular

    hamate

    lunate

    capitate

    scaphoid

    trapezoid

    trapezium

  metacarpal bones

  phalanges

    proximal phalanx

    middle phalanx

    distal phalanx

4. Complete Parts B and C of the laboratory report.

Figure 14.5 Label the bones and groups of bones in this posterior view of the left hand.

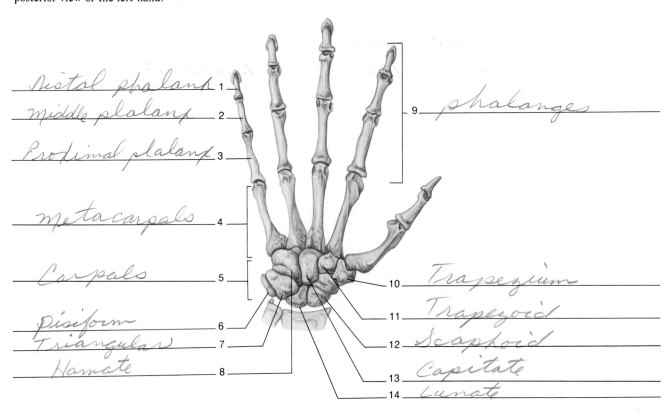

distal phalanx 1

middle phalanx 2

proximal phalanx 3

metacarpals 4

carpals 5

pisiform 6

Triangular 7

Hamate 8

9 _phalanges_

10 _Trapezium_

11 _Trapezoid_

12 _Scaphoid_

13 _Capitate_

14 _Lunate_

Name _____

Date _____

Section _____

Pectoral Girdle and Upper Limb

Part A

Complete the following statements:

1. The pectoral girdle is an incomplete ring because it is open in the back between the _scapulae_ .

2. The medial ends of the clavicles articulate with the _manubrium_

3. The lateral ends of the clavicles articulate with the _scapulae_ . _acromion process_

4. A clavicle is structurally weak as a result of its _elongated double curve_ .

5. The _spine_ divides the scapula into unequal portions.

6. The tip of the shoulder is due to the _acromion process_ of the scapula.

7. At the head of the scapula, the _coracoid process_ curves forward and downward below the clavicle.

8. The glenoid cavity of the scapula articulates with the _head_ of the humerus.

Part B

Match the bones in column A with the features in column B. Place the letter of your choice in the space provided.

Column A		Column B
A. carpals	_A_ 1.	capitate
B. humerus	_B_ 2.	capitulum
C. metacarpals	_B_ 3.	coronoid fossa
D. phalanges	_B_ 4.	deltoid tuberosity
E. radius	_B_ 5.	greater tubercle
F. ulna	_B_ 6.	intertubercular groove
	_A_ 7.	lunate
	_B_ 8.	olecranon fossa
	_F_ 9.	radial notch
	_E_ 10.	radial tuberosity
	_A_ 11.	trapezium
	_A_ 12.	triangular
	_B_ 13.	trochlea
	_F_ 14.	trochlear notch
	_E_ 15.	ulnar notch

Figure 14.6 Identify the features indicated on this X-ray film of the elbow.

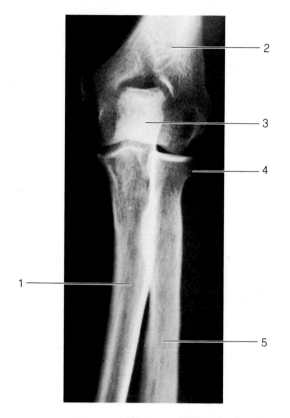

Figure 14.7 Identify the features indicated on this X-ray film of the shoulder.

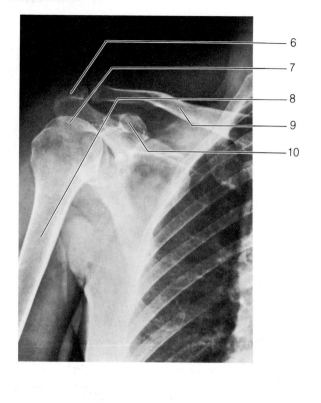

Part C

Identify the features indicated in the X-ray films of figures 14.6 and 14.7.

1. _ulna_
2. _tumera_
3. _electron proc_
4. _radius_
5. _ra_
6. _acrom proc_
7. _humm_
8. _hum_
9. _clavicle_
10. _coracoid_

Pelvic Girdle and Lower Limb

The pelvic girdle includes two coxal bones that articulate with each other anteriorly at the symphysis pubis and with the sacrum posteriorly. Together, the pelvic girdle, sacrum, and coccyx comprise the pelvis. The pelvis, in turn, provides support for the trunk of the body and provides attachments for the legs.

The bones of the lower limb form the framework of the leg, ankle, instep, and toes. Each limb includes a femur, a tibia, a fibula, and several tarsals, metatarsals, and phalanges.

Purpose of the Exercise

To examine the bones of the pelvic girdle and lower limb and to identify the major features of these parts.

Learning Objectives

After completing this exercise, you should be able to:

1. Locate and identify the bones of the pelvic girdle and their major features.
2. Locate and identify the bones of the lower limb and their major features.

Materials Needed:

human skeleton, articulated

For Demonstration:

male and female pelves

Procedure A—The Pelvic Girdle

1. Review the section entitled "Pelvic Girdle" in chapter 7 of the textbook.
2. As a review activity, label figures 15.1 and 15.2.

3. Examine the bones of the pelvic girdle and locate the following:

coxal bone

    acetabulum

    ilium

    iliac crest

    sacroiliac joint

    anterior superior iliac spine

    posterior superior iliac spine

    greater sciatic notch

    ischium

    ischial tuberosity

    ischial spine

    pubis

    symphysis pubis

    pubic arch

    obturator foramen

4. Complete Part A of Laboratory Report 15 on page 103.

Demonstration

Examine the male and female pelves. Look for major differences between them. Note especially the flare of the iliac bones, the angle of the pubic arch, the distance between the ischial spines and ischial tuberosities, and the curve of the sacrum. In what ways are the differences you observed related to the function of the female pelvis as a birth canal? _____

_____

_____
_____

Figure 15.1 Label the posterior features of the pelvic girdle.

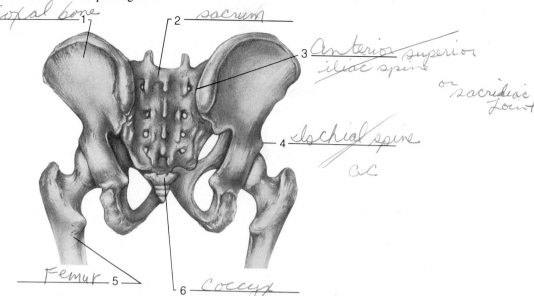

1 _Coxal bone_
2 _sacrum_
3 _Anterior superior iliac spine or sacroiliac joint_
4 _ischial spine CC_
5 _Femur_
6 _Coccyx_

Procedure B—The Lower Limb

1. Review the section entitled "Lower Limb" in chapter 7 of the textbook.
2. As a review activity, label figures 15.3, 15.4, and 15.5.
3. Examine the bones of the lower limb and locate each of the following:

femur

  head

  fovea capitis

  neck

  greater trochanter

  lesser trochanter

  linea aspera

  lateral condyle

  medial condyle

  lateral epicondyle

  medial epicondyle

patella

tibia

  medial condyle

  lateral condyle

  tibial tuberosity

  anterior crest

  medial malleolus

fibula

  head

  lateral malleolus

foot

  tarsal bones

    talus

    calcaneus

    navicular

    cuboid

    lateral cuneiform

    intermediate cuneiform

    medial cuneiform

  metatarsal bones

  phalanges

    proximal phalanx

    middle phalanx

    distal phalanx

4. Complete Parts B and C of the laboratory report.

Figure 15.2 Label (*a*) the lateral and (*b*) the medial features of the right coxal bone.

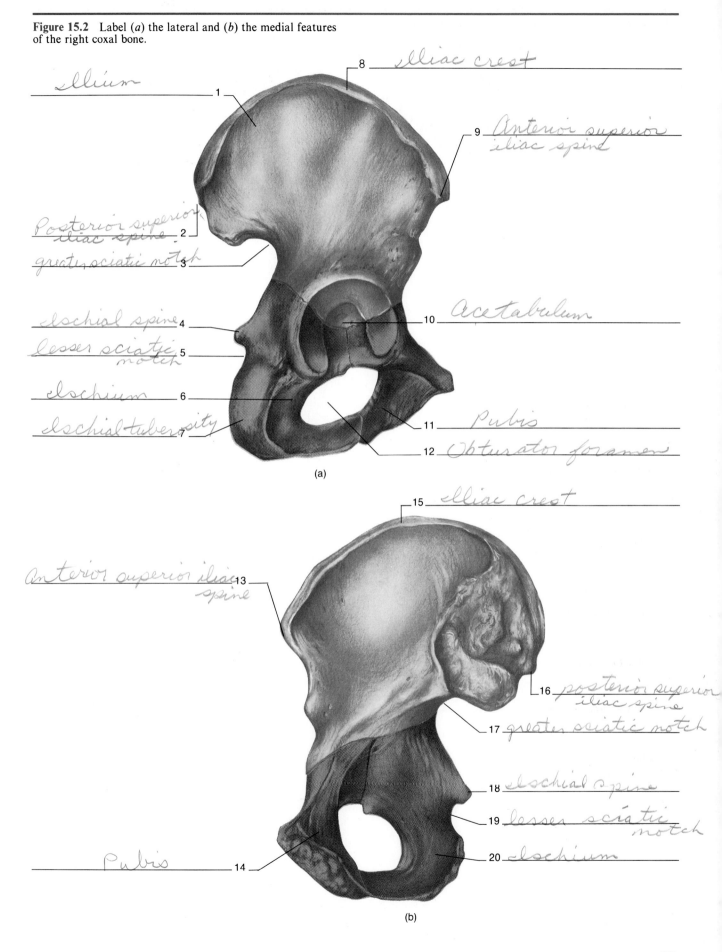

ilium 1

iliac crest 8

9 *Anterior superior iliac spine*

Posterior superior iliac spine 2

greater sciatic notch 3

ischial spine 4

lesser sciatic notch 5

ischium 6

ischial tuberosity 7

10 *Acetabulum*

11 *Pubis*

12 *Obturator foramen*

(a)

15 *iliac crest*

Anterior superior iliac spine 13

16 *posterior superior iliac spine*

17 *greater sciatic notch*

18 *ischial spine*

19 *lesser sciatic notch*

20 *ischium*

Pubis 14

(b)

Figure 15.3 Label the features of (*a*) the anterior surface and (*b*) the posterior surface of the left femur.

fovea capitis 1

Neck 2

lesser trochanter 3

greater trochanter 7

head 8

Neck 9

lesser trochanter 10

linea aspera 11

medial epicondyle 4
lateral epicondyle 5
patella surface 6
lateral condyle 12
medial condyle 13

(a)

(b)

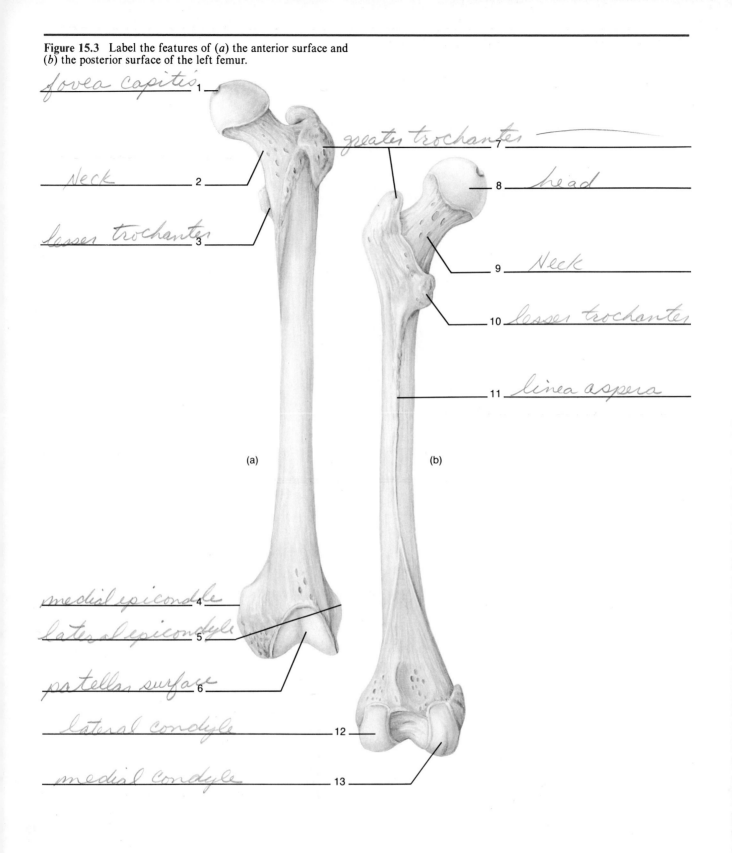

Figure 15.4 Label the anterior features of the left tibia and fibula.

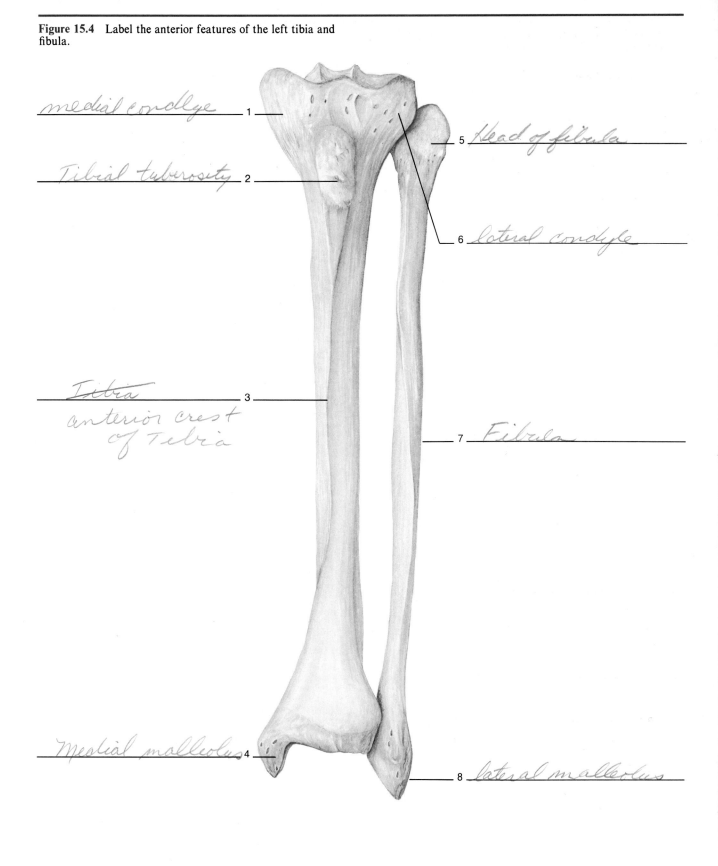

medial condlye _____ 1

Tibial tuberosity _____ 2

Itbia _____ 3
anterior crest
of Tebia

Medial malleolus _____ 4

5 _____ Head of fibula

6 _____ lateral condyle

7 _____ Fibula

8 _____ lateral malleolus

Figure 15.5 Label the features of (*a*) the medial surface and (*b*) the superior surface of the left foot.

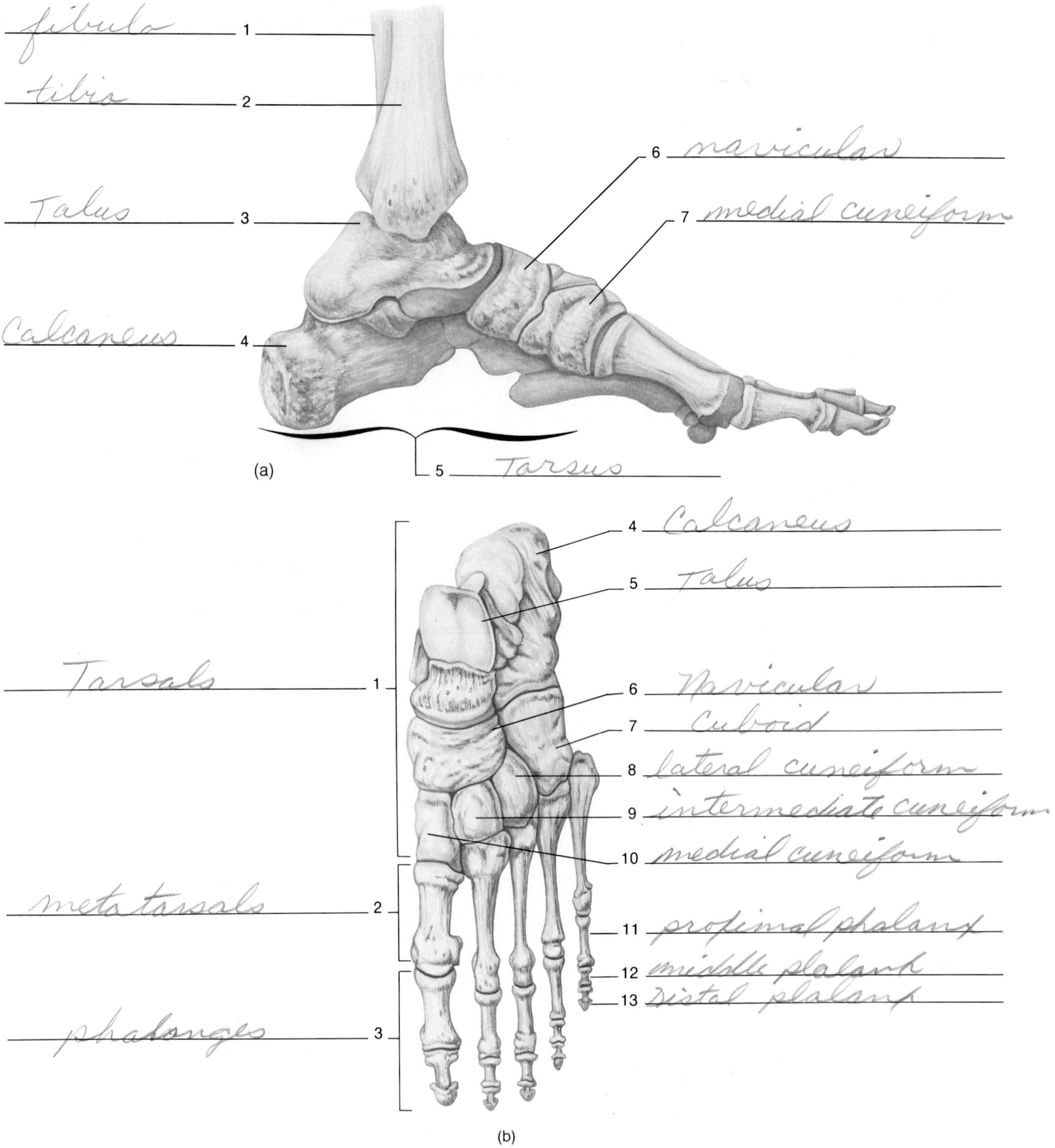

fibula _____ 1

tibia _____ 2

Talus _____ 3

Calcaneus _____ 4

6 _____ navicular

7 _____ medial cuneiform

(a)

5 _____ Tarsus

Tarsals _____ 1

metatarsals _____ 2

phalanges _____ 3

4 _____ Calcaneus

5 _____ Talus

6 _____ Navicular

7 _____ Cuboid

8 _____ lateral cuneiform

9 _____ intermediate cuneiform

10 _____ medial cuneiform

11 _____ proximal phalanx

12 _____ middle phalanx

13 _____ distal phalanx

(b)

Pelvic Girdle and Lower Limb

Part A

Complete the following statements:

1. The pelvic girdle consists of two _coxal bones_.
2. The head of the femur articulates with the _acetabulum_ of the hipbone.
3. The _ilium_ is the largest portion of the coxal bone.
4. The distance between the _ischial spine_ represents the shortest diameter of the pelvic outlet.
5. The pubic bones come together anteriorly to form the joint called _symphysis pubis_.
6. The _iliac crest_ is the portion of the coxal bone that causes the prominence of the hip.
7. When a person sits, the _tuberosity_ of the ischium supports the weight of the body.
8. The angle formed by the pubic bones below the symphysis pubis is called the _pubic arch_.
9. _obturator foramen_ is the largest foramen in the skeleton.
10. The ilium joins the sacrum at the _sacroiliac joint_.

Part B

Match the bones in column A with the features in column B. Place the letter of your choice in the space provided.

Column A		Column B
A. femur	_E_ 1.	middle phalanx
B. fibula	_A_ 2.	lesser trochanter
C. metatarsals	_G_ 3.	medial malleolus
D. patella	_A_ 4.	fovea capitis
E. phalanges	_E_ 5.	calcaneus
F. tarsals	_F_ 6.	lateral cuneiform
G. tibia	_G_ 7.	tibial tuberosity
	_F_ 8.	talus
	_A_ 9.	linea aspera
	_B_ 10.	lateral malleolus

Part C

Identify the features indicated in the X-ray films of figures 15.6, 15.7, and 15.8.

1. _tub forc___
2. _sinus_
3. _____
4. _sacrum___
5. _acatabulum___
6. _neck a femur head_
7. _femur_
8. _lateral lip_
9. _ales_
10. _later fibia_
11. _tibia_
12. _fibia_
13. _tibia_
14. _patelas_
15. _calcanius_
16. _navicula_
17. _cunform_
18. _metarsal_
19. _1 prox ph_
20. _1 dist pha_

Figure 15.6   Identify the features indicated on this X-ray film of the pelvic girdle.

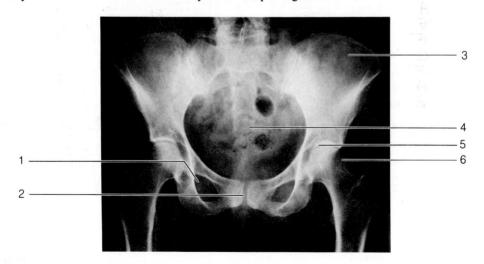

Figure 15.7 Identify the features indicated on this X-ray film of the knee.

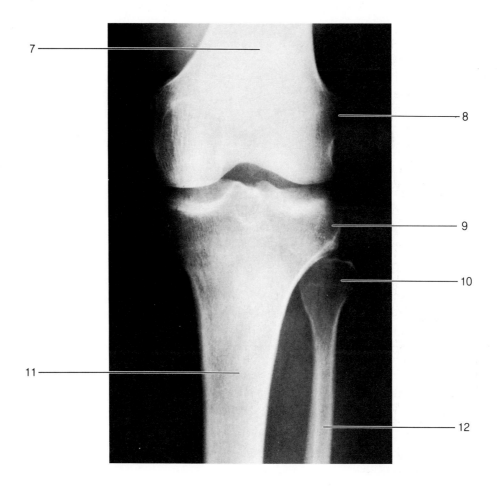

7

8

9

10

11

12

Figure 15.8 Identify the features indicated on this X-ray film of the left foot.

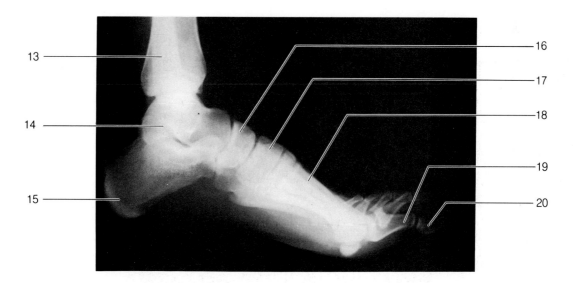

13

14

15

16

17

18

19

20

Laboratory Exercise 16

The Joints

Joints are junctions between bones. Although they vary considerably in structure, they can be classified according to the type of tissue that binds the bones together. Thus, three groups of joints can be identified: (1) fibrous joints, (2) cartilaginous joints, and (3) synovial joints.

Movements occurring at freely movable joints are due to the contractions of skeletal muscles. In each case, the type of movement depends on the kind of joint involved and the way in which the muscles are attached to the bones on either side of the joint.

Purpose of the Exercise

To examine examples of the three types of joints, to identify the major features of these joints, and to review the types of movements produced at synovial joints.

Learning Objectives

After completing this exercise, you should be able to:

1. Distinguish between fibrous, cartilaginous, and synovial joints.
2. Identify examples of each type of joint.
3. Identify the major features of each type of joint.
4. Locate and identify examples of each of the six types of synovial joints.
5. Identify the types of movements that occur at synovial joints.
6. Describe the structure of the shoulder, elbow, hip, and knee joints.

Materials Needed:
human skull
human skeleton, articulated
models of synovial joints (shoulder, elbow, hip, and knee)

For Demonstration:

fresh animal joint
X-ray films of major joints

Procedure A—Types of Joints

1. Review the section entitled "Classification of Joints" in chapter 8 of the textbook.
2. Examine the human skull and articulated skeleton to locate examples of the following types of joints:

   fibrous joints

      syndesmosis

      suture

      gomphosis

   cartilaginous joints

      synchondrosis

      symphysis

   synovial joints

3. Complete Part A of Laboratory Report 16 on page 113.
4. Review the sections entitled "General Structure of a Synovial Joint" and "Types of Synovial Joints" in chapter 8.
5. As a review activity, label figures 16.1, 16.2, and 16.3.
6. Locate examples of the following types of synovial joints in the skeleton. At the same time, examine the corresponding joints in the models and in your own skeleton. Experiment with each joint to experience its range of movements.

   ball-and-socket joint

   condyloid joint

   gliding joint

   hinge joint

   pivot joint

   saddle joint

7. Complete Parts B, C, and D of the laboratory report.

Demonstration

Examine the fresh synovial animal joint. Locate the fibrous connective tissue that forms the joint capsule and the hyaline cartilage that forms the articular cartilage on the ends of the bones. Locate the synovial membrane on the inside of the joint capsule. Does the joint have any semilunar cartilages (menisci)? _____

What is the function of such cartilages? _____
_____
_____

Figure 16.1 Label this generalized structure of a synovial joint.

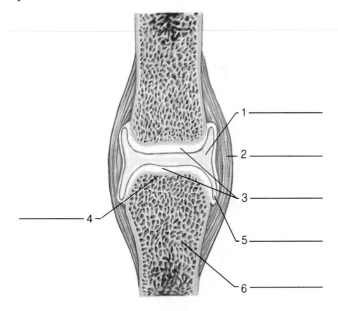

Procedure B—Joint Movements

1. Review the section entitled "Types of Joint Movements" in chapter 8 of the textbook.
2. Move various parts of your own body to demonstrate the following joint movements:

   flexion

   extension

   hyperextension

   dorsiflexion

   plantar flexion

   abduction

   adduction

   rotation

   circumduction

   supination

   pronation

   eversion

   inversion

   protraction

   retraction

   elevation

   depression

3. Have your laboratory partner do some of the preceding movements and see if you can identify correctly the movements made.
4. Complete Part E of the laboratory report.

Figure 16.2 Label the parts of the knee joint.

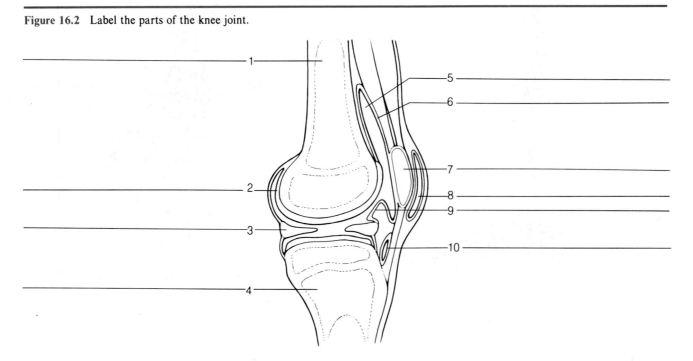

Figure 16.3 Label the types of synovial joints numbered in these diagrams.

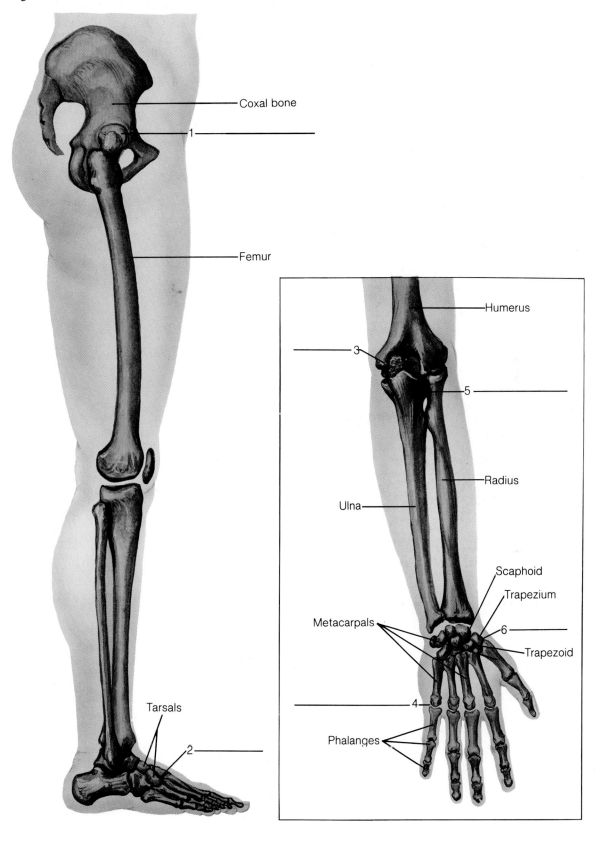

Coxal bone

1

Femur

Tarsals

2

Humerus

3

5

Radius

Ulna

Scaphoid

Trapezium

Metacarpals

6

Trapezoid

4

Phalanges

Figure 16.4 Label the features of (*a*) the shoulder joint and (*b*) the elbow joint.

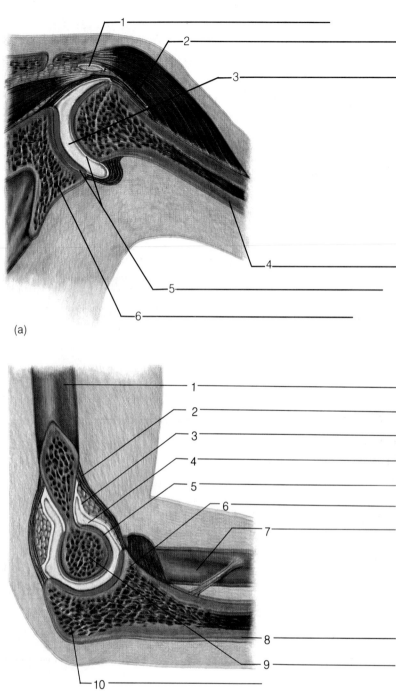

(a)

(b)

Figure 16.5 Label the features of the hip joint.

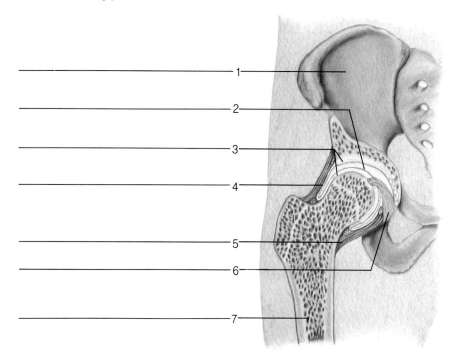

1 _____

2 _____

3 _____

4 _____

5 _____

6 _____

7 _____

Procedure C—Examples of Synovial Joints

1. Review the section entitled "Examples of Synovial Joints" in chapter 8 of the textbook.
2. As a review activity, label figures 16.4 and 16.5.
3. Examine the models of the shoulder, elbow, hip, and knee joints. Locate the following features:

   shoulder joint

       coracohumeral ligament

       glenohumeral ligament

       transverse humeral ligament

       glenoidal labrum

   elbow joint

       ulnar collateral ligament

       radial collateral ligament

   hip joint

       iliofemoral ligament

       pubofemoral ligament

       ischiofemoral ligament

   knee joint

       patellar ligament

       oblique popliteal ligament

       arcuate popliteal ligament

       tibial collateral ligament

       fibular collateral ligament

       anterior cruciate ligament

       posterior cruciate ligament

       medial meniscus

       lateral meniscus

4. Complete Part F of the laboratory report.

Optional Activity

Use colored pencils to color the joints in figures 16.1, 16.2, 16.4, and 16.5. This activity should help you to visualize the features of joints. You can check your work by referring to the corresponding figures in the textbook, which are presented in full color.

Demonstration

Study the available X-ray films of joints by holding the films in front of a light source. Identify the bones incorporated in the joint. Also identify other major features of the structure.

laboratory report 16

Name _____

Date _____

Section _____

The Joints

Part A

Match the terms in column A with the descriptions in column B. Place the letter of your choice in the space provided.

Column A	Column B
A. gomphosis	___ 1. joint between flat bones united by a thin layer of connective tissue
B. suture	___ 2. articular cartilages at joint attached by pad of fibrocartilage
C. symphysis	___ 3. temporary joint in which bones are united by bands of hyaline cartilage
D. synchondrosis	___ 4. immovable joint in which bones are united by interosseous ligament
E. syndesmosis	___ 5. joint formed by union of cone-shaped bony process in bony socket

Part B

Complete the following statements:

1. _____ is the most common type of joint in the human body.

2. Articular cartilage is composed of _____ tissue.

3. The outer layer of the joint capsule is attached to the _____ of each bone at the joint.

4. Bundles of strong, collagenous fibers that bind bones to bones at joints are called _____ .

5. _____ acts as a joint lubricant in freely movable joints.

6. The pads of fibrocartilage between the articulating surfaces of the knee joint are called

_____ .

7. The fluid-filled sacs sometimes associated with freely movable joints are called _____ .

8. The inner lining of a bursa is formed by _____ .

Part C

Identify the types of joints illustrated in figure 16.6.

1. _____   6. _____

2. _____   7. _____

3. _____   8. _____

4. _____   9. _____

5. _____   10. _____

Figure 16.6 Identify the types of joints illustrated (*a–h*).

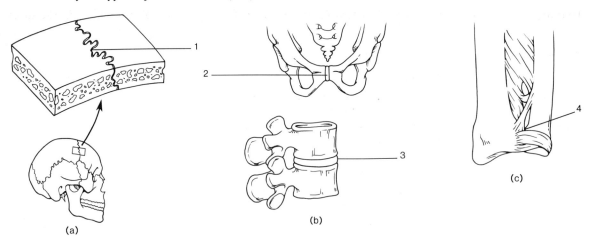

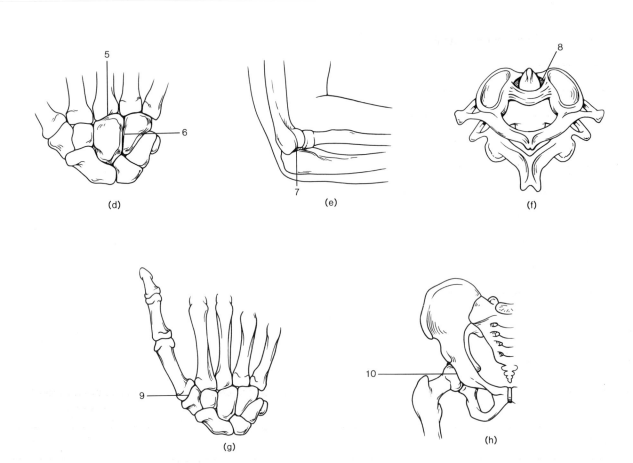

Part D

Match the types of joints in column A with the examples in column B. Place the letter of your choice in the space provided.

Column A

A. ball-and-socket
B. condyloid
C. gliding
D. hinge
E. pivot
F. saddle

Column B

—— 1. hip joint
—— 2. metacarpal-phalanx
—— 3. proximal radius-ulna
—— 4. elbow joint
—— 5. phalanx-phalanx
—— 6. shoulder joint
—— 7. knee joint
—— 8. carpal-metacarpal of the thumb
—— 9. carpal-carpal
—— 10. tarsal-tarsal

Part E

Identify the types of joint movements illustrated in figure 16.7.

1. _____
2. _____
3. _____
4. _____
5. _____
6. _____
7. _____
8. _____
9. _____
10. _____

Part F

Complete the following chart:

Name of Joint	Type of Joint	Bones Included	Types of Movement Possible
Shoulder joint			
Elbow joint			
Hip joint			
Knee joint			

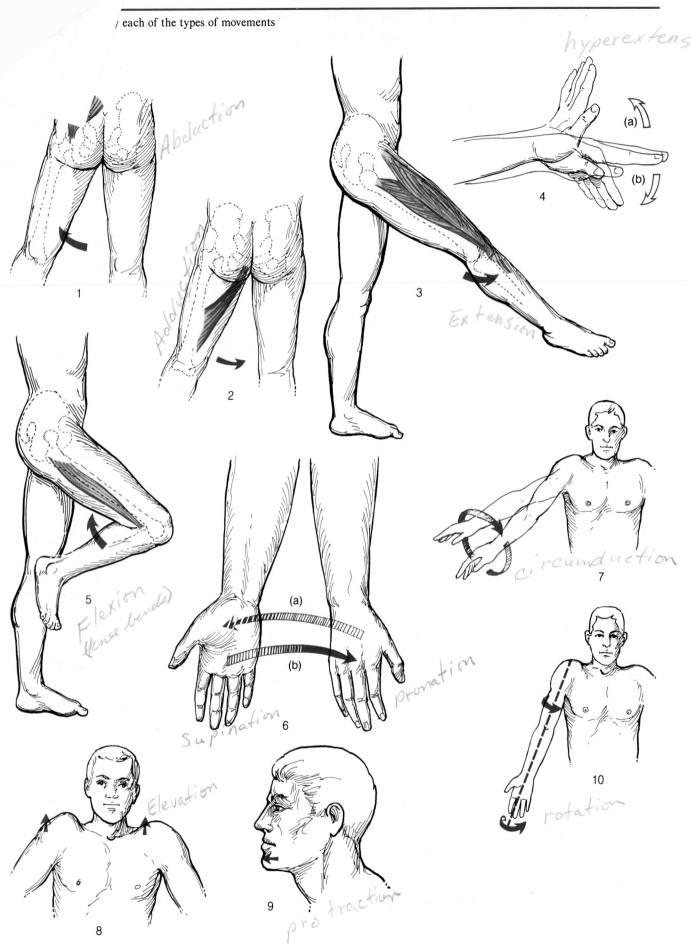

hyperextension

Abduction

Aolduction

Extension

Flexion
(knee bends)

Supination

Pronation

circumduction

rotation

Elevation

protraction

1
2
3
4
5
6
7
8
9
10
(a)
(b)

Skeletal Muscle Structure

A skeletal muscle represents an organ of the muscular system and is composed of several kinds of tissues. These tissues include skeletal muscle tissue, nerve tissue, blood, and various connective tissues.

Each skeletal muscle is surrounded by connective tissue, and the connective tissue often extends beyond the end of a muscle, providing an attachment to other muscles or to bones. Connective tissue also extends into the structure of a muscle and separates it into compartments.

Purpose of the Exercise

To review the structure of a skeletal muscle.

Learning Objectives

After completing this exercise, you should be able to:

1. Describe how connective tissue is associated with muscle tissue within a skeletal muscle.
2. Name and locate the major parts of a skeletal muscle fiber on a model.
3. Distinguish between the origin and insertion of a muscle.
4. Describe the general actions of prime movers, synergists, and antagonists.

Materials Needed:
   compound microscope
   prepared microscope slide of skeletal muscle tissue
   manikin
   model of skeletal muscle fiber

For Demonstration:

   fresh round beefsteak

Procedure

1. Review the section entitled "Skeletal Muscle Tissue" in chapter 5 of the textbook.
2. Reexamine the microscopic structure of skeletal muscle by observing a prepared microscope slide of this tissue. Use the illustration of skeletal muscle tissue in chapter 5 to locate the following features:

   skeletal muscle fiber

   nuclei

   striations

3. Review the section entitled "Structure of a Skeletal Muscle" in chapter 9 of the textbook.
4. As a review activity, label figures 17.1 and 17.2.
5. Examine the manikin, and locate examples of fascia, tendons, and aponeuroses. Locate examples of tendons in your own body.
6. Complete Part A of Laboratory Report 17 on page 121.

Demonstration

Examine the fresh beefsteak. It represents a cross section through the beef thigh muscles. Note the white lines of connective tissue that separate the individual skeletal muscles. Also note how the connective tissue extends into the structure of a muscle and separates it into small compartments of muscle tissue. Identify as many layers of fascia as you can.

---

7. Examine the model of the skeletal muscle fiber and locate the following:

sarcolemma	I band
sarcoplasm	H zone
myofibril	M line
myosin filament	Z line
actin filament	sarcoplasmic reticulum
sarcomere	cisternae
A band	transverse tubules

Figure 17.1 Label the parts of a skeletal muscle shown in this illustration (*a–d*).

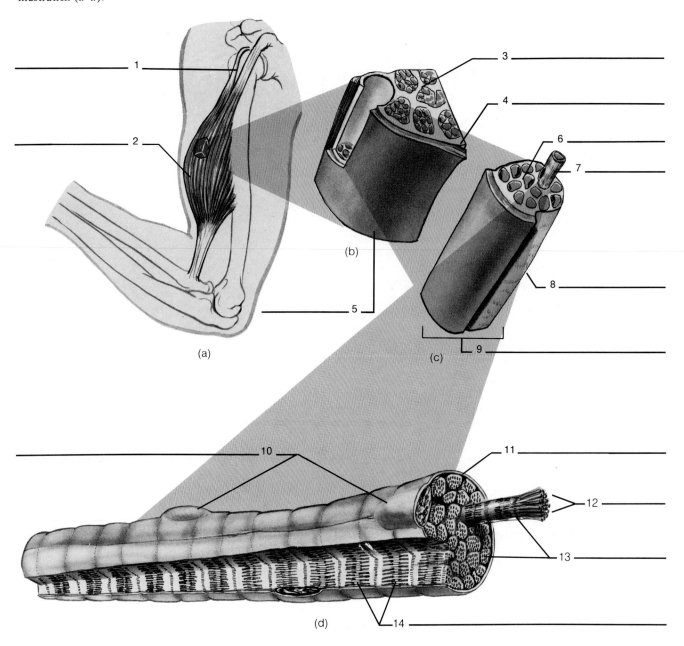

(a)

(b)

(c)

(d)

8. Complete Part B of the laboratory report.
9. Review the section entitled "Skeletal Muscle Actions" in chapter 9 of the textbook.
10. Provide labels for figure 17.3.
11. Locate the following muscles and their origins and insertions in the manikin and in your own body:

    biceps brachii

    pectoralis major

    latissimus dorsi

12. Make various movements with your arm at the shoulder and elbow. For each movement, determine the location of the muscles functioning as prime movers and as antagonists. (Remember, when a prime mover contracts, its antagonist must relax.)
13. Complete Part C of the laboratory report.

Figure 17.2 Supply the labels for this illustration of a skeletal muscle fiber.

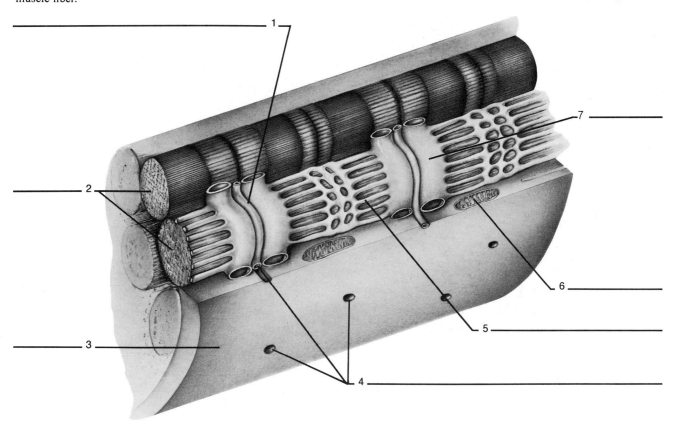

Figure 17.3 Label the parts indicated in this illustration of the upper arm.

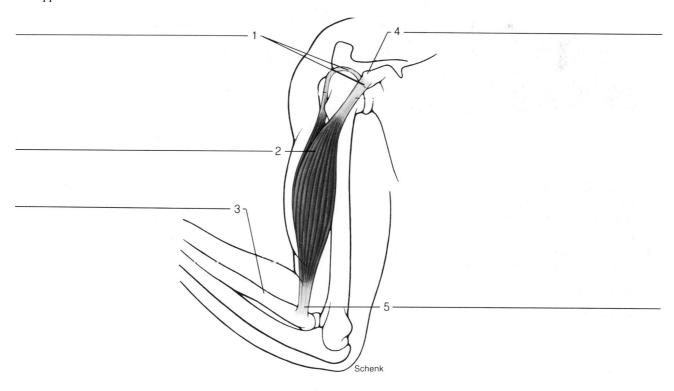

Schenk

laboratory report 17

Name _____

Date _____

Section _____

Skeletal Muscle Structure

Part A

Match the terms in column A with the definitions in column B. Place the letter of your choice in the space provided.

Column A

A. aponeuroses
B. cisterna
C. endomysium
D. epimysium
E. fascia
F. fascicle
G. myosin
H. perimysium
I. sarcomere
J. sarcolemma
K. sarcoplasm
L. sarcoplasmic reticulum
M. tendon
N. transverse tubule

Column B

____ 1. membranous channel extending inward from muscle fiber membrane
____ 2. cytoplasm of a muscle fiber
____ 3. network of connective tissue that extends throughout the muscular system
____ 4. layer of connective tissue that separates a muscle into small bundles
____ 5. enlarged portion of sarcoplasmic reticulum on either side of a transverse tubule
____ 6. broad sheet of connective tissue that attaches coverings of adjacent tissues
____ 7. cell membrane of a muscle fiber
____ 8. layer of connective tissue that surrounds a skeletal muscle
____ 9. unit of alternating light and dark striations
____ 10. layer of connective tissue that surrounds an individual muscle fiber
____ 11. cellular organelle in muscle fiber corresponding to the endoplasmic reticulum
____ 12. cordlike part that attaches a muscle to a bone
____ 13. protein found within a myofibril
____ 14. a bundle of muscle fibers

Part B

Provide the labels for the electron micrograph in figure 17.4.

1. _____

2. _____

3. _____

4. _____

5. _____

121

Figure 17.4 Identify the bands of the striations in this transmission electron micrograph of myofibrils (×20,000).

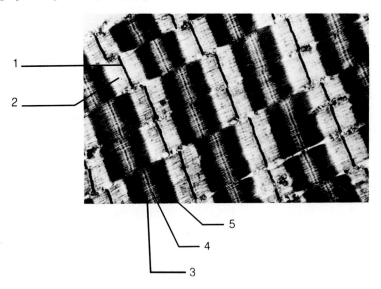

Part C

Complete the following statements:

1. _____ is a muscle in the upper arm that has two origins.

2. _____ pulls the arm toward the back.

3. _____ moves the arm across the chest.

4. _____ acts as the insertion of the biceps brachii.

5. _____ acts to bend the arm at the elbow.

6. _____ acts as the origin of the pectoralis major during chinning exercises.

7. _____ acts as the insertion of the pectoralis major when the arm is moved across the chest.

8. _____ acts as the origin of the latissimus dorsi during chinning exercises.

9. _____ acts as the insertion of the latissimus dorsi during chinning exercises.

Demonstration of Physiological Recording Equipment

To study the characteristics of certain physiological events such as muscle contractions, it often is necessary to use a recording device, such as a *kymograph* or a *Physiograph*. These devices are relatively simple to use, and they can provide accurate recordings of various physiological changes.

Purpose of the Exercise

To become familiar with the equipment available in your laboratory to record muscle contractions.

Learning Objectives

After completing this exercise, you should be able to:

1. Identify the major parts of the recording apparatus.
2. Explain how the apparatus can be used to record muscle contractions.

Materials Needed:
For Procedure A—The Kymograph:

kymograph recording system
electronic stimulator (or inductorium)
frog muscles

For Procedure B—The Physiograph:

Physiograph
myograph and stand
frog muscle

Procedure A—The Kymograph

1. Observe the kymograph, and, at the same time, study figure 18.1 to learn the names of its major parts.
2. Note that the kymograph consists of a cylindrical *drum* around which a sheet of paper is wrapped. The drum is mounted on a motor-driven *shaft,* and the speed of the motor can be varied. Thus, the drum can be rotated rapidly if rapid physiological events are being recorded or rotated slowly for events that occur more slowly.

    A *stylus* that can mark on the paper is attached to a *movable lever,* and the lever, in turn, is connected to an isolated muscle. The origin of the muscle is fixed in position by a *clamp,* while its insertion is hooked to the muscle lever. The muscle also is connected by wires to an *electronic stimulator* (or inductorium). The stimulator can deliver single or multiple electrical shocks to the muscle, and it can be adjusted so that the intensity (voltage), duration (milliseconds), and frequency (stimuli per second) can be varied. Another stylus, on the *signal marker,* records the time each stimulus is given to the muscle. As the muscle responds, the duration and relative length of its contraction are recorded by the stylus on the muscle lever.
3. Watch carefully while the laboratory instructor demonstrates the operation of the kymograph to record a frog muscle contraction.

Figure 18.1 Kymograph set up to record frog muscle contractions.

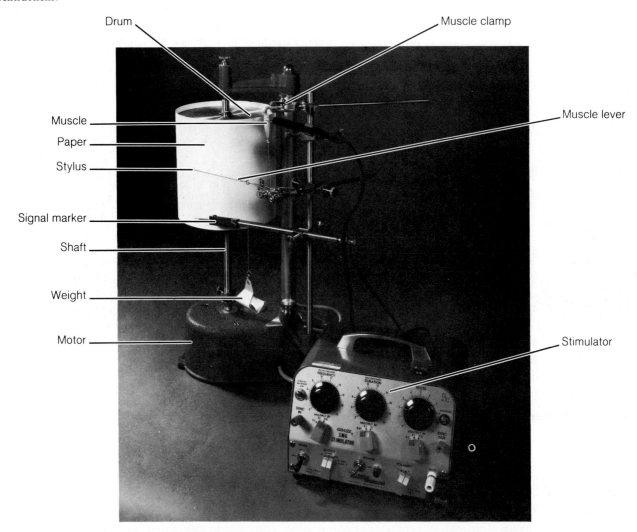

Drum

Muscle clamp

Muscle

Paper

Stylus

Muscle lever

Signal marker

Shaft

Weight

Motor

Stimulator

Procedure B—The Physiograph

1. Observe the Physiograph and, at the same time, study figures 18.2 and 18.3 to learn the names of its major parts.
2. Note that the recording system of the Physiograph includes a transducer, an amplifier, and a recording pen. The *transducer* is a sensing device that can respond to some kind of physiological change by sending an electrical signal to the amplifier. The *amplifier* increases the strength of the electrical signal and relays it to an electric motor that moves the recording pen. As the pen moves, a line is drawn on paper.

   To record a frog muscle contraction, a transducer called a *myograph* is used. The origin of the muscle is held in a fixed position, and its insertion is attached to a small lever in the myograph by a thread. The myograph, in turn, is connected to the amplifier by a transducer cable. The muscle also is connected by wires to the electronic stimulator, which is part of the Physiograph. This stimulator can be adjusted to deliver single or multiple electrical shocks to the muscle, and the intensity (voltage), duration (milliseconds), and frequency (stimuli per second) can be varied.

   The speed at which the paper moves under the recording pen can be controlled. A second pen, driven by a timer, marks time units on the paper and indicates when the stimulator is activated. As the muscle responds to stimuli, the recording pen records the duration and relative length of each muscle contraction.
3. Watch carefully while the laboratory instructor operates the Physiograph to record a frog muscle contraction.

Figure 18.2 Major parts of a Physiograph.

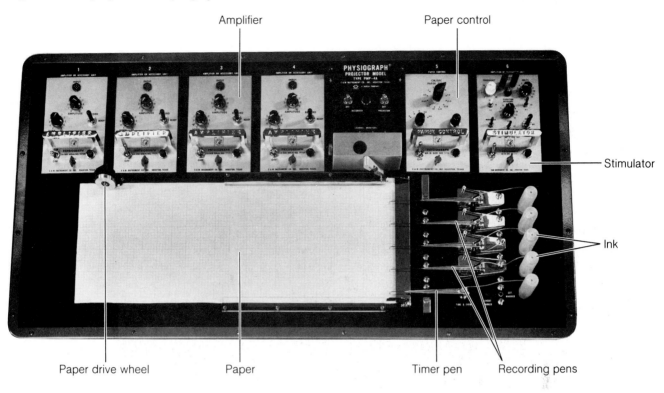

Amplifier

Paper control

Stimulator

Ink

Paper drive wheel

Paper

Timer pen

Recording pens

Figure 18.3 Myograph attached to frog muscle.

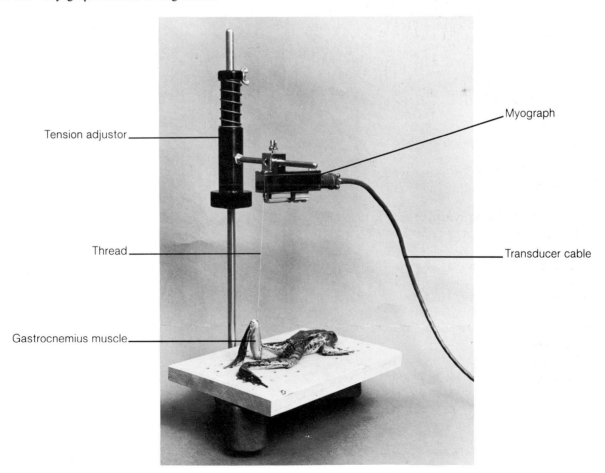

Tension adjustor

Myograph

Thread

Transducer cable

Gastrocnemius muscle

Laboratory Exercise 19

Skeletal Muscle Contraction

To observe the phenomenon of skeletal muscle contractions, muscles can be removed from live, anesthetized frogs. These muscles can be attached to recording systems and stimulated by electrical shocks of varying strength, duration, and frequency. Recordings obtained from such procedures can be used to study the basic characteristics of skeletal muscle contractions.

Purpose of the Exercise

To observe and record the responses of an isolated frog muscle to electrical stimulation of varying strength and frequency.

Learning Objectives

After completing this exercise, you should be able to:

1. Make use of a recording system and stimulator to record frog muscle responses to electrical stimulation.
2. Determine the threshold level of electrical stimulation in frog muscle.
3. Determine the intensity of stimulation needed for maximal muscle contraction.
4. Record a single muscle twitch and identify its phases.
5. Record the response of a muscle to increasing frequency of stimulation, and identify the patterns of tetany and fatigue.

Materials Needed:
   recording system (kymograph, Physiograph, etc.)
   stimulator and connecting wires
   live frog
   dissecting tray
   dissecting instruments
   probe for pithing
   heavy thread
   frog Ringer's solution

Procedure A—Textbook Review

1. Review the section entitled "Muscular Responses" in chapter 9 of the textbook.
2. Complete Part A of Laboratory Report 19 on page 130.

Procedure B—Recording System

1. Set up the recording system and stimulator to record the contractions of a frog muscle according to the directions provided by the laboratory instructor.
2. Obtain a live frog, and prepare its calf muscle (gastrocnemius) as described in Procedure C.

Procedure C—Muscle Preparation

1. Anesthetize the live frog by pithing so that it will have no feelings when its muscle is removed. To do this:
   a. Hold the frog securely in one hand so that its legs are extended downward.
   b. Position the frog's head between your thumb and index finger.
   c. Bend the frog's head forward at an angle of about 90° by pressing on its snout with your index finger (figure 19.1).
   d. Use a sharp probe to locate the foramen magnum between the occipital condyles in the midline between the frog's tympanic membranes.
   e. Insert the probe through the skin and into the foramen magnum, then quickly move the probe from side to side to separate the brain from the spinal cord.
   f. Slide the probe forward into the braincase, and continue to move the probe from side to side to destroy the brain.

Figure 19.1 Hold the frog's head between your thumb and index finger to pith (*a*) its brain and (*b*) its spinal cord.

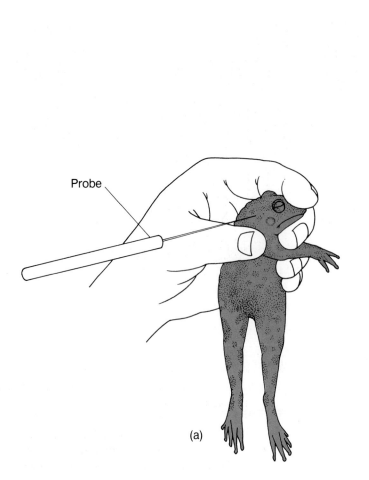

Probe

(a)

Probe

(b)

g. Remove the probe from the braincase, and insert it into the spinal cord through the same opening in the skin.

h. Move the probe up and down the spinal cord to destroy it. If the frog has been pithed correctly, its legs will be extended and relaxed. Also, the eyes will not respond when touched with a probe.

2. Remove the frog's gastrocnemius muscle by proceeding as follows:

a. Place the pithed frog in a dissecting tray.

b. Use scissors to cut through the skin completely around the leg in the thigh.

c. Pull the skin downward and off the leg.

d. Locate the gastrocnemius muscle in the calf and the Achilles tendon at its distal end.

e. Separate the Achilles tendon from the underlying tissue, using forceps.

f. Tie a thread firmly around the tendon (figure 19.2).

g. When the thread is secure, free the distal end of the tendon by cutting it with scissors.

h. Attach the frog muscle to the recording system in the manner suggested by your laboratory instructor (figures 18.1 and 18.3).

i. Insert the ends of the stimulator wires into the muscle so that one wire is located on either side of the belly of the muscle.

Keep the frog muscle moist at all times by dripping frog Ringer's solution on it. When the muscle is not being used, cover it with some paper toweling that has been saturated with frog Ringer's solution.

Before you begin operating the recording system and stimulator, have the laboratory instructor inspect your setup.

Figure 19.2 (*a*) Separate the Achilles tendon from the underlying tissue; (*b*) tie a thread around the tendon, and cut its distal attachment.

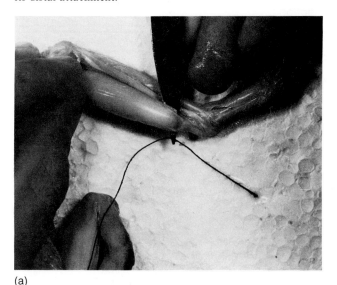

(a)

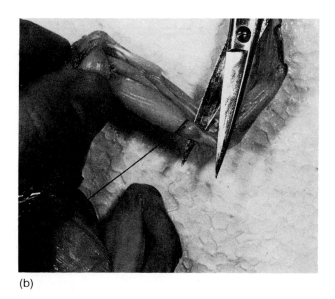

(b)

Procedure D—Threshold Stimulation

1. To determine the threshold or minimal strength (voltage) of electrical stimulation needed to elicit a contraction in the frog muscle:
   a. Set the stimulus duration to a minimum (about 0.1 milliseconds).
   b. Set the voltage to a minimum (about 0.1 volts).
   c. Set the stimulator so that it will administer single stimuli.
2. Administer a single stimulus to the muscle and watch to see if it responds. If no response is observed, increase the voltage to the next higher setting and repeat the procedure until the muscle responds by contracting.
3. After determining the threshold level of stimulation, continue to increase the voltage in increments of 1 or 2 volts until a maximal muscle contraction is obtained.
4. Complete Part B of the laboratory report.

Procedure E—Single Muscle Twitch

1. To record a single muscle twitch, set the voltage for a maximal muscle contraction as determined in Procedure D.
2. Set the paper speed at maximum, and with the paper moving, administer a single electrical stimulus to the frog muscle.
3. Repeat this procedure to obtain several recordings of single muscle twitches.
4. Complete Part C of the laboratory report.

Procedure F—Sustained Contraction

1. To record a sustained muscle contraction:
   a. Set the stimulator for continuous stimulation.
   b. Set the voltage for maximal muscle contraction as determined in Procedure D.
   c. Set the frequency of stimulation at a minimum.
   d. Set the paper speed at about 0.05 cm/sec.
   e. With the paper moving, administer electrical stimulation and slowly increase the frequency of stimulation until the muscle sustains a contraction (tetany).
   f. Continue to stimulate the muscle at the frequency that produces sustained contractions until the muscle fatigues and relaxes.
2. Every fifteen seconds for the next several minutes, stimulate the muscle to see how long it takes to recover from the fatigue.
3. Complete Part D of the laboratory report.

Optional Activity

To demonstrate the staircase effect (treppe), obtain a fresh frog gastrocnemius muscle and attach it to the recording system as before. Set the paper control for slow speed, and set the stimulator voltage to produce a maximal muscle contraction. Stimulate the muscle once each second for several seconds. How do you explain the differences in the lengths of successive muscle contractions? _____
_____
_____
_____

laboratory report 19

Skeletal Muscle Contraction

Part A

Complete the following statements:

1. The minimal intensity of stimulation necessary to trigger a muscle contraction is called the _____ .

2. When a muscle fiber is stimulated sufficiently to respond, it contracts completely. This phenomenon is called the _____ response.

3. A(an) _____ consists of a single motor neuron and all of the muscle fibers with which the neuron is associated.

4. As the intensity of stimulation increases, the number of motor units activated also increases. This phenomenon is called _____ .

5. When all possible motor units are activated, the muscle contracts with _____ tension.

6. The recording of a muscle contraction is called a(an) _____ .

7. When a muscle contracts and immediately relaxes, its action is called a(an) _____ .

8. The time between stimulation and response is called the _____ .

9. The time following an action potential, during which the muscle remains unresponsive to stimulation, is called the _____ .

10. The staircase effect may be caused by an increasing concentration of _____ that occurs in a rested muscle when it is subjected to a series of stimuli.

11. When a muscle is exposed to a series of high-frequency stimuli and cannot relax before the next stimulus of the series arrives, a(an) _____ contraction results.

12. A forceful, sustained contraction is called _____ .

13. The most useful skeletal muscle contractions are _____ contractions.

14. Even when a muscle is at rest, some contraction is occurring in its fibers. This phenomenon is called _____ .

15. When a person loses consciousness, the body collapses because the skeletal muscles lose _____ .

Part B

Complete the following:

1. What was the threshold voltage for stimulation of the frog gastrocnemius muscle? _____

2. What voltage produced maximal contraction of this muscle? _____

3. Do you think other frog muscles would respond in a similar way to these voltages of stimulation? _____
   Why do you think so? _____

   _____

   _____

Part C

Complete the following:

1. Fasten a recording of two single muscle twitches in the space below.

2. On a muscle twitch recording, label the *latent period, period of contraction, period of relaxation,* and indicate the time it took for each of these phases to occur. (See figure 9.15 in the textbook.)

3. What differences, if any, do you note in the two myograms of a single muscle twitch? How do you explain these differences? _____

   _____

   _____

Part D

Complete the following:

1. Fasten a recording of a sustained contraction in the space below.

2. On the sustained contraction recording, indicate when the muscle twitches began to combine (summate), and label the period of tetany and the period of fatigue.

3. At what frequency of stimulation did tetany occur? _____

4. How long did it take for the tetanic muscle to fatigue? _____

5. Is the length of muscle contraction at the beginning of tetany the same or different from the length of the single muscle contractions before tetany occurred? _____ How do you explain this? _____
    _____
    _____

6. How long did it take for the fatigued muscle to become responsive again? _____

Laboratory Exercise 20

Muscles of the Face and Head

The skeletal muscles of the face and head include the muscles of facial expression that lie just beneath the skin, the muscles of mastication that are attached to the mandible, and the muscles that move the head.

Purpose of the Exercise

To review the actions, origins, and insertions of the muscles of the face and head.

Learning Objectives

After completing this exercise, you should be able to:

1. Locate and identify the muscles of facial expression, the muscles of mastication, and the muscles that move the head.
2. Describe and demonstrate the action of each of these muscles.
3. Identify the origin and insertion of each of these muscles in a human skeleton.

Materials Needed:
    manikin
    human skull
    articulated human skeleton

Procedure

1. Review the sections entitled "Muscles of Facial Expression," "Muscles of Mastication," and "Muscles That Move the Head" in chapter 9 of the textbook.

2. As a review activity, label figures 20.1, 20.2, and 20.3.
3. Locate the following muscles in the manikin and in your own body whenever possible:

    epicranius

    orbicularis oculi

    orbicularis oris

    buccinator

    zygomaticus

    platysma

    masseter

    temporalis

    medial pterygoid

    lateral pterygoid

    sternocleidomastoid

    splenius capitis

    semispinalis capitis

    longissimus capitis

4. Demonstrate the action of these muscles in your own body.
5. Locate the origins and insertions of these muscles in the human skull and skeleton.
6. Complete Parts A, B, and C of Laboratory Report 20 on pages 137 and 138.

Figure 20.1 Label the muscles of expression and mastication.

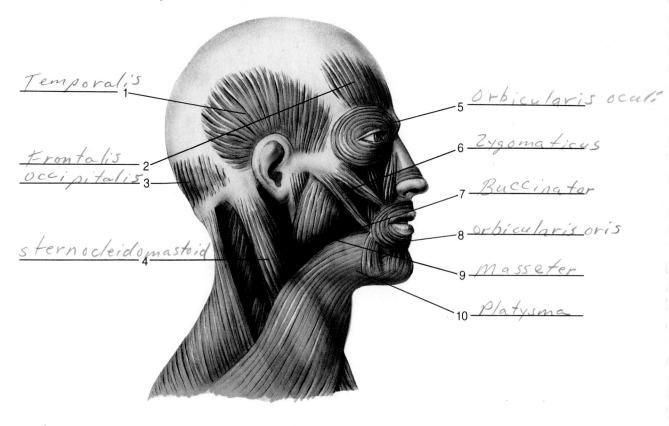

Temporalis — 1

Frontalis — 2
occipitalis — 3

sternocleidomastoid — 4

Orbicularis oculi — 5

Zygomaticus — 6

Buccinator — 7

orbicularis oris — 8

Masseter — 9

Platysma — 10

Figure 20.2 Label these muscles of mastication.

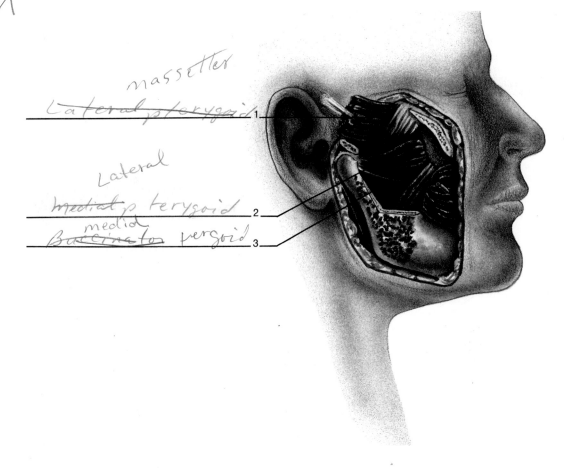

masseter
Lateral pterygoid — 1

Lateral
medial pterygoid — 2
medial
Buccinator pterygoid — 3

Figure 20.3 Label the deep muscles in the back of the neck that help to move the head.

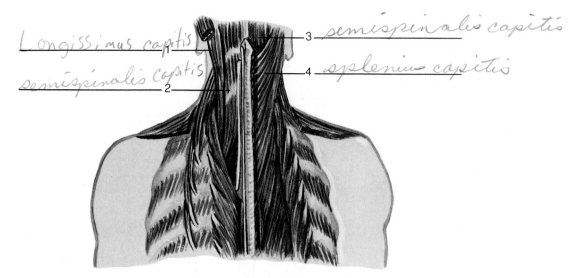

Longissimus capitis 1

semispinalis capitis 2

3 semispinalis capitis

4 splenius capitis

Name _____

Date _____

Section _____

Muscles of the Face and Head

Part A

Complete the following statements:

1. When the _____zygomaticus_____ contracts, the corner of the mouth is drawn upward.

2. The _____buccinator_____ acts to compress the wall of the cheeks when air is blown out of the mouth.

3. The _____orbicularis oris_____ causes the lips to pucker.

4. The _____masseter~~lateral pterygoid~~_____ and _____platysma_____ help to lower the mandible.

5. The temporalis acts to _____raise jaw_____ .

6. The _____medial_____ pterygoid can close the jaw.

7. The _____lateral_____ pterygoid can pull the jaw sideways.

8. The _____sternocleidomastoid_____ can raise the sternum during forceful inhalation.

9. The _____sternocleidomastoid_____ can bend the head toward the chest.

10. The _____splenius capitis_____ can bend the head to one side or bring it into an upright position.

Part B

Name the muscle indicated by the following combinations of origin and insertion.

Origin	Insertion	Muscle
1. occipital bone	skin and muscle around eye	*Epicranus*
2. zygomatic bone	orbicularis oris	*zygo*
3. zygomatic arch	lateral surface of mandible	*maceter*
4. sphenoid bone	anterior surface of mandibular condyle	*lateral ptar*
5. anterior surface of sternum and upper clavicle	mastoid process of temporal bone	*sterclidomastoid*
6. outer surfaces of mandible and maxilla	orbicularis oris	*buccinetor*
7. fascia in upper chest	lower border of mandible	*patyma*
8. temporal bone	coronoid process and anterior ramus of mandible	*temps*
9. spinous processes of cervical and thoracic vertebrae	mastoid process of temporal bone	*spin capitis*
10. processes of cervical and thoracic vertebrae	occipital bone	*sem n capti*

Part C

Identify the muscles of various facial expressions in the photographs of figure 20.4.

a. *Epicran eye — zygo mouth*

b. *pytimus mouth batt zygo mouth*

c. *oblider oris mouth obiculo otic*

Figure 20.4 Identify the muscles of facial expression being contracted by the model in each of these photographs.

(a) (b) (c)

Muscles of the Chest, Shoulder, and Arm

The muscles of the chest and shoulder are responsible for moving the scapula and upper arm, while those within the arm act to move joints in the elbow, wrist, hand, and fingers.

Purpose of the Exercise

To review the actions, origins, and insertions of the muscles in the chest, shoulder, and arm.

Learning Objectives

After completing this exercise, you should be able to:

1. Locate and identify the muscles of the chest, shoulder, and arm.
2. Describe and demonstrate the action of each of these muscles.
3. Identify the origin and insertion of each of these muscles.

Materials Needed:
    manikin
    articulated human skeleton
    muscular models of the arm and hand

Procedure

1. Review the sections entitled "Muscles That Move the Pectoral Girdle," "Muscles That Move the Upper Arm," "Muscles That Move the Forearm," and "Muscles That Move the Wrist, Hand, and Fingers" in chapter 9 of the textbook.
2. As a review activity, label figures 21.1, 21.2, 21.3, and 21.4.

3. Locate the following muscles in the manikin and models of the arm and hand. Also, locate in your own body as many of the muscles as you can.

trapezius

rhomboideus major

levator scapulae

serratus anterior

pectoralis minor

coracobrachialis

pectoralis major

teres major

latissimus dorsi

supraspinatus

deltoid

subscapularis

infraspinatus

teres minor

biceps brachii

brachialis

brachioradialis

triceps brachii

supinator

pronator teres

pronator quadratus

Figure 21.1 Label the muscles of the posterior shoulder.

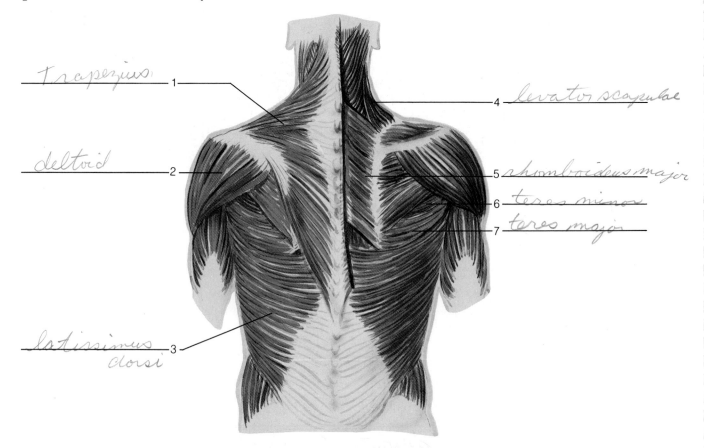

Trapezius 1

deltoid 2

latissimus dorsi 3

4 levator scapulae

5 rhomboideus major

6 teres minor

7 teres major

flexor carpi radialis

flexor carpi ulnaris

palmaris longus

flexor digitorum profundus

flexor digitorum superficialis

extensor carpi radialis longus

extensor carpi radialis brevis

extensor carpi ulnaris

extensor digitorum

4. Demonstrate the action of these muscles in your own body.
5. Locate the origins and insertions of these muscles in the human skeleton.
6. Complete Parts A, B, C, and D of Laboratory Report 21 on pages 145–48.

Figure 21.2 Label the muscles of the anterior chest.

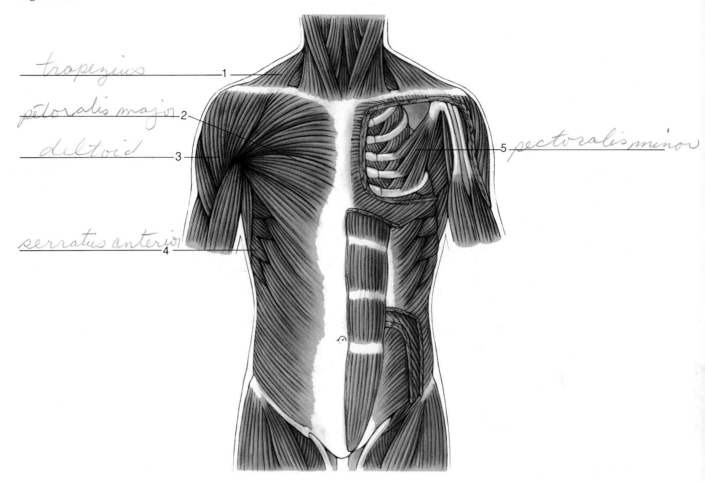

trapezius ——— 1

pectoralis major — 2

deltoid ——— 3

serratus anterior — 4

5 — pectoralis minor

Figure 21.3 Label (*a*) the muscles of the posterior shoulder and upper arm and (*b*) the muscles of the anterior shoulder and upper arm.

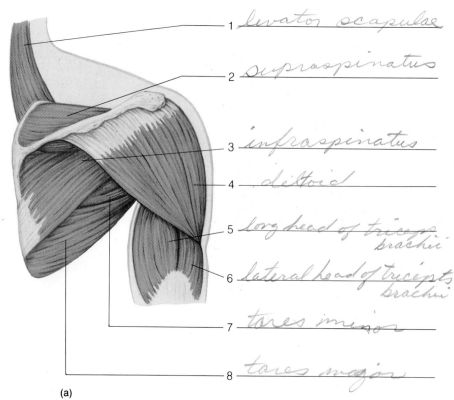

1 — *levator scapulae*

2 — *supraspinatus*

3 — *infraspinatus*

4 — *deltoid*

5 — *long head of triceps brachii*

6 — *lateral head of triceps brachii*

7 — *teres minor*

8 — *teres major*

(a)

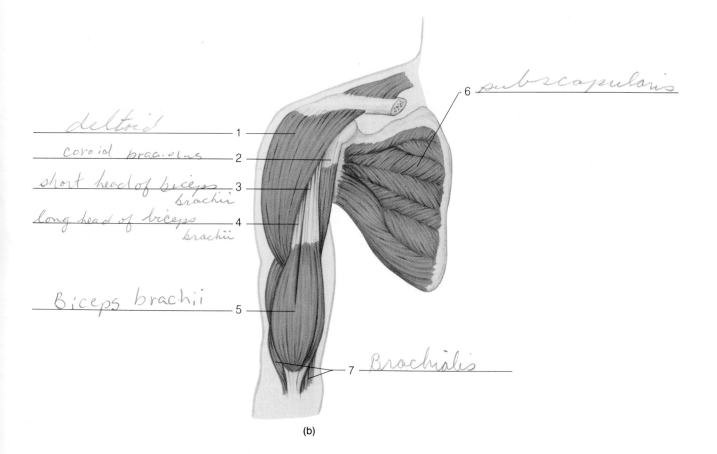

deltoid — 1

coroid bracioलns — 2

short head of biceps brachii — 3

long head of biceps brachii — 4

Biceps brachii — 5

6 — *subscapularis*

7 — *Brachialis*

(b)

Figure 21.4 Label (*a*) the muscles of the anterior forearm and (*b*) the muscles of the posterior forearm.

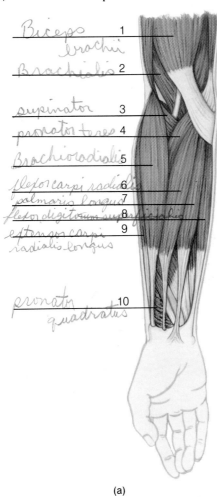

Biceps
brachii — 1
Brachialis — 2

supinator — 3
pronator teres — 4
Brachioradialis — 5
flexor carpi radialis — 6
palmaris longus — 7
flexor digitorum superficialis — 8
extensor carpi
radialis longus — 9

cario alarge

pronator
quadratus — 10

(a)

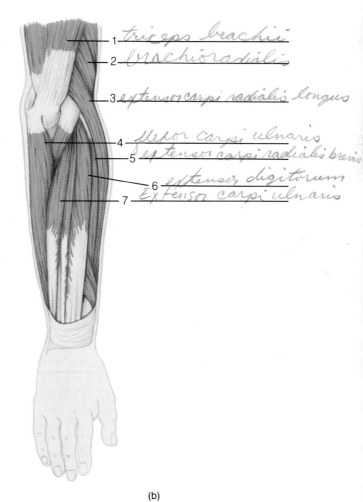

1 — triceps brachii
2 — brachioradialis
3 — extensor carpi radialis longus
4 — flexor carpi ulnaris
5 — extensor carpi radialis brevis
6 — extensor digitorum
7 — Extensor carpi ulnaris

(b)

Muscles of the Chest, Shoulder, and Arm

Part A

Match the muscles in column A with the actions in column B. Place the letter of your choice in the space provided.

Column A		Column B
A. brachialis	C 1.	abducts upper arm
B. coracobrachialis	H 2.	pulls upper arm forward and across chest
C. deltoid	E 3.	flexes and adducts wrist
D. extensor carpi ulnaris	K 4.	raises scapula and adducts it
E. flexor carpi ulnaris	J 5.	turns palm downward
F. flexor digitorum profundus	I 6.	raises ribs in forceful inhalation
G. infraspinatus	M 7.	rotates forearm laterally
H. pectoralis major	L 8.	pulls scapula downward and anteriorly
I. pectoralis minor	A 9.	flexes elbow
J. pronator teres	B 10.	flexes and adducts upper arm
K. rhomboideus major	O 11.	extends elbow
L. serratus anterior	N 12.	adducts and rotates upper arm medially
M. supinator	D 13.	extends and adducts wrist
N. teres major	G 14.	rotates upper arm laterally
O. triceps brachii	F 15.	flexes distal joints of fingers

Part B

Identify the features indicated in the cross section of the forearm illustrated in figure 21.5.

1. _____radius_____
2. _____radius_____
3. _____
4. _____parad teres_____
5. _____radialus_____
6. _____ex to_____
7. _____
8. _____ex_____
9. _____c_____
10. _____

Figure 21.5 Identify the features indicated in this cross section of the forearm.

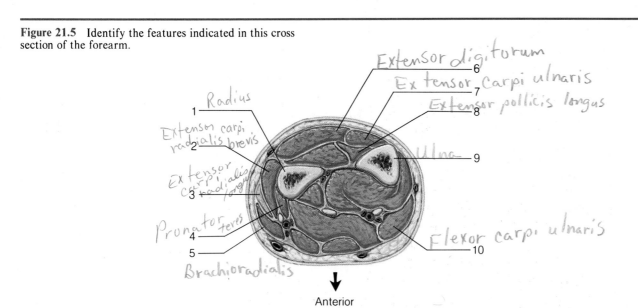

Extensor digitorum
Extensor carpi ulnaris
Extensor pollicis longus

Radius
1

Extensor carpi
radialis brevis
2

Extensor
carpi
radialis
3 longus

Pronator teres
4

5

Brachioradialis

Ulna
9

Flexor carpi ulnaris
10

Anterior

Part C

Name the muscle indicated by the following combinations of origin and insertion.

	Origin	Insertion	Muscle
1.	spines of upper thoracic vertebrae	medial border of scapula	_____
2.	outer surfaces of upper ribs	ventral surface of scapula	_____
3.	sternal ends of upper ribs	coracoid process of scapula	_____
4.	coracoid process of scapula	shaft of humerus	_____
5.	lateral border of scapula	intertubercular groove of humerus	_____
6.	anterior surface of scapula	lesser tubercle of humerus	Subscap
7.	lateral border of scapula	greater tubercle of humerus	subsca minor
8.	anterior shaft of humerus	coronoid process of ulna	_____
9.	medial epicondyle of humerus	lateral surface of radius	_____
10.	anterior distal end of ulna	anterior distal end of radius	pronator quad
11.	distal lateral end of humerus	lateral surface of radius	_____
12.	medial epicondyle of humerus	base of second and third metacarpals	_____
13.	medial epicondyle of humerus	fascia of palm	_____
14.	distal end of humerus	base of second metacarpal	ext long
15.	lateral epicondyle of humerus	base of fifth metacarpal	ex long

Part D

Identify the muscles indicated in figure 21.6.

1. pectoralis major
2. Serratus anterior
3. biceps brachii
4. External oblique
5. trapezius
6. deltoid
7. rectus abdominis
8. Trapezius
9. Teres major
10. biceps brachii
11. triceps brachii
12. deltoid

13. Infraspinatus
14. Latissimus dorsi
15. deltoid
16. pectoralis major
17. Biceps brachii
18. serratus anterior
19. Flexor carpi radialis Brachial
20. Extensor carpi radialis longus
21. Brachioradialis Ex digitorum
22. Trapezius
23. Triceps Brachii
24. Latissimus dorsi

Figure 21.6 Identify the muscles that appear as body surface features in these photographs (a–c).

(a)

Figure 21.6—*Continued*

8

9

10

11

12

13

14

(b)

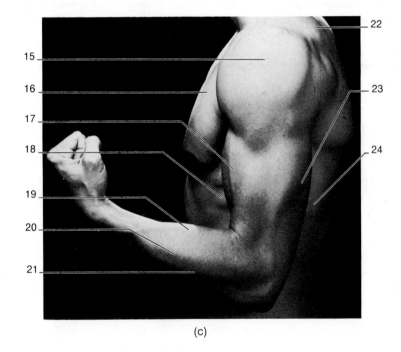

15

16

17

18

19

20

21

22

23

24

(c)

Muscles of the Abdominal Wall and Pelvic Outlet

The anterior and lateral walls of the abdomen contain broad, flattened muscles arranged in layers. These muscles connect the rib cage and vertebral column to the pelvic girdle.

The muscles of the pelvic outlet are arranged in two muscular sheets: (1) a deeper pelvic diaphragm that forms the floor of the pelvic cavity and (2) a urogenital diaphragm that fills the space within the pubic arch.

Purpose of the Exercise

To review the actions, origins, and insertions of the muscles of the abdominal wall and pelvic outlet.

Learning Objectives

After completing this exercise, you should be able to:

1. Locate and identify the muscles of the abdominal wall and pelvic outlet.
2. Describe the action of each of these muscles.
3. Identify the origin and insertion of each of these muscles.

Materials Needed:
manikin
articulated human skeleton
muscular models of male and female pelves

Procedure

1. Review the sections entitled "Muscles of the Abdominal Wall" and "Muscles of the Pelvic Outlet" in chapter 9 of the textbook.
2. As a review activity, label figures 22.1, 22.2, and 22.3.
3. Locate the following muscles in the manikin:

   external oblique

   internal oblique

   transversus abdominis

   rectus abdominis

4. Demonstrate the actions of these muscles in your own body.
5. Locate the origin and insertion of each of these muscles in the human skeleton.
6. Complete Part A of Laboratory Report 22 on page 152.
7. Locate the following muscles in the models of the male and female pelves:

   levator ani

   coccygeus

   superficial transversus perinei

   bulbospongiosus

   ischiocavernosus

8. Locate the origin and insertion of each of these muscles in the human skeleton.
9. Complete Part B of the laboratory report.

Figure 22.1 Label the muscles of the abdominal wall.

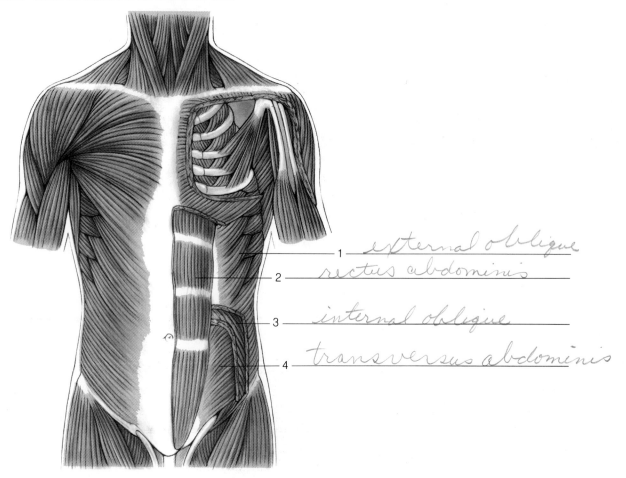

1 _external oblique_

2 _rectus abdominis_

3 _internal oblique_

4 _transversus abdominis_

Figure 22.2 Label the muscles of the male pelvic outlet.

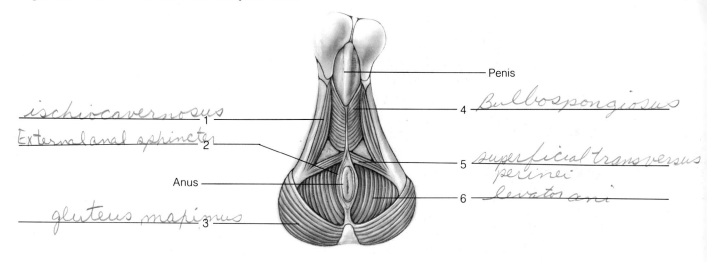

ischiocavernosus 1

External anal sphincter 2

Anus —

gluteus maximus 3

— Penis

4 — Bulbospongiosus

5 — superficial transversus perinei

6 — levator ani

Figure 22.3 Label the muscles of the female pelvic outlet.

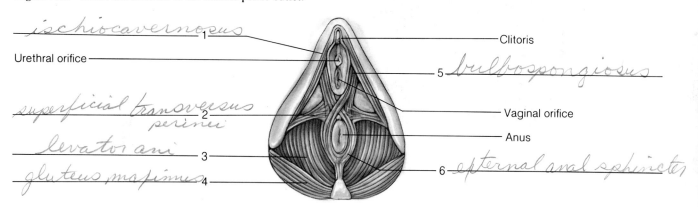

ischiocavernosus 1

Urethral orifice —

superficial transversus perinei 2

levator ani 3

gluteus maximus 4

— Clitoris

5 — bulbospongiosus

— Vaginal orifice

— Anus

6 — external anal sphincter

Name _____

Date _____

Section _____

Muscles of the Abdominal Wall and Pelvic Outlet

Part A

Complete the following statements:

1. A band of tough connective tissue in the midline of the anterior abdominal wall called the _____ serves as a muscle attachment.

2. The _____ connects the ribs and sternum to the pubic bones.

3. The _____ forms the third layer (deepest layer) of the abdominal wall muscles.

4. The function of the external oblique muscle is to _____.

5. The action of the rectus abdominis is to _____.

Part B

Complete the following statements:

1. The levator ani and coccygeus together form the _____.

2. The levator ani provides a sphincterlike action in the _____.

3. The action of the coccygeus is to _____.

4. The _____ surrounds the base of the penis.

5. In females, the bulbospongiosus acts to _____.

6. The ischiocavernosus extends from the margin of the pubic arch to the _____.

7. In the female, the _____ muscles are separated by the vagina.

8. The function of the superficial transversus perinei is to _____.

9. The coccygeus extends from the coccyx and sacrum to the _____.

10. The _____ assists in closing the urethra.

Muscles of the Thigh and Leg

The muscles that move the thigh are attached to the femur and to some part of the pelvic girdle. Those attached anteriorly act primarily to flex the thigh, while those attached posteriorly act to extend, abduct, or rotate it.

The muscles that move the lower leg connect the tibia or fibula to the femur or to the pelvic girdle. They function to flex or extend the knee. Other muscles, located in the lower leg, act to move the ankle, foot, and toes.

Purpose of the Exercise

To review the actions, origins, and insertions of the muscles that move the thigh, leg, ankle, foot, and toes.

Learning Objectives

After completing this exercise, you should be able to:

1. Locate and identify the muscles that move the thigh, leg, ankle, foot, and toes.
2. Describe and demonstrate the actions of each of these muscles.
3. Identify the origin and insertion of each of these muscles.

Materials Needed:
manikin
articulated human skeleton
muscular models of the leg and foot

Procedure

1. Review the sections entitled "Muscles That Move the Thigh," "Muscles That Move the Lower Leg," and "Muscles That Move the Ankle, Foot, and Toes" in chapter 9 of the textbook.
2. As a review activity, label figures 23.1, 23.2, 23.3, 23.4, 23.5, and 23.6.
3. Locate the following muscles in the manikin and leg and foot models. Also locate as many of them as possible in your own body.

psoas major

iliacus

gluteus maximus

gluteus medius

gluteus minimus

tensor fasciae latae

adductor longus

adductor magnus

gracilis

biceps femoris

semitendinosus

semimembranosus

sartorius

Figure 23.1 Label the muscles of the anterior right thigh.

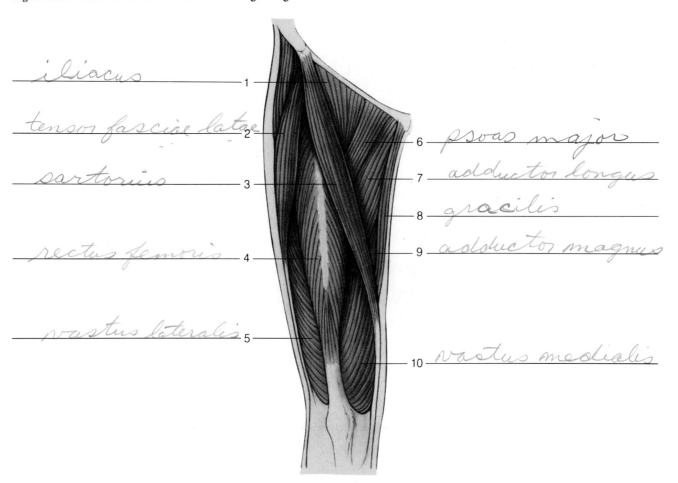

iliacus 1

tensor fasciae latae 2

sartorius 3

rectus femoris 4

vastus lateralis 5

psoas major 6

adductor longus 7

gracilis 8

adductor magnus 9

vastus medialis 10

quadriceps femoris

tibialis anterior

peroneus tertius

extensor digitorum longus

gastrocnemius

soleus

flexor digitorum longus

tibialis posterior

peroneus longus

4. Demonstrate the action of each of these muscles in your own body.
5. Locate the origin and insertion of each of these muscles in the human skeleton.
6. Complete Parts A, B, C, and D of Laboratory Report 23 on pages 158–60.

Figure 23.2 Label the muscles of the lateral right thigh.

gluteus medius 1

tensor fasciae latae 4

gluteus maximus 2

sartorius 5

rectus femoris 6

Biceps femoris 3

vastus lateralis 7

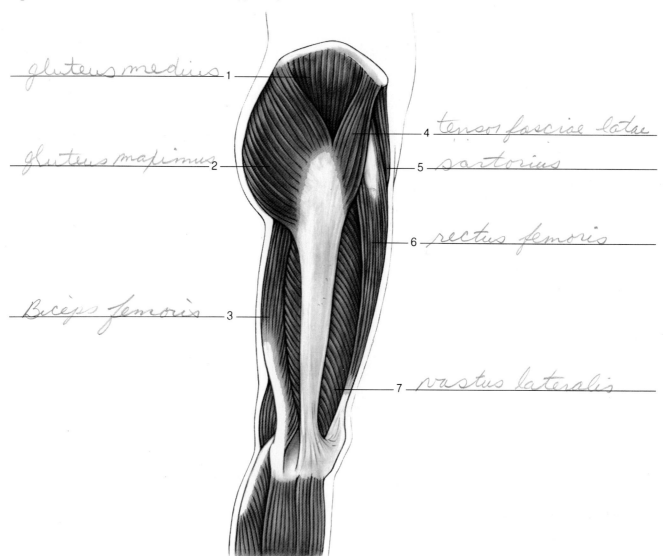

Figure 23.3 Label the muscles of the posterior right thigh.

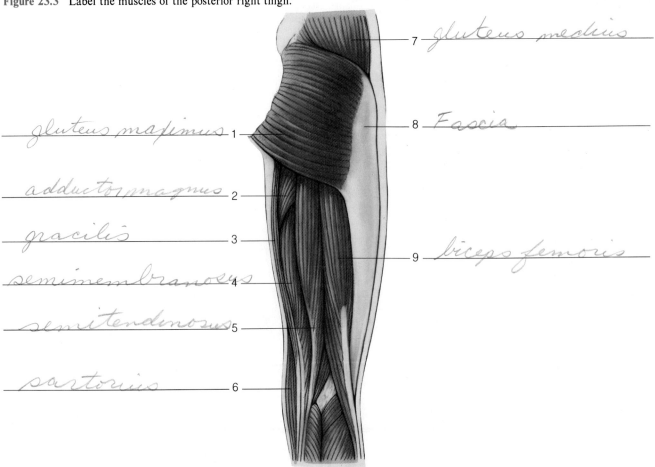

gluteus maximus 1

adductor magnus 2

gracilis 3

semimembranosus 4

semitendinosus 5

sartorius 6

7 _gluteus medius_

8 _Fascia_

9 _biceps femoris_

Figure 23.4 Label the muscles of the anterior right lower leg.

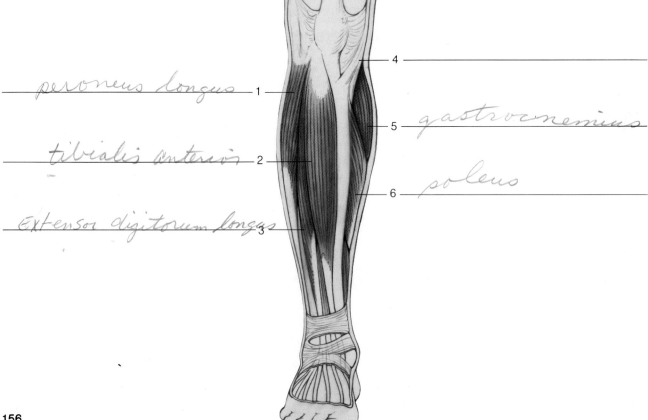

peroneus longus 1

tibialis anterior 2

Extensor digitorum longus 3

4

5 _gastrocnemius_

6 _soleus_

156

Figure 23.5 Label the muscles of the lateral right lower leg.

gastrocnemius — 1

peroneus longus — 2

soleus — 3

4 — tibialis anterior

5 — Extensor digitorum longus

6 —

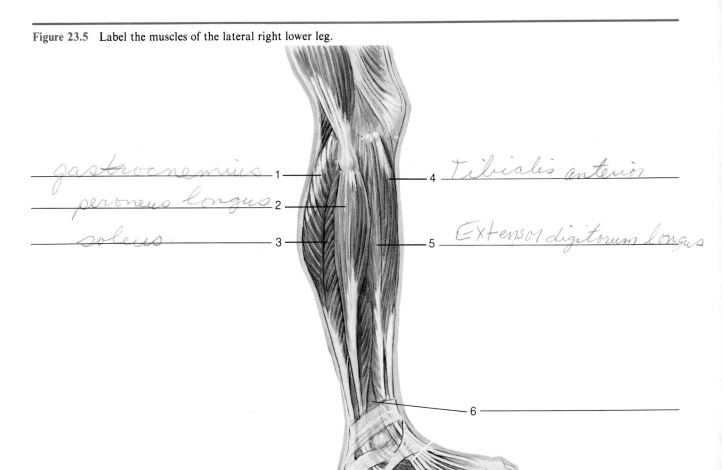

Figure 23.6 Label the muscles of the medial right lower leg.

tibialis anterior — 1

2 — gastrocnemius

3 — soleus

4 —

5 —

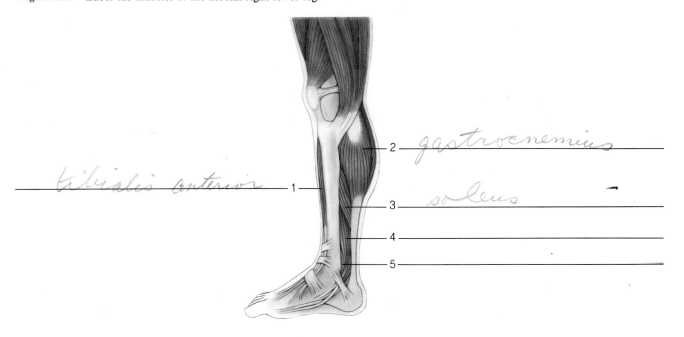

Muscles of the Thigh and Leg

Part A

Match the muscles in column A with the actions in column B. Place the letter of your choice in the space provided.

Column A   **Column B**

A. biceps femoris   ____ 1. adducts thigh and rotates leg medially
B. gluteus medius   ____ 2. plantar flexion and eversion of foot
C. gracilis   ____ 3. flexes thigh
D. peroneus longus   ____ 4. abducts thigh and rotates it laterally
E. peroneus tertius   ____ 5. dorsal flexion and eversion of foot
F. psoas major   ____ 6. abducts thigh and rotates it medially
G. quadriceps femoris   ____ 7. plantar flexion and inversion of foot
H. sartorius   ____ 8. flexes and rotates leg laterally
I. tibialis anterior   ____ 9. extends the knee
J. tibialis posterior   ____ 10. dorsal flexion and inversion of foot

Part B

Name the muscle indicated by the following combinations of origin and insertion.

	Origin	**Insertion**	**Muscle**
1.	lateral surface of ilium	greater trochanter of femur	_____
2.	ischial tuberosity	posterior surface of femur	_____
3.	spine of ilium	medial surface of tibia	_____
4.	lateral and medial condyles of femur	posterior surface of calcaneus	_____
5.	anterior iliac crest	fascia of the thigh	_____
6.	greater trochanter and posterior surface of femur	tibial tuberosity	_____
7.	ischial tuberosity	medial surface of tibia	_____
8.	medial surface of femur	tibial tuberosity	_____
9.	posterior surface of tibia	distal phalanges of four lateral toes	_____
10.	lateral condyle and lateral surface of tibia	cuneiform and first metatarsal	_____

Figure 23.7 Identify the features indicated in this cross section of the lower leg.

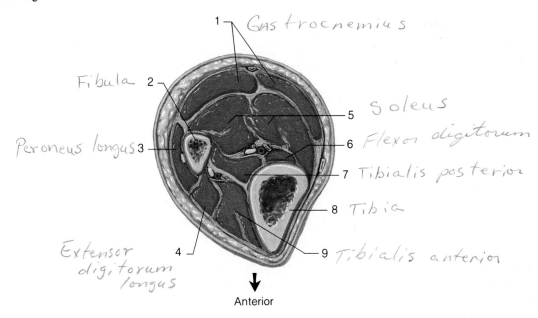

1 Gastrocnemius

Fibula 2

Peroneus longus 3

5 Soleus

6 Flexor digitorum

7 Tibialis posterior

8 Tibia

Extensor digitorum longus 4

9 Tibialis anterior

↓
Anterior

Part C

Identify the features indicated in the cross section of the lower leg illustrated in figure 23.7.

1. _____   6. _____

2. _____   7. _____

3. _____   8. _____

4. _____   9. _____

5. _____

Part D

Identify the muscles indicated in figure 23.8.

1. _____   9. _____

2. _____   10. _____

3. _____   11. _____

4. _____   12. _____

5. _____   13. _____

6. _____   14. _____

7. _____   15. _____

8. _____

Figure 23.8 Identify the muscles that appear as leg surface features in these photographs (a–c).

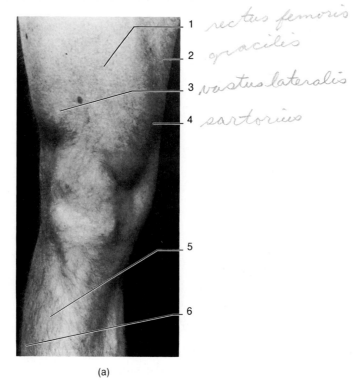

1 *rectus femoris*
2 *gracilis*
3 *vastus lateralis*
4 *sartorius*
5
6

(a)

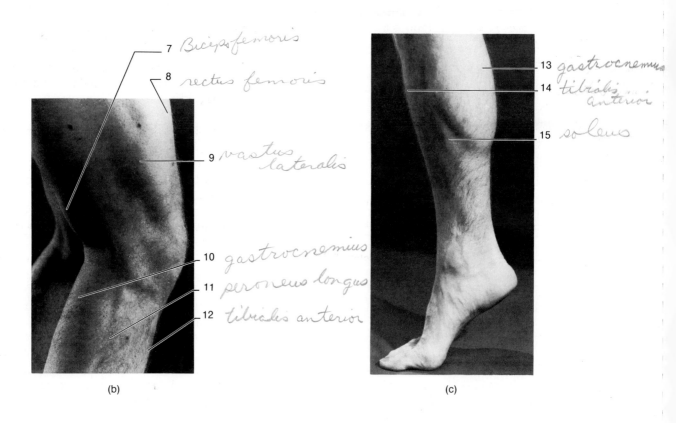

7 *Biceps femoris*
8 *rectus femoris*
9 *vastus lateralis*
10 *gastrocnemius*
11 *peroneus longus*
12 *tibialis anterior*

(b)

13 *gastrocnemius*
14 *tibialis anterior*
15 *soleus*

(c)

Cat Dissection: Musculature

Although the primary aim of this exercise is to become more familiar with the human musculature, human cadavers are not always available for dissection. Instead, embalmed cats often are used for dissection because they are relatively small and can be purchased from biological suppliers. Also, as mammals, cats have many features in common with humans, including similar skeletal muscles (with similar names).

On the other hand, cats make use of four limbs for support while humans use only two limbs. Because the musculature of each type of organism is adapted to provide for its special needs, comparisons of the muscles of cats and humans may not be precise.

Purpose of the Exercise

To observe the musculature of the cat and to compare it with that of the human.

Learning Objectives

After completing this exercise, you should be able to:

1. Name and locate the major skeletal muscles of the cat.
2. Name and locate the corresponding muscles of the human.
3. Name the origins, insertions, and actions of the muscles designated by the laboratory instructor.

Materials Needed:

embalmed cat	identification tag
dissecting tray	disposable gloves
dissecting instruments	human manikin
large plastic bag	human arm and leg models

Procedure A—Skin Removal

1. Cats usually are embalmed with a mixture of alcohol, formaldehyde, and glycerol, which prevents microorganisms from causing the tissues to decompose. However, because fumes from this embalming fluid may be annoying and may irritate skin, be sure to work in a well-ventilated room and wear disposable gloves to protect your hands.
2. Obtain an embalmed cat, a dissecting tray, a set of dissecting instruments, a large plastic bag, and an identification tag.
3. If the cat is in a storage bag, dispose of any excess embalming fluid in the bag, as directed by the laboratory instructor. However, you may want to save some of the embalming fluid to keep the cat moist until you have completed your work later in the year.
4. If the cat is too wet with embalming fluid, blot the cat dry with paper towels.
5. Remove the skin from the cat. To do this:
   a. Place the cat in the dissecting tray with its ventral side down.
   b. Use a sharp scalpel to make a short, shallow incision through the skin in the dorsal midline at the base of the cat's neck.
   c. Insert the pointed blade of a dissecting scissors through the opening in the skin and cut along the dorsal midline, following the tips of the vertebral spinous processes to the base of the tail (figure 24.1).
   d. Use the scissors to make an incision encircling the region of the tail, anus, and genital organs. Do not remove the skin from this area.
   e. Use bone shears or a bone saw to sever the tail at its base and discard it.
   f. From the incision at the base of the tail, cut along the lateral surface of each hind limb to the paw, and encircle the ankle. Also clip off the claws from each paw to prevent them from tearing the plastic storage bag.

Figure 24.1 Incisions to be made for removing the skin of the cat.

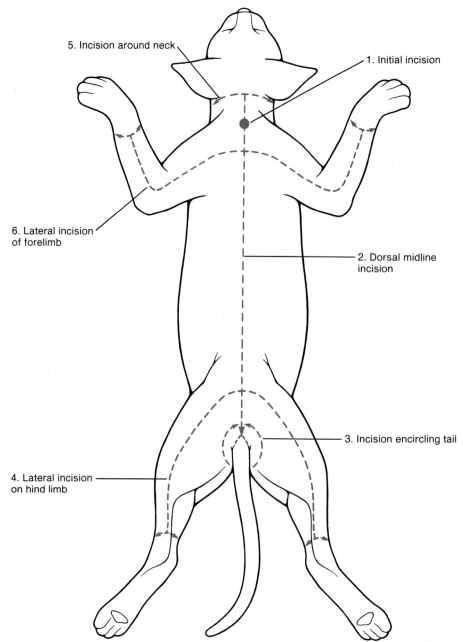

5. Incision around neck

1. Initial incision

6. Lateral incision of forelimb

2. Dorsal midline incision

3. Incision encircling tail

4. Lateral incision on hind limb

g. Make an incision from the initial cut at the base of the neck to encircle the neck.

h. From the incision around the neck, cut along the lateral surface of each forelimb to the paw and encircle the wrist (figure 24.1).

i. Grasp the skin on either side of the dorsal midline incision, and carefully pull the skin laterally away from the body. At the same time, use your fingers or the blunt end of the scalpel handle to help separate the skin from the underlying connective tissue. As you pull the skin away, you may note a thin sheet of muscle tissue attached to it. This muscle is called the *cutaneous maximus* and it functions to move the cat's skin. Humans lack the cutaneous maximus, but a similar sheet of muscle (platysma) is present in the neck.

j. As you remove the skin, work toward the ventral surface, then work toward the head, and finally work toward the tail. Pull the skin over each limb as if you were removing a glove.

k. If the cat is a female, note the mammary glands on the ventral surface of the thorax and abdomen. (See reference plate 1.) These glands will adhere to the skin and can be removed from it and discarded.

Chart 24.1   Muscles of the head and neck

Muscle	Origin	Insertion	Action
Sternomastoid	Manubrium of sternum	Mastoid process of temporal bone	Turns and depresses head
Sternohyoid	Costal cartilage	Hyoid bone	Depresses hyoid bone
Digastric	Mastoid process and occipital bone	Ventral border of mandible	Depresses mandible
Mylohyoid	Inner surface of mandible	Hyoid bone	Raises floor of mouth
Masseter	Zygomatic arch	Coronoid fossa of mandible	Elevates mandible
Hyoglossus	Hyoid bone	Tongue	Pulls tongue back
Styloglossus	Styloid process	Tip of tongue	Pulls tongue back
Geniohyoid	Inner surface of mandible	Hyoid bone	Pulls hyoid bone forward
Cleidomastoid	Clavicle	Mastoid process	Turns head to side
Thyrohyoid	Thyroid cartilage	Hyoid bone	Raises larynx
Sternothyroid	Sternum	Thyroid cartilage	Pulls larynx back

l. After the skin has been pulled away, carefully remove as much of the remaining connective tissue as possible to expose the underlying skeletal muscles. The muscles should appear light brown and fibrous.

6. After skinning the cat:
   a. Discard the tissues you have removed as directed by the laboratory instructor.
   b. Wrap the skin around the cat to help keep its body moist, and place it in a plastic storage bag.
   c. Write your name in pencil on an identification tag, and tie the tag to the storage bag so that you can identify your specimen.
7. Wash and dry the dissecting tray and instruments, and rinse and dry the laboratory table.
8. Wash your hands thoroughly with soap and water.

Procedure B—Skeletal Muscle Dissection

1. The purpose of a skeletal muscle dissection is to separate the individual muscles from any surrounding tissues and thus expose the muscles for observation. To do a muscle dissection:
   a. Use the appropriate figure as a guide and locate the muscle to be dissected in the specimen.
   b. Use a blunt probe or a finger to separate the muscle from the surrounding connective tissue along its natural borders. The muscle should separate smoothly. If the border appears ragged, you probably have torn the muscle fibers.
2. If it is necessary to cut through (transect) a superficial muscle to observe a deeper one, use scissors to transect the muscle about halfway between its origin and insertion. Then, lift aside the cut ends, leaving their attachments intact.
3. The following procedures will instruct you to dissect some of the larger and more easily identified muscles of the cat. In each case, the procedure will include the names of the muscles to be dissected, figures illustrating their locations, and charts listing the origins, insertions, and actions of the muscles.

(More detailed dissection instructions can be obtained by consulting a guide or atlas for cat dissection.) As you dissect each muscle, study the figures and charts in chapter 9 of the textbook and identify any corresponding (homologous) muscles of the human body. Also locate these homologous muscles in the manikin or models of the arm and leg.

Procedure C—Muscles of the Head and Neck

1. Place the cat in the dissecting tray with its ventral side up.
2. Remove the skin and underlying connective tissues from one side of the neck, forward to the chin and up to the ear.
3. Study figure 24.2 and reference plate 2, then locate and dissect the following muscles:

sternomastoid	mylohyoid
sternohyoid	masseter
digastric	

4. To find the deep muscles of the cat's neck and chin, carefully remove the sternomastoid, sternohyoid, and mylohyoid muscles from the right side.
5. Study figure 24.3, and locate the following muscles:

hyoglossus	cleidomastoid
styloglossus	thyrohyoid
geniohyoid	sternothyroid

6. See chart 24.1 for the origins, insertions, and actions of these head and neck muscles.
7. Complete Part A of Laboratory Report 24 on page 174.

Figure 24.2 Muscles on the ventral surface of the cat's head and neck.

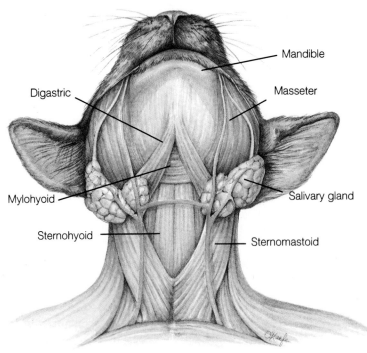

Figure 24.3 Deep muscles of the cat's neck and chin regions.

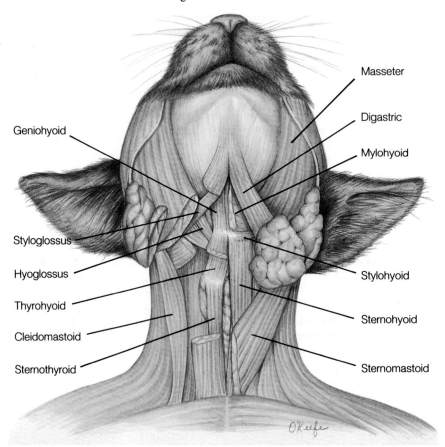

Figure 24.4 Muscles of the cat's thorax and abdomen.

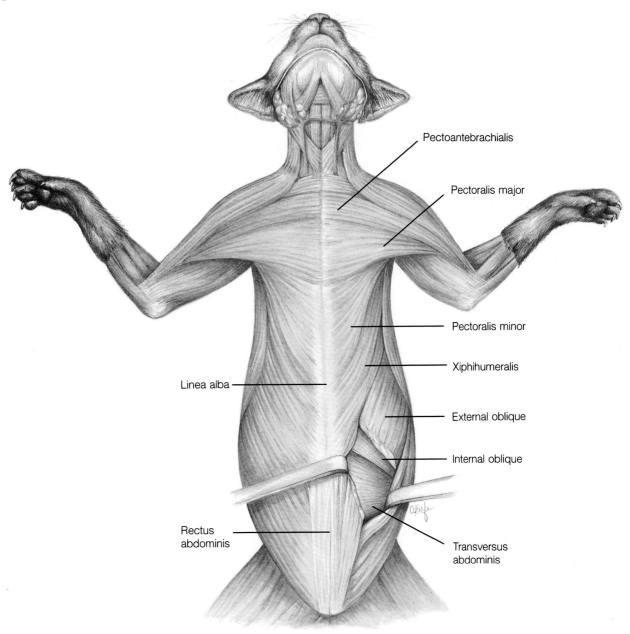

Pectoantebrachialis

Pectoralis major

Pectoralis minor

Xiphihumeralis

Linea alba

External oblique

Internal oblique

Rectus
abdominis

Transversus
abdominis

Procedure D—Muscles of the Thorax

1. Place the cat in the dissecting tray with its ventral side up.
2. Remove any remaining fat and connective tissue to expose the muscles in the walls of the thorax and abdomen.
3. Study figure 24.4 and reference plate 3, then locate and dissect the following pectoral muscles:

   pectoantebrachialis

   pectoralis major

   pectoralis minor

   xiphihumeralis

4. To find the deep thoracic muscles, transect the pectoralis minor, pectoralis major, and pectoantebrachialis muscles, and move their cut edges to the sides. Remove the underlying fat and connective tissues to expose the following muscles, as shown in figure 24.5:

   scalenus

   serratus ventralis

   levator scapulae

   transversus costarum

5. See chart 24.2 for the origins, insertions, and actions of these muscles.

Figure 24.5 Deep muscles of the cat's thorax.

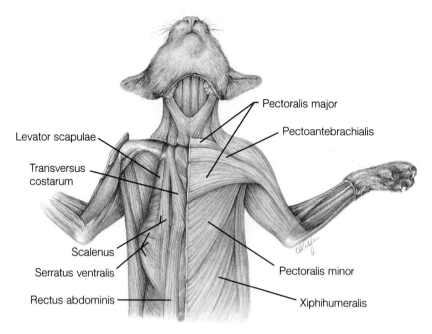

Pectoralis major

Pectoantebrachialis

Levator scapulae

Transversus costarum

Scalenus

Serratus ventralis

Rectus abdominis

Pectoralis minor

Xiphihumeralis

Chart 24.2   Muscles of the thorax

Muscle	Origin	Insertion	Action
Pectoantebrachialis	Manubrium of sternum	Fascia of forelimb	Adducts forelimb
Pectoralis major	Manubrium of sternum and costal cartilages	Greater tubercle of humerus	Adducts and rotates forelimb
Pectoralis minor	Sternum	Proximal end of humerus	Adducts and rotates forelimb
Xiphihumeralis	Xiphoid process of sternum	Proximal end of humerus	Adducts forelimb
Scalenus	Ribs	Cervical vertebrae	Flexes neck
Serratus ventralis	Ribs	Scapula	Pulls scapula ventrally and posteriorly
Levator scapulae	Cervical vertebrae	Scapula	Pulls scapula ventrally and anteriorly
Transversus costarum	Sternum	First rib	Pulls sternum toward head

Procedure E—Muscles of the Abdominal Wall

1. Study figure 24.4 and reference plate 4.
2. Locate the *external oblique muscle* in the abdominal wall.
3. Make a shallow, longitudinal incision through the external oblique. Lift up the cut edge and expose the *internal oblique muscle* beneath (figure 24.4). Note that the fibers of the internal oblique run at a right angle to those of the external oblique.
4. Make a longitudinal incision through the internal oblique. Lift up the cut edge and expose the *transversus abdominis*.
5. Expose the *rectus abdominis muscle* on one side of the midventral line. This muscle lies beneath an aponeurosis.

6. See chart 24.3 for the origins, insertions, and actions of these muscles.
7. Complete Part B of the laboratory report.

Procedure F—Muscles of the Shoulder and Back

1. Place the cat in the dissecting tray with its dorsal side up.
2. Remove any remaining fat and connective tissue to expose the muscles of the shoulder and back.
3. Study figure 24.6 and reference plate 5, then locate and dissect the following superficial muscles:

clavotrapezius   acromiodeltoid

acromiotrapezius   spinodeltoid

spinotrapezius   latissimus dorsi

clavobrachialis

Chart 24.3 Muscles of the abdominal wall

Muscle	Origin	Insertion	Action
External oblique	Ribs and fascia of back	Linea alba	Compresses abdominal wall
Internal oblique	Fascia of back	Linea alba	Compresses abdominal wall
Transversus abdominis	Lower ribs	Linea alba	Compresses abdominal wall
Rectus abdominis	Symphysis pubis	Sternum and costal cartilages	Compresses abdominal wall and flexes trunk

Figure 24.6 Superficial muscles of the cat's shoulder and back (*right side*); deep muscles of the shoulder and back (*left side*).

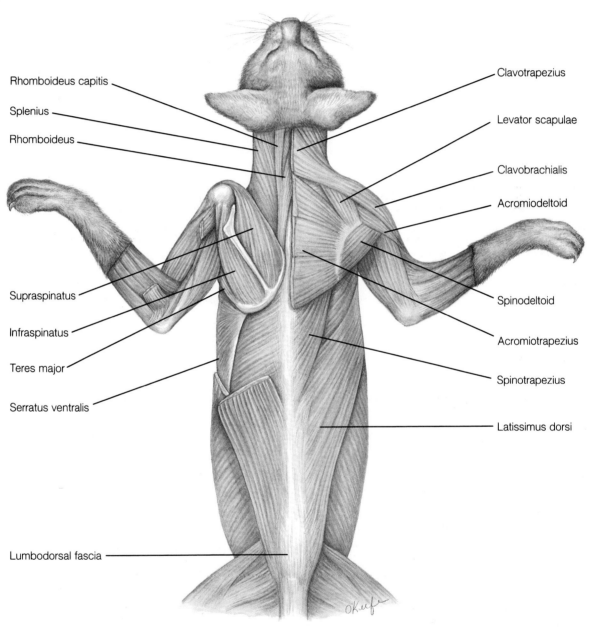

Rhomboideus capitis

Splenius

Rhomboideus

Supraspinatus

Infraspinatus

Teres major

Serratus ventralis

Lumbodorsal fascia

Clavotrapezius

Levator scapulae

Clavobrachialis

Acromiodeltoid

Spinodeltoid

Acromiotrapezius

Spinotrapezius

Latissimus dorsi

Chart 24.4 Muscles of the shoulder and back

Muscle	Origin	Insertion	Action
Clavotrapezius	Occipital bone of skull and spines of cervical vertebrae	Clavicle	Pulls clavicle upward
Acromiotrapezius	Spines of cervical and thoracic vertebrae	Spine and acromion process of scapula	Pulls scapula upward
Spinotrapezius	Spines of thoracic vertebrae	Spine of scapula	Pulls scapula upward and back
Clavobrachialis	Clavicle	Ulna near semilunar notch	Flexes forelimb
Acromiodeltoid	Acromion process of scapula	Proximal end of humerus	Flexes and rotates forelimb
Spinodeltoid	Spine of scapula	Proximal end of humerus	Flexes and rotates forelimb
Latissimus dorsi	Thoracic and lumbar vertebrae	Proximal end of humerus	Pulls forelimb upward and back
Supraspinatus	Fossa above spine of scapula	Greater tubercle of humerus	Extends forelimb
Infraspinatus	Fossa below spine of scapula	Greater tubercle of humerus	Rotates forelimb
Teres major	Posterior border of scapula	Proximal end of humerus	Rotates forelimb
Rhomboideus	Spines of cervical and thoracic vertebrae	Medial border of scapula	Pulls scapula back
Rhomboideus capitis	Occipital bone of skull	Medial border of scapula	Pulls scapula forward
Splenius	Fascia of neck	Occipital bone	Turns and raises head

4. Using scissors, transect the latissimus dorsi and the group of trapezius muscles. Lift aside their cut edges and remove any underlying fat and connective tissue. Study figure 24.6 and reference plate 6, then locate and dissect the following deep muscles of the shoulder and back:

supraspinatus

infraspinatus

teres major

rhomboideus

rhomboideus capitis

splenius

5. See chart 24.4 for the origins, insertions, and actions of these muscles in the shoulder and back.
6. Complete Part C of the laboratory report.

Procedure G—Muscles of the Forelimb

1. Place the cat in the dissecting tray with its ventral side up.
2. Remove any remaining fat and connective tissue from a forelimb to expose the muscles.
3. Study figure 24.7(a), then locate and dissect the following muscles from the medial surface of the upper forelimb:

epitrochlearis

triceps brachii

4. Transect the pectoral muscles (pectoantebrachialis, pectoralis major, and pectoralis minor) and lift aside their cut edges.
5. Study figure 24.7(b) and reference plate 7, then locate and dissect the following muscles:

coracobrachialis

biceps brachii

long head of the triceps brachii

medial head of the triceps brachii

6. Study figure 24.8(a) and reference plate 8, which show the lateral surface of the upper forelimb.
7. Locate the lateral head of the triceps muscle, transect it, and lift aside its cut edges.
8. Study figure 24.8(b). Locate and dissect the following deeper muscles:

anconeus

brachialis

9. Separate the muscles in the lower forelimb into an *extensor group* (figure 24.8a) and a *flexor group* (figure 24.7a). Generally, the extensors are located on the posterior surface of the lower forelimb and the flexors are located on the anterior surface.
10. See chart 24.5 for the origins, insertions, and actions of these muscles of the forelimb.
11. Complete Part D of the laboratory report.

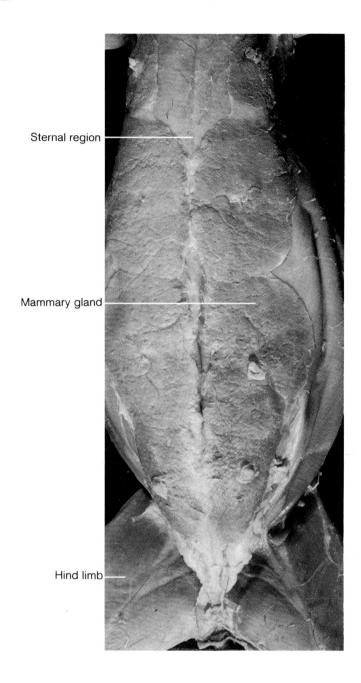

Sternal region

Mammary gland

Hind limb

Reference plate 2 Deep muscles on the right ventral side of the cat's neck.

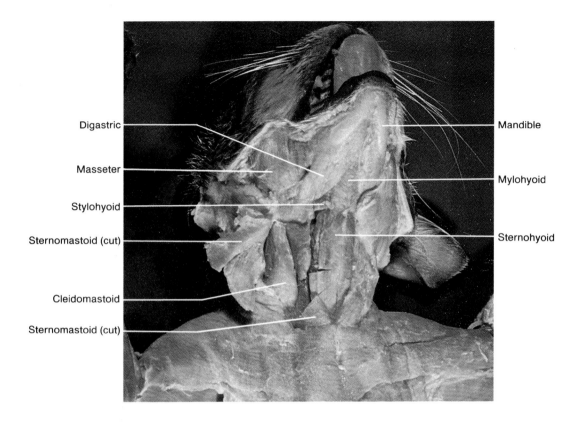

Digastric

Masseter

Stylohyoid

Sternomastoid (cut)

Cleidomastoid

Sternomastoid (cut)

Mandible

Mylohyoid

Sternohyoid

Reference plate 3 Superficial muscles of the cat's thorax.

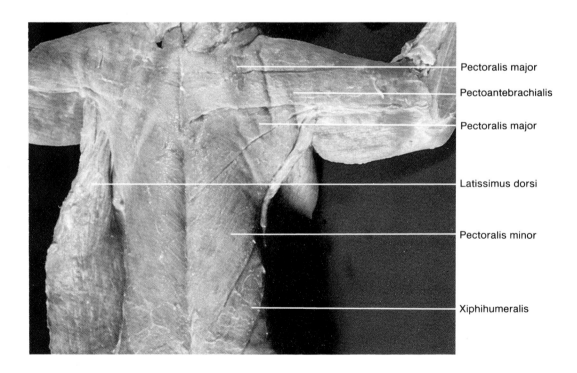

Pectoralis major

Pectoantebrachialis

Pectoralis major

Latissimus dorsi

Pectoralis minor

Xiphihumeralis

Reference plate 4 Superficial muscles of the cat's thorax and abdomen.

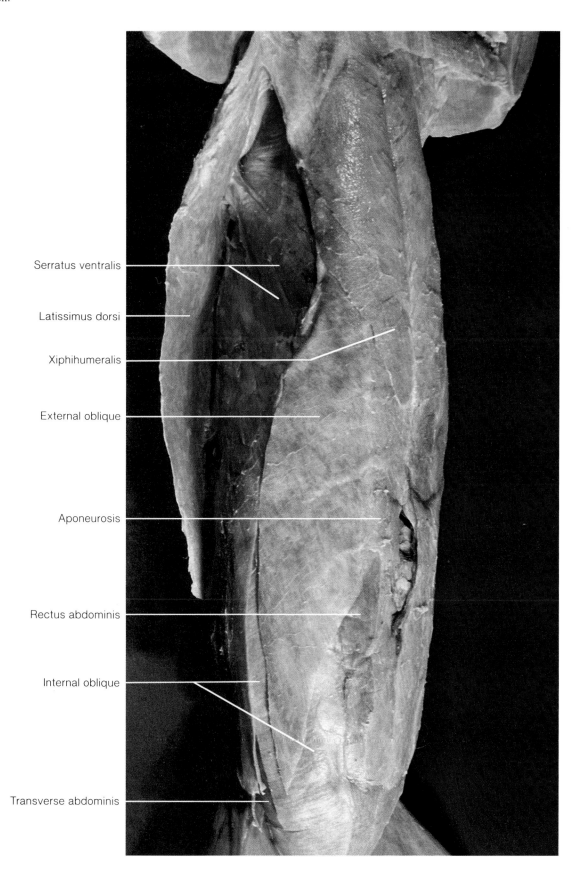

Serratus ventralis

Latissimus dorsi

Xiphihumeralis

External oblique

Aponeurosis

Rectus abdominis

Internal oblique

Transverse abdominis

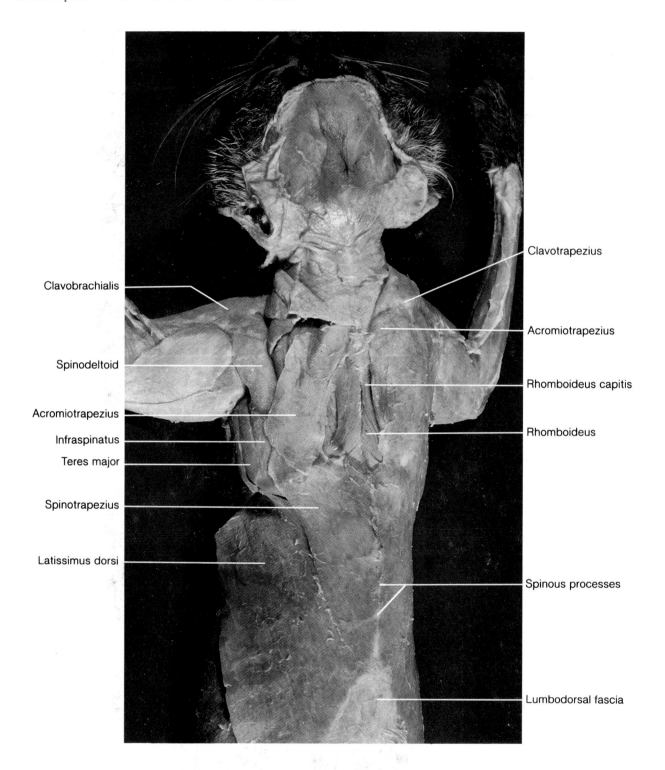

Clavobrachialis

Spinodeltoid

Acromiotrapezius

Infraspinatus

Teres major

Spinotrapezius

Latissimus dorsi

Clavotrapezius

Acromiotrapezius

Rhomboideus capitis

Rhomboideus

Spinous processes

Lumbodorsal fascia

Reference plate 6 Deep muscles of the cat's back.

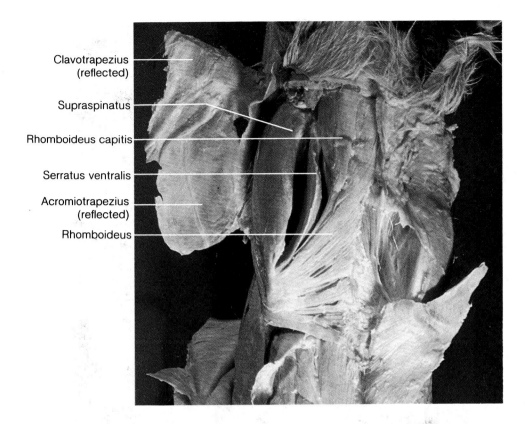

Clavotrapezius (reflected)

Supraspinatus

Rhomboideus capitis

Serratus ventralis

Acromiotrapezius (reflected)

Rhomboideus

Reference plate 7 Deep muscles of the cat's right medial forelimb.

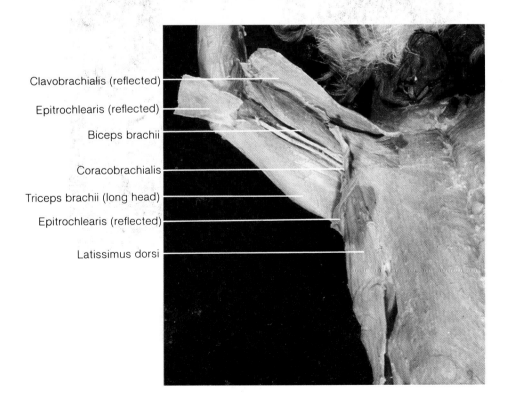

Clavobrachialis (reflected)

Epitrochlearis (reflected)

Biceps brachii

Coracobrachialis

Triceps brachii (long head)

Epitrochlearis (reflected)

Latissimus dorsi

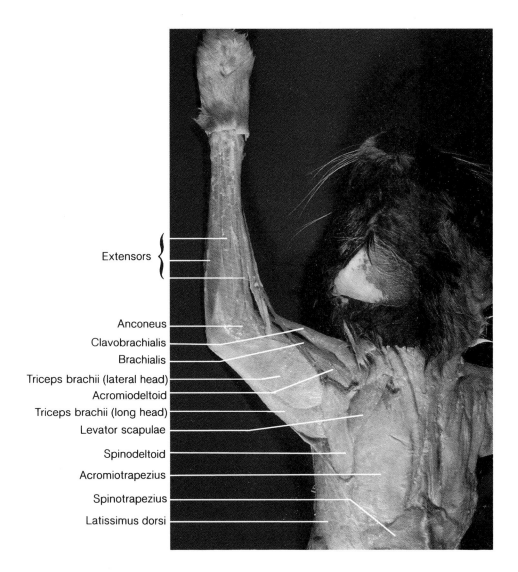

Extensors {

Anconeus

Clavobrachialis

Brachialis

Triceps brachii (lateral head)

Acromiodeltoid

Triceps brachii (long head)

Levator scapulae

Spinodeltoid

Acromiotrapezius

Spinotrapezius

Latissimus dorsi

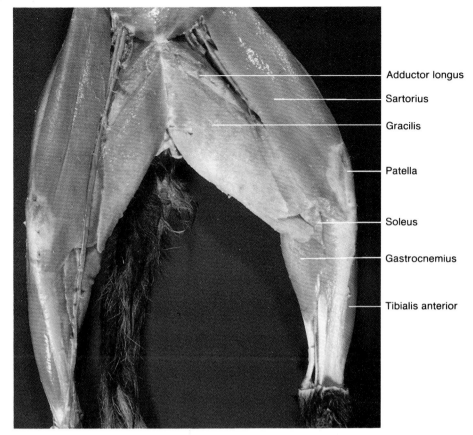

Adductor longus

Sartorius

Gracilis

Patella

Soleus

Gastrocnemius

Tibialis anterior

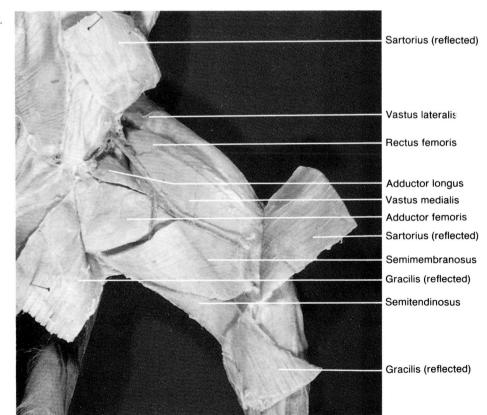

Sartorius (reflected)

Vastus lateralis

Rectus femoris

Adductor longus

Vastus medialis

Adductor femoris

Sartorius (reflected)

Semimembranosus

Gracilis (reflected)

Semitendinosus

Gracilis (reflected)

Reference plate 11 Muscles of the cat's hind limb, lateral view.

Sartorius
Gluteus medius
Tensor fasciae latae
Gluteus maximus
Caudofemoralis
Biceps femoris
Semitendinosus
Gastrocnemius
Extensor
Soleus
Peroneus

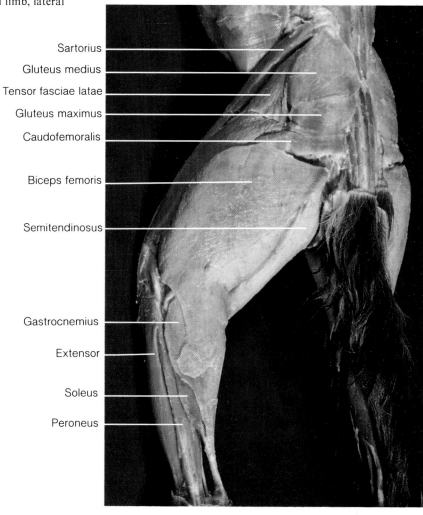

Reference plate 12 Deep muscles of the cat's hind limb, lateral view.

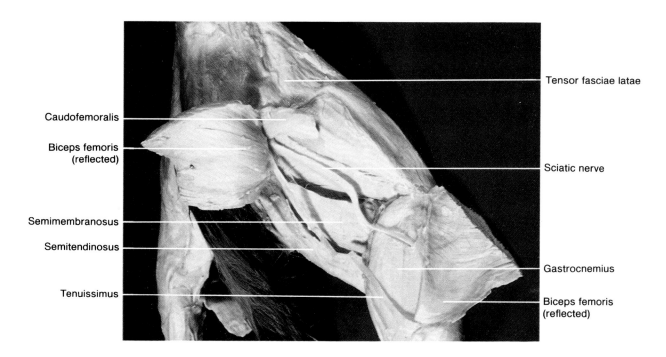

Caudofemoralis
Biceps femoris
(reflected)
Semimembranosus
Semitendinosus
Tenuissimus

Tensor fasciae latae
Sciatic nerve
Gastrocnemius
Biceps femoris
(reflected)

Reference plate 13 Organs of the cat's thoracic and abdominal cavities.

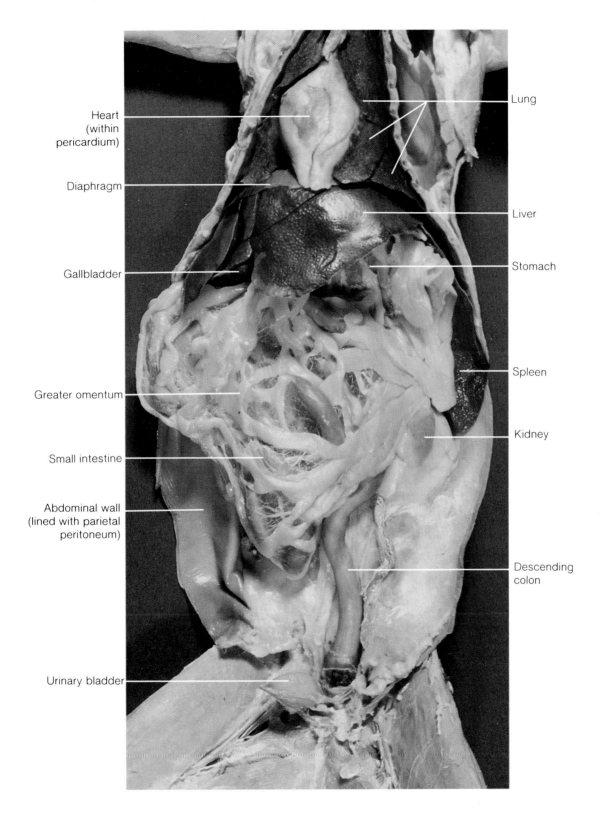

Heart (within pericardium)

Diaphragm

Gallbladder

Greater omentum

Small intestine

Abdominal wall (lined with parietal peritoneum)

Urinary bladder

Lung

Liver

Stomach

Spleen

Kidney

Descending colon

Reference plate 14 Abdominal organs of the cat with the greater omentum removed.

Left lateral lobe of liver

Caudate lobe of liver

Gallbladder

Left median lobe of liver

Common bile duct

Right median lobe of liver

Jejunum

Kidney

Ileum

Right horn of uterus

Lesser omentum

Stomach

Duodenum

Pancreas

Spleen

Colon

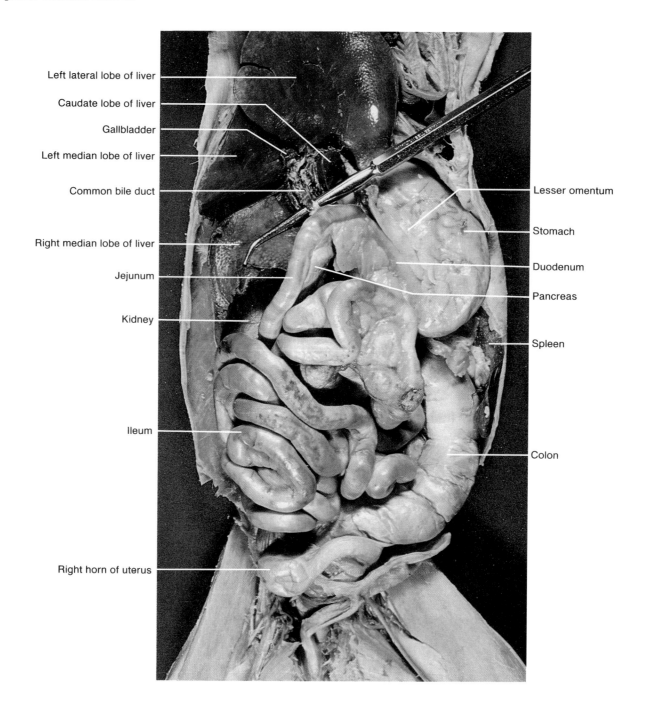

Reference plate 15 Features of the cat's neck and thoracic cavity.

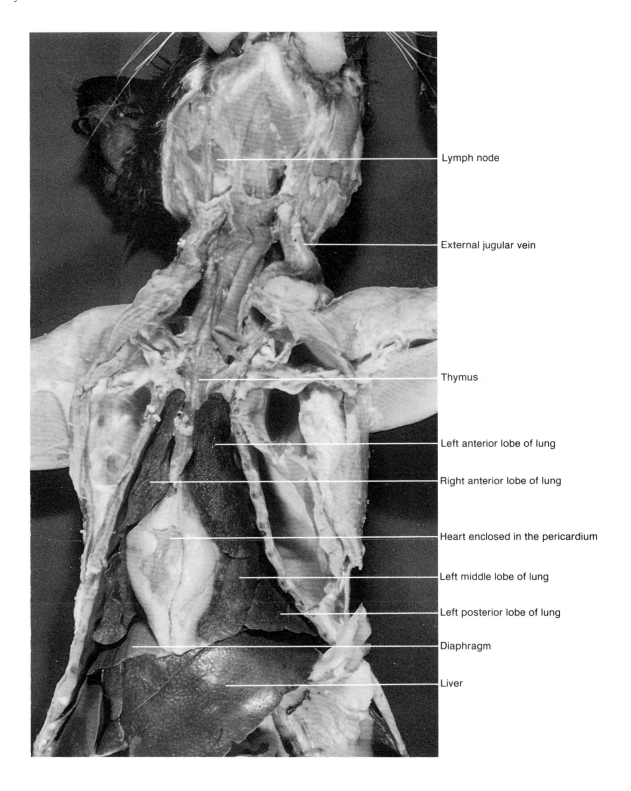

Lymph node

External jugular vein

Thymus

Left anterior lobe of lung

Right anterior lobe of lung

Heart enclosed in the pericardium

Left middle lobe of lung

Left posterior lobe of lung

Diaphragm

Liver

Reference plate 16 Features of the cat's neck and oral cavity.

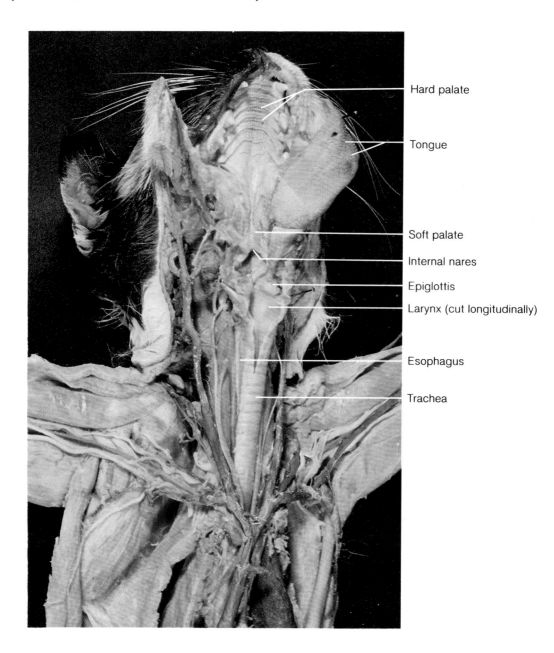

Hard palate

Tongue

Soft palate

Internal nares

Epiglottis

Larynx (cut longitudinally)

Esophagus

Trachea

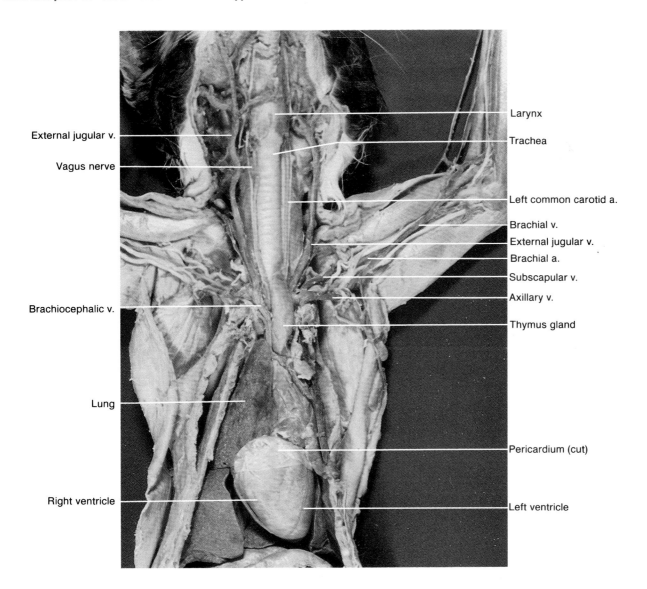

External jugular v.

Vagus nerve

Brachiocephalic v.

Lung

Right ventricle

Larynx

Trachea

Left common carotid a.

Brachial v.

External jugular v.

Brachial a.

Subscapular v.

Axillary v.

Thymus gland

Pericardium (cut)

Left ventricle

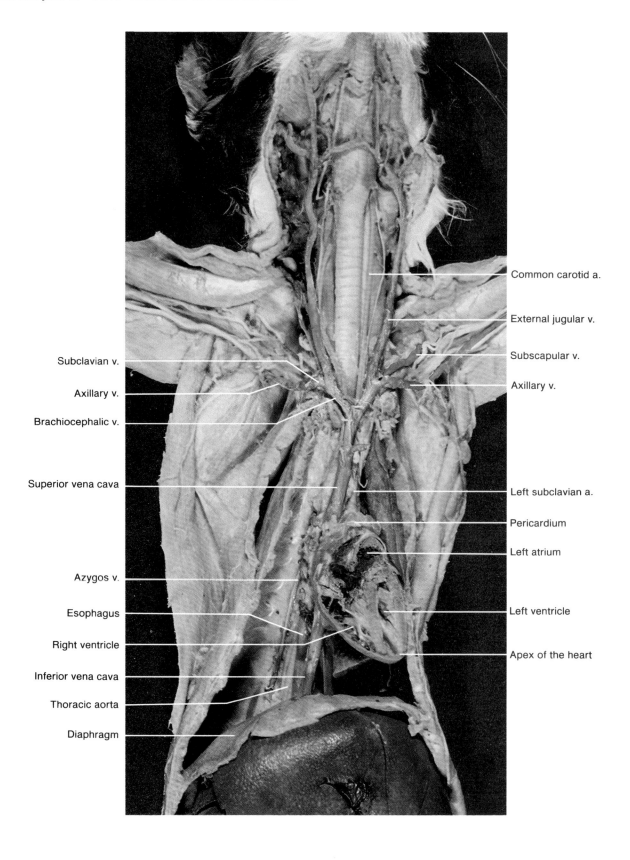

Common carotid a.

External jugular v.

Subscapular v.

Axillary v.

Subclavian v.

Axillary v.

Brachiocephalic v.

Superior vena cava

Left subclavian a.

Pericardium

Left atrium

Azygos v.

Left ventricle

Esophagus

Right ventricle

Apex of the heart

Inferior vena cava

Thoracic aorta

Diaphragm

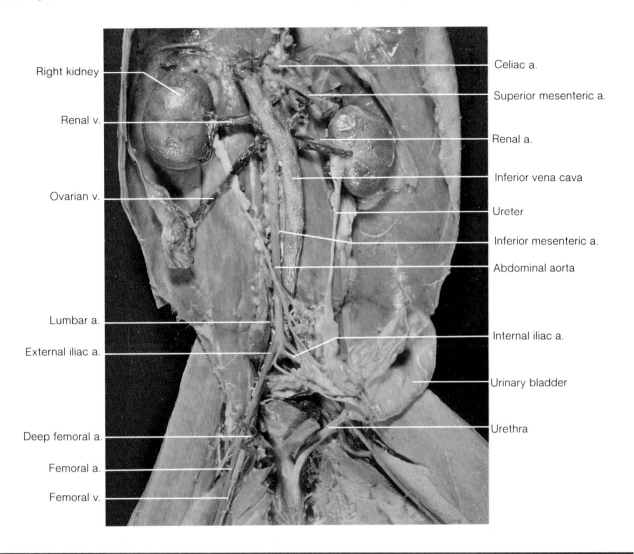

Right kidney

Renal v.

Ovarian v.

Lumbar a.

External iliac a.

Deep femoral a.

Femoral a.

Femoral v.

Celiac a.

Superior mesenteric a.

Renal a.

Inferior vena cava

Ureter

Inferior mesenteric a.

Abdominal aorta

Internal iliac a.

Urinary bladder

Urethra

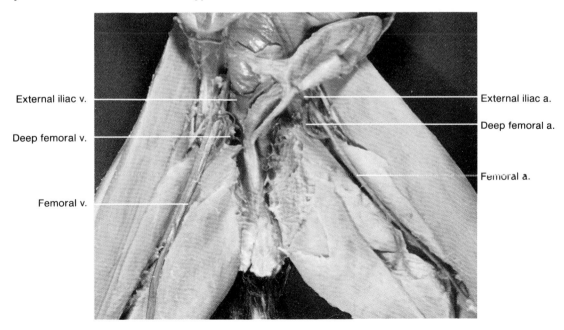

External iliac v.

Deep femoral v.

Femoral v.

External iliac a.

Deep femoral a.

Femoral a.

Reference plate 21 Female reproductive system of the cat.

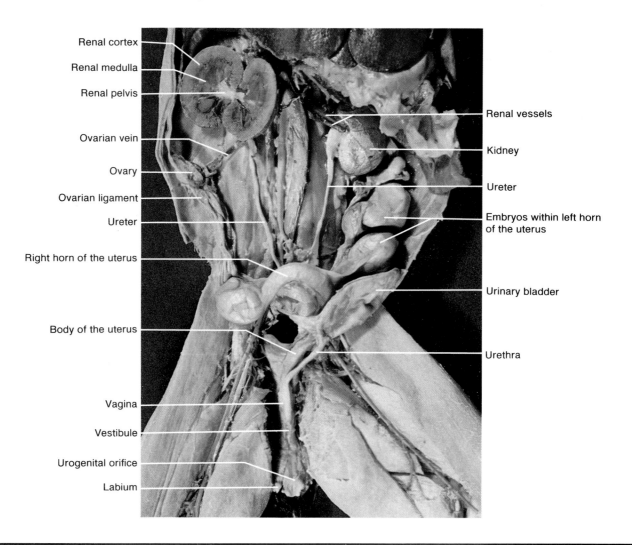

Renal cortex
Renal medulla
Renal pelvis
Ovarian vein
Ovary
Ovarian ligament
Ureter
Right horn of the uterus
Body of the uterus
Vagina
Vestibule
Urogenital orifice
Labium

Renal vessels
Kidney
Ureter
Embryos within left horn of the uterus
Urinary bladder
Urethra

Reference plate 22 Male reproductive system of the cat.

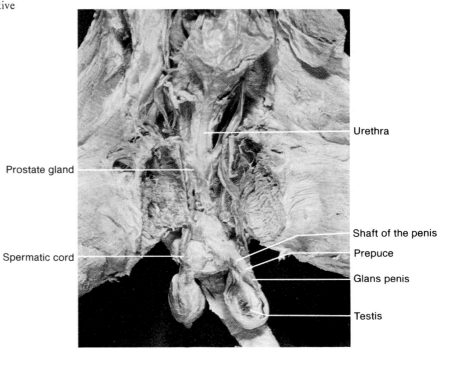

Prostate gland
Spermatic cord

Urethra
Shaft of the penis
Prepuce
Glans penis
Testis

Figure 24.7 Superficial muscles of the cat's medial forelimb (*a*); deep muscles of the medial forelimb (*b*).

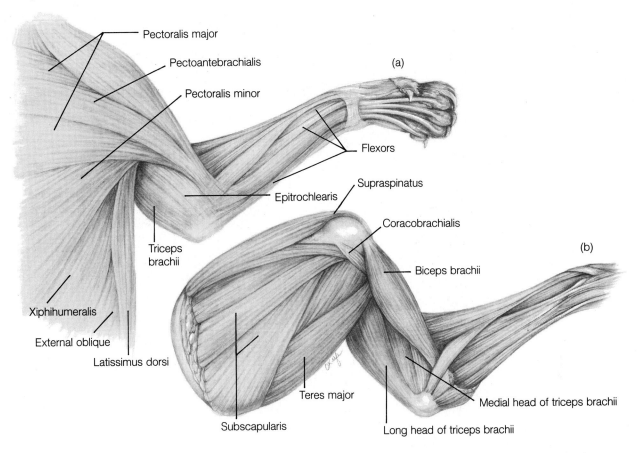

Chart 24.5 Muscles of the forelimb

Muscle	Origin	Insertion	Action
Coracobrachialis	Coracoid process of scapula	Proximal end of humerus	Adducts forelimb
Epitrochlearis	Fascia of latissimus dorsi	Olecranon process of ulna	Extends forelimb
Biceps brachii	Border of glenoid cavity of scapula	Radial tuberosity of radius	Flexes forelimb
Triceps brachii			
Lateral head	Deltoid tuberosity of humerus	Olecranon process of ulna	Extends forelimb
Long head	Border of glenoid cavity of scapula	Olecranon process of ulna	Extends forelimb
Medial head	Shaft of humerus	Olecranon process of ulna	Extends forelimb
Anconeus	Lateral epicondyle of humerus	Olecranon process of ulna	Extends forelimb
Brachialis	Lateral surface of humerus	Proximal end of ulna	Flexes forelimb

Figure 24.8 Superficial muscles of the cat's lateral forelimb (*a*); deep muscles of the lateral forelimb (*b*).

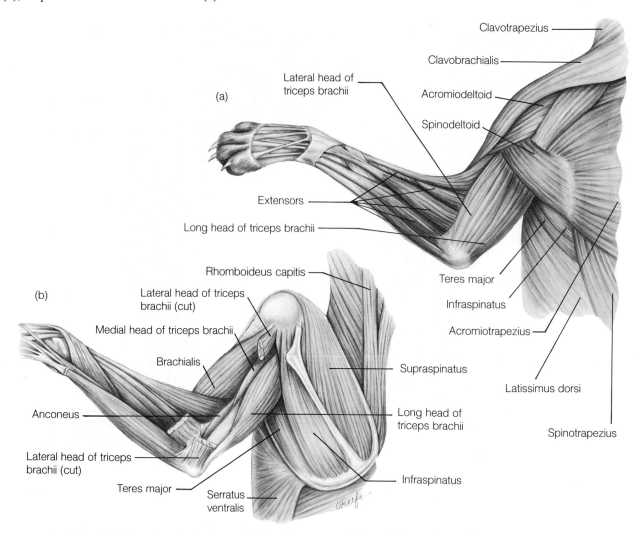

(a)

Clavotrapezius

Clavobrachialis

Acromiodeltoid

Spinodeltoid

Lateral head of triceps brachii

Extensors

Long head of triceps brachii

Teres major

Infraspinatus

Acromiotrapezius

Latissimus dorsi

Spinotrapezius

(b)

Rhomboideus capitis

Lateral head of triceps brachii (cut)

Medial head of triceps brachii

Brachialis

Anconeus

Lateral head of triceps brachii (cut)

Teres major

Serratus ventralis

Supraspinatus

Long head of triceps brachii

Infraspinatus

Procedure H—Muscles of the Hip and Hind Limb

1. Place the cat in the dissecting tray with its ventral side up.
2. Remove any remaining fat and connective tissue from the hip and hind limb to expose the muscles.
3. Study figure 24.9(a) and reference plate 9, then locate and dissect the following muscles from the medial surface of the thigh:

   sartorius

   gracilis

4. Using scissors, transect the sartorius and gracilis and lift aside their cut edges to observe the deeper muscles of the thigh.

5. Study figure 24.9(b) and reference plate 10, then locate and dissect the following muscles:

   rectus femoris

   vastus medialis

   adductor longus

   adductor femoris

   semimembranosus

   semitendinosus

   tensor fasciae latae

6. Transect the tensor fasciae latae and rectus femoris muscles and turn their ends aside. Locate the *vastus intermedius* and *vastus lateralis* muscles beneath. (See figure 24.10.) (*Note:* In some specimens, the vastus intermedius, vastus lateralis, and vastus medialis muscles are closely united by connective tissue and are difficult to separate.)

Figure 24.9 Superficial muscles of the cat's medial hind limb
(*a*); deep muscles of the medial hind limb (*b*).

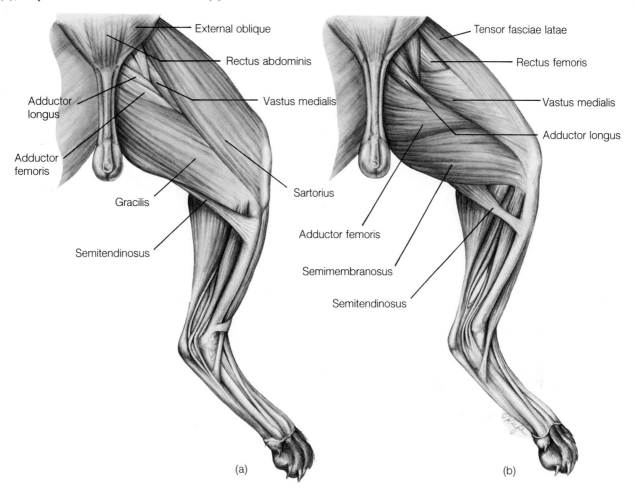

External oblique

Rectus abdominis

Adductor longus

Vastus medialis

Adductor femoris

Gracilis

Sartorius

Semitendinosus

(a)

Tensor fasciae latae

Rectus femoris

Vastus medialis

Adductor longus

Adductor femoris

Semimembranosus

Semitendinosus

(b)

7. Study figure 24.11(a) and reference plate 11, then locate and dissect the following muscles from the lateral surface of the hip and thigh:

   biceps femoris

   caudofemoralis

   gluteus maximus

8. Using scissors, transect the tensor fasciae latae and biceps femoris and lift aside their cut edges to observe the deeper muscles of the thigh.

9. Locate and dissect the following muscles:

   tenuissimus

   gluteus medius

   (*Note:* The tenuissimus muscle and the sciatic nerve may adhere tightly to the dorsal surface of the biceps femoris muscle. Take care to avoid cutting these structures.)

10. On the lateral surface of the lower hind limb (figure 24.11b and reference plate 12), locate and dissect the following muscles:

    gastrocnemius

    soleus

    tibialis anterior

    peroneus group

11. See chart 24.6 for the origins, insertions, and actions of these muscles of the hip and hind limb.

12. Complete Part E of the laboratory report.

Figure 24.10 Other deep muscles of the cat's thigh, medial view.

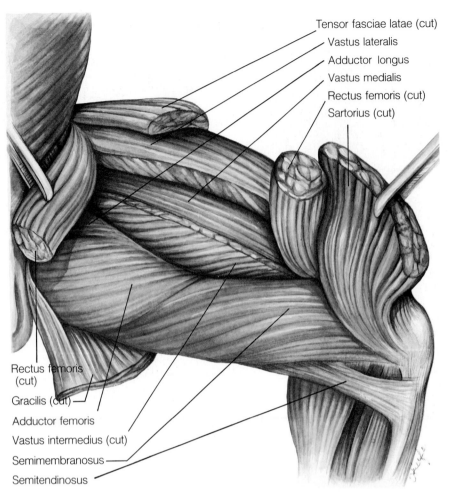

Tensor fasciae latae (cut)
Vastus lateralis
Adductor longus
Vastus medialis
Rectus femoris (cut)
Sartorius (cut)

Rectus femoris (cut)
Gracilis (cut)
Adductor femoris
Vastus intermedius (cut)
Semimembranosus
Semitendinosus

Chart 24.6 Muscles of the hip and hind limb

Muscle	Origin	Insertion	Action
Sartorius	Crest of ilium	Fascia of knee and proximal end of tibia	Rotates and extends hind limb
Gracilis	Symphysis pubis and ischium	Proximal end of tibia	Adducts hind limb
Quadriceps femoris			
Vastus lateralis	Shaft of femur and greater trochanter	Patella	Extends hind limb
Vastus intermedius	Shaft of femur	Patella	Extends hind limb
Rectus femoris	Ilium	Patella	Extends hind limb
Vastus medialis	Shaft of femur	Patella	Extends hind limb
Adductor longus	Pubis	Proximal end of femur	Adducts hind limb
Adductor femoris	Pubis and ischium	Shaft of femur	Adducts hind limb
Semimembranosus	Ischial tuberosity	Medial epicondyle of femur	Extends thigh
Tensor fasciae latae	Ilium	Fascia of thigh	Extends thigh

Figure 24.11 Superficial muscles of the cat's lateral hind limb (*a*); deep muscles of the lateral hind limb (*b*).

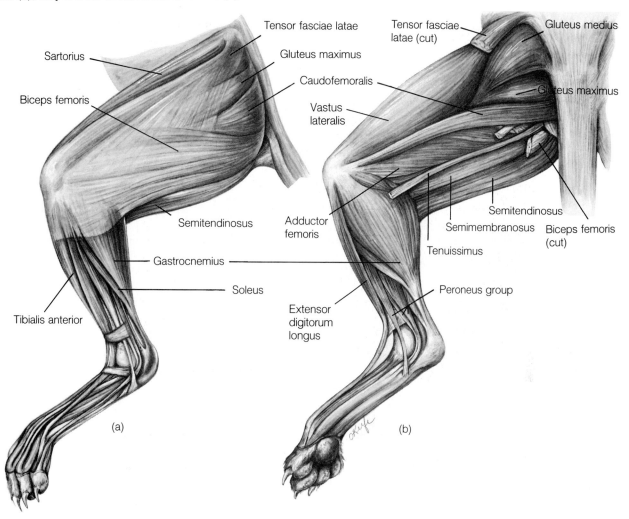

Chart 24.6—*Continued*

Muscle	Origin	Insertion	Action
Biceps femoris	Ischial tuberosity	Patella and tibia	Abducts thigh and flexes lower hind limb
Tenuissimus	Second caudal vertebra	Tibia and fascia of biceps femoris	Abducts thigh and flexes lower hind limb
Semitendinosus	Ischial tuberosity	Crest of tibia	Flexes lower hind limb
Caudofemoralis	Caudal vertebrae	Patella	Abducts thigh and extends lower hind limb
Gluteus maximus	Sacral and caudal vertebrae	Greater trochanter of femur	Abducts thigh
Gluteus medius	Ilium and sacral and caudal vertebrae	Greater trochanter of femur	Abducts thigh
Gastrocnemius	Lateral and medial epicondyles of femur	Calcaneus	Extends foot
Soleus	Proximal end of fibula	Calcaneus	Extends foot
Tibialis anterior	Proximal end of tibia and fibula	First metatarsal	Flexes foot
Peroneus group	Shaft of fibula	Metatarsals	Flexes foot

Name _____

Date _____

Section _____

Cat Dissection: Musculature

Part A

Complete the following statements:

1. The _____ muscle of the human is homologous to the sternomastoid muscle of the cat.

2. The _____ muscle elevates the mandible in the human and in the cat.

3. Two muscles of the cat that are inserted on the hyoid bone are the _____ and the _____ .

Part B

Complete the following:

Name two muscles that are found in the thoracic wall of the cat but are absent in the human.

1. _____
2. _____

Name two muscles that are found in the thoracic wall of the cat and the human.

3. _____
4. _____

Name four muscles that are found in the abdominal wall of the cat and the human.

5. _____
6. _____
7. _____
8. _____

Part C

Complete the following:

Name three muscles of the cat that together correspond to the trapezius muscle in the human.

1. _____
2. _____
3. _____

Name three muscles of the cat that together correspond to the deltoid muscle in the human.

4. _____

5. _____

6. _____

Name the muscle in the cat and in the human that occupies the fossa above the spine of the scapula.

7. _____

Name the muscle in the cat and in the human that occupies the fossa below the spine of the scapula.

8. _____

Name two muscles found in the cat and in the human that can rotate the forelimb.

9. _____

10. _____

Part D

Complete the following:

Name two muscles found in the cat and in the human that can flex the forelimb.

1. _____

2. _____

Name a muscle found in the cat but absent in the human that can extend the forelimb.

3. _____

Name a muscle that has three heads, can extend the forelimb, and is found in the cat and in the human.

4. _____

Part E

Each of the muscles listed in figure 24.12 is found both in the cat and in the human. Identify each of the numbered muscles in the figure by placing its name in the space next to its number.

Figure 24.12 Identify the numbered muscles that occur in both the cat and the human.

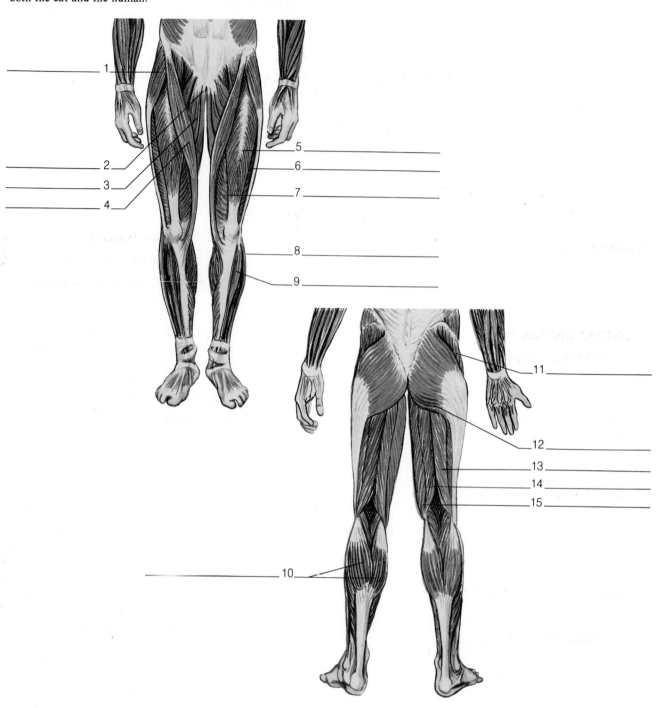

Laboratory Exercise 25

Nerve Tissue

Nerve tissue, which occurs in the brain, spinal cord, and peripheral nerves, contains neurons and neuroglial cells. The neurons are the basic structural and functional units of the nervous system. The neuroglial cells perform various supportive and protective functions.

Purpose of the Exercise

To review the characteristics of nerve tissue and to observe neurons, neuroglial cells, and various features of peripheral nerves.

Learning Objectives

After completing this exercise, you should be able to:

1. Describe the general characteristics of nerve tissue.
2. Distinguish between neurons and neuroglial cells.
3. Identify the major parts of a neuron and a peripheral nerve.

Materials Needed:
   compound microscope
   prepared microscope slides of the following:
      spinal cord (smear)
      dorsal root ganglion (section)
      neuroglial cells (astrocytes)
      peripheral nerve (cross section and longitudinal section)

For Optional Activity:

   prepared microscope slide of Purkinje cells, Golgi

Procedure

1. Review the sections entitled "Nerve Tissue" and "Classification of Neurons and Nerve Fibers" in chapter 10 of the textbook.
2. As a review activity, label figures 25.1 and 25.2.
3. Complete Parts A and B of Laboratory Report 25 on page 181.
4. Obtain a prepared microscope slide of a spinal cord smear. Using low-power magnification, search the slide and locate the relatively large, deeply stained cell bodies of motor neurons (multipolar neurons).
5. Observe a single motor neuron, using high-power magnification, and note the following features:

   cell body

   nucleus

   nucleolus

   Nissl bodies (granular structures in the cytoplasm)

   neurofibrils (threadlike structures extending into the nerve fibers)

   nerve fibers (axon and dendrites)

   You also may note small, darkly stained nuclei of neuroglial cells around the motor neuron.
6. Sketch and label a single motor neuron in the space provided in Part C of the laboratory report.
7. Obtain a prepared microscope slide of a dorsal root ganglion. Search the slide and locate a cluster of sensory neuron cell bodies. You also may note bundles of nerve fibers passing among groups of neuron cell bodies. (See figure 25.3.)

Figure 25.1 Label this diagram of a motor neuron.

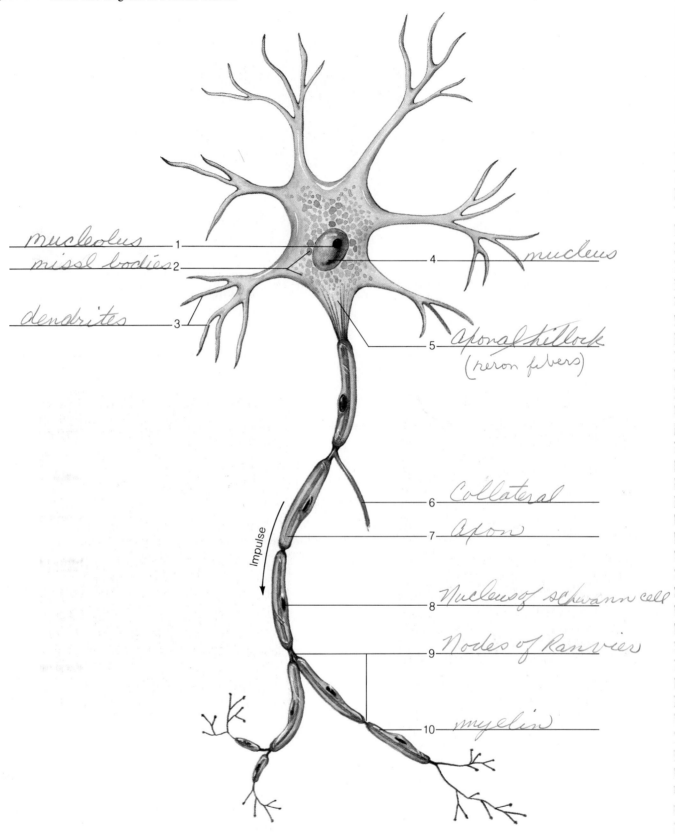

nucleolus _____ 1

nissl bodies _____ 2

dendrites _____ 3

4 _____ nucleus

5 _____ axonal hillock
(axon fibers)

Impulse

6 _____ Collateral

7 _____ axon

8 _____ Nucleus of schwann cell

9 _____ Nodes of Ranvier

10 _____ myelin

Figure 25.2 Label this portion of a myelinated nerve fiber.

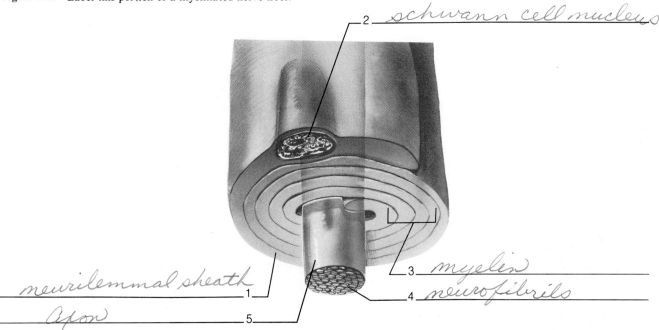

2 _schwann cell nucleus_

neurilemmal sheath 1

axon 5

3 _myelin_

4 _neurofibrils_

Figure 25.3 Micrograph of a dorsal root ganglion (×50).

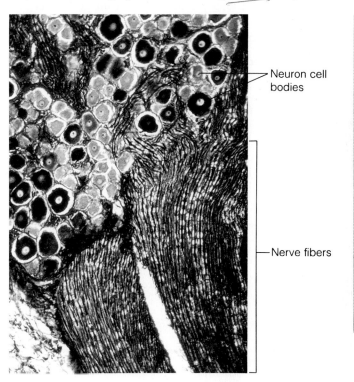

Neuron cell bodies

Nerve fibers

Figure 25.4 Micrograph of astrocytes (×250).

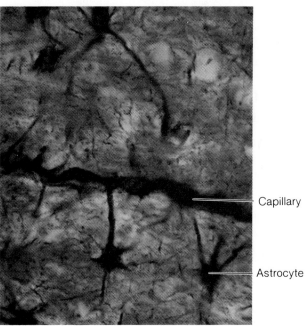

Capillary

Astrocyte

8. Sketch and label a single sensory neuron cell body in the space provided in Part C of the laboratory report.

9. Obtain a prepared microscope slide of neuroglial cells. Search the slide and locate some darkly stained astrocytes with numerous long, slender processes. (See figure 25.4.) (Why do you think these cells are sometimes called "spider cells"?)

10. Sketch a single neuroglial cell in the space provided in Part C of the laboratory report.

11. Obtain a prepared microscope slide of a peripheral nerve. Locate the cross section of the nerve, and note the many round nerve fibers inside. Also note the dense layer of connective tissue (perineurium) that encircles the nerve fibers and holds them together in a bundle. The individual nerve fibers are surrounded by a layer of more delicate connective tissue (endoneurium). (See figure 25.5.)

Figure 25.5 Cross section of a peripheral nerve (×400).

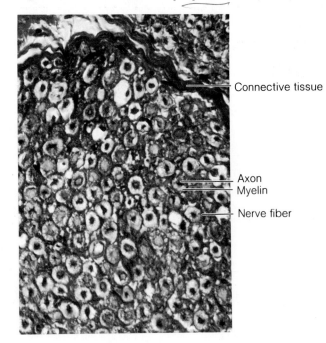

Connective tissue

Axon
Myelin

Nerve fiber

Figure 25.6 Longitudinal section of a peripheral nerve (×250).

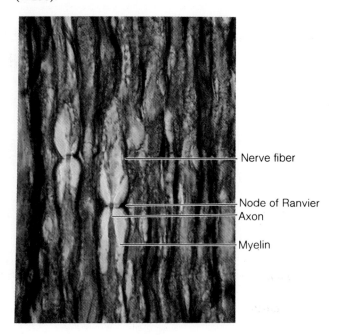

Nerve fiber

Node of Ranvier
Axon

Myelin

12. Observe a single nerve fiber, using high-power magnification, and note the following features:

central axon

myelin around the axon (actually, most of the myelin may have been dissolved and lost during the slide preparation)

neurilemma

13. Sketch and label a single nerve fiber (cross section) in the space provided in Part D of the laboratory report.

14. Locate the longitudinal section of the peripheral nerve on the slide. (See figure 25.6.) Note the following:

central axons

myelin sheaths

neurilemmal sheaths

nodes of Ranvier

15. Sketch and label a single nerve fiber (longitudinal section) in the space provided in Part D of the laboratory report.

Optional Activity

Obtain a prepared microscope slide of Purkinje cells. To locate these neurons, search the slide for large, flask-shaped cell bodies. Note that each cell body has one or two large, thick dendrites that give rise to branching networks of fibers. These cells are found in a particular region of the brain (cerebellar cortex).

Nerve Tissue

Part A

Match the terms in column A with the descriptions in column B. Place the letter of your choice in the space provided.

Column A			**Column B**
A. astrocyte	_G_	1.	sheath that encloses myelin
B. axon	_H_	2.	corresponds to endoplasmic reticulum in other cells
C. collateral	_F_	3.	network of fine threads within nerve fiber
D. dendrite	_E_	4.	substance composed of lipid-protein
E. myelin	_D_	5.	nerve fiber with many tiny thornlike spines
F. neurofibrils	_C_	6.	branch of an axon
G. neurilemma	_A_	7.	star-shaped neuroglial cell
H. Nissl body	_B_	8.	nerve fiber arising from a slight elevation of the cell body

Part B

Match the terms in column A with the descriptions in column B. Place the letter of your choice in the space provided.

Column A			**Column B**
A. effector	_C_	1.	transmits impulse from sensory to motor neuron
B. ganglion	_F_	2.	transmits impulse out of the brain or spinal cord
C. interneuron	_E_	3.	transmits impulse into brain or spinal cord
D. microglia	_G_	4.	myelin-forming neuroglial cell
E. sensory neuron	_D_	5.	phagocytic neuroglial cell
F. motor neuron	_A_	6.	part capable of responding to motor impulse
G. oligodendrocyte	_B_	7.	specialized mass of neuron cell bodies outside the brain or spinal cord

Part C

1. Sketch and label a single motor neuron.

2. Sketch and label a single sensory neuron cell body.

3. Sketch a single neuroglial cell.

Part D

1. Sketch and label a single nerve fiber (cross section).

2. Sketch and label a single nerve fiber (longitudinal section).

Nerve Impulse Stimulation

A nerve cell usually is polarized due to an unequal distribution of ions on either side of its membrane. When such a polarized membrane is stimulated at or above its threshold intensity, a wave of action potentials is triggered to move in all directions away from the site of stimulation. This wave constitutes a nerve impulse, and if it reaches a muscle, the muscle may respond by contracting.

Purpose of the Exercise

To review the characteristics of a nerve impulse and to investigate the effects of certain stimuli on a nerve.

Learning Objectives

After completing this exercise, you should be able to:

1. Describe the events that lead to the stimulation of a nerve impulse.
2. List four types of factors that can stimulate a nerve impulse.
3. Test the effects of various factors on a nerve-muscle preparation.

Materials Needed:

live frog	glass plate
dissecting tray	ring stand and ring
dissecting instruments	microscope slides
frog Ringer's solution	Bunsen burner
electronic stimulator	ice
filter paper	1% HCl
glass rod	1% NaCl

For Optional Activity:

2% Novocain solution (procaine hydrochloride)

Procedure

1. Review the section entitled "Cell Membrane Potential" in chapter 10 of the textbook.
2. Complete Part A of Laboratory Report 26 on page 187.

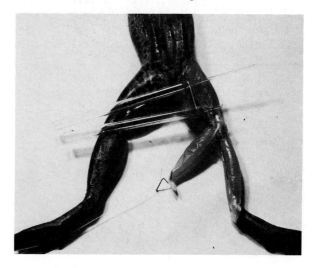

Figure 26.1 The sciatic nerve appears as a silvery white thread between the muscles of the thigh.

3. Obtain a live frog and pith its brain and spinal cord as described in Procedure C of Laboratory Exercise 19 on pages 127–28.
4. Place the pithed frog in a dissecting tray and remove the skin from its hind limb, beginning at the waist, as described in Procedure C of Laboratory Exercise 19. (As the skin is removed, keep the exposed tissues moist by flooding them with frog Ringer's solution.)
5. Expose the frog's sciatic nerve. To do this:
   a. Use a glass rod to separate the gastrocnemius muscle from the adjacent muscles.
   b. Locate the Achilles tendon at the distal end of the gastrocnemius, and cut it with scissors.
   c. Place the frog ventral side down, and separate the muscles of the thigh to locate the sciatic nerve. The nerve will look like a silvery white thread passing through the thigh, dorsal to the femur (figure 26.1).
   d. Dissect the nerve to its origin in the spinal cord.
   e. Use scissors to cut the nerve at its origin, and carefully snip off all of the branch nerves in the thigh, leaving only its connection to the gastrocnemius muscle.

f. Use a scalpel to free the proximal end of the gastrocnemius.

g. Carefully remove the nerve and attached muscle, and transfer the preparation to a glass plate supported on the ring of a ring stand.

h. Use a glass rod to position the preparation so that the sciatic nerve is hanging over the edge of the glass plate. (Be sure to keep the preparation moistened with frog Ringer's solution at all times.)

6. Determine the threshold voltage and the voltage needed for maximal muscle contraction by using the electronic stimulator, as described in Procedure D of Laboratory Exercise 19.

7. Expose the cut end of the sciatic nerve to each of the following, and observe the response of the gastrocnemius muscle.

a. Pinch the end of the nerve firmly between two glass microscope slides.

b. Touch the cut end with a glass rod at room temperature.

c. Touch the cut end with a glass rod that has been cooled in ice water for five minutes.

d. Touch the cut end with a glass rod that has been heated in the flame of a Bunsen burner.

e. Dip the cut end in 1% HCl.

f. Dip the cut end in 1% NaCl.

8. Complete Part B of the laboratory report.

Optional Activity

Test the effect of Novocain on a frog sciatic nerve. To do this:

1. Place a nerve-muscle preparation on a glass plate supported by the ring of a ring stand, as before.

2. Use the electronic stimulator to determine the voltage needed for maximal muscle contraction.

3. Saturate a small piece of filter paper with 2% Novocain solution, and wrap the paper around the midsection of the sciatic nerve.

4. At two-minute intervals, stimulate the nerve, using the voltage needed for maximal contraction until the muscle fails to respond.

5. Remove the filter paper, and flood the nerve with frog Ringer's solution.

6. At two-minute intervals, stimulate the nerve until the muscle responds again. How long did it take for the nerve to recover from the effect of the Novocain?

laboratory report 26

Name _____

Date _____

Section _____

Nerve Impulse Stimulation

Part A

Complete the following statements:

1. _____ ions tend to pass through cell membranes more easily than sodium ions.

2. When a nerve cell is at rest, there is a relatively greater concentration of _____ ions outside of its membrane.

3. When sodium ions are actively transported outward through a nerve cell membrane, _____ ions are transported inward.

4. The difference in electrical charge between the inside and the outside of a nerve cell membrane is called the _____ .

5. If the resting potential becomes less negative in response to stimulation, the cell membrane is said to be _____ .

6. As a result of an additive phenomenon called _____ , the threshold potential of a membrane may be reached.

7. Following depolarization, potassium ions diffuse outward and cause the cell membrane to become _____ .

8. An action potential is a rapid sequence of changes involving _____ and _____ .

9. The moment following the passage of an action potential during which an ordinary stimulus will not trigger another impulse is called the _____ .

10. Muscle fiber contraction and nerve impulse conduction are similar in that both responses are _____ .

11. Myelin contains a high proportion of _____ .

12. Nodes of Ranvier occur between adjacent _____ .

13. The type of conduction in which an impulse seems to jump from node to node is called _____ .

14. The greater the diameter of a nerve fiber, the _____ the impulse travels.

Part B

1. What was the threshold voltage for the frog sciatic nerve? _____

2. What was the voltage needed for maximal contraction of the gastrocnemius muscle? _____

3. Complete the following chart:

Factor Tested	Muscle Response	Effect on Nerve
Pinching		
Glass rod (room temperature)		
Glass rod (cooled)		
Glass rod (heated)		
1% HCl		
1% NaCl		

4. Write a statement to summarize the results of these tests.

_____

_____

_____

_____

The Reflex Arc and Reflexes

A reflex arc represents the simplest type of nerve pathway found in the nervous system. This pathway begins with a receptor at the end of a sensory nerve fiber. The sensory fiber leads into the central nervous system and may connect with one or more interneurons. Some of these interneurons, in turn, communicate with motor neurons, whose fibers lead outward to effectors.

Thus, when a sensory receptor is stimulated by some kind of change occurring inside or outside the body, nerve impulses may pass through a reflex arc, and as a result, effectors may respond. Such an automatic, unconscious response is called a reflex.

Purpose of the Exercise

To review the characteristics of reflex arcs and reflex behavior and to demonstrate some of the reflexes that occur in the human body.

Learning Objectives

After completing this exercise, you should be able to:

1. Describe a reflex arc.
2. Distinguish between a reflex arc and a reflex.
3. Identify and demonstrate several reflex actions that occur in humans.

Materials Needed:
rubber percussion hammer

Procedure

1. Review the section entitled "Nerve Pathways" in chapter 10 of the textbook.
2. As a review activity, label figure 27.1.
3. Complete Part A of Laboratory Report 27 on page 191.

4. Work with a laboratory partner to demonstrate each of the reflexes listed below (figure 27.2). After each demonstration, record your observations in the chart provided in Part B of the laboratory report.
   a. *Knee-jerk reflex.* Have your laboratory partner sit on a table (or sturdy chair) with legs relaxed and hanging freely over the edge without touching the floor. Strike your partner's patellar ligament (just below the patella) with the blunt side of a rubber percussion hammer.
   b. *Ankle-jerk reflex.* Have your partner kneel on the table with back toward you and with feet over the edge and relaxed. Strike the Achilles tendon (just above its insertion on the calcaneus) with the blunt side of the rubber hammer.
   c. *Biceps-jerk reflex.* Have your partner place a bare arm bent at the elbow on the table. Press your index finger on the inside of the elbow over the tendon of the biceps brachii, and strike your finger with the blunt side of the rubber hammer. Watch the biceps brachii for a response.
   d. *Triceps-jerk reflex.* Have your partner lie on the back with an arm bent across the abdomen. Strike the tendon of the triceps brachii near its insertion at the tip of the elbow. Watch the triceps brachii for a response.
   e. *Babinski reflex.* Have your partner remove a shoe and sock and lie on the back. Draw the tip of the rubber hammer, applying firm pressure, over the sole from the heel to the base of the large toe. Watch the toes for a response.
5. Complete steps 2 and 3 of Part B of the laboratory report.

Figure 27.1 Label this diagram of a withdrawal reflex by placing the correct numbers in the spaces provided.

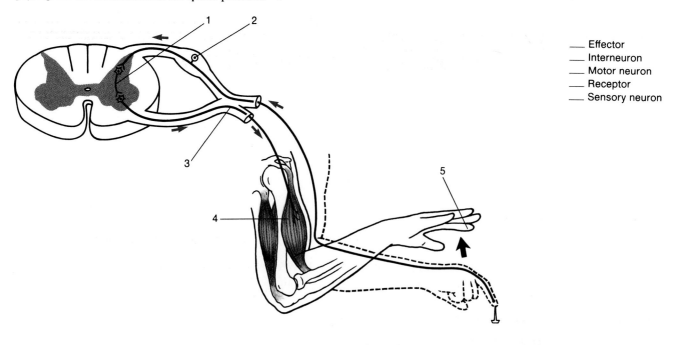

_____ Effector
_____ Interneuron
_____ Motor neuron
_____ Receptor
_____ Sensory neuron

Figure 27.2 Demonstrate each of the following reflexes: (a) knee-jerk reflex; (b) ankle-jerk reflex; (c) biceps-jerk reflex; (d) triceps-jerk reflex; and (e) Babinski reflex.

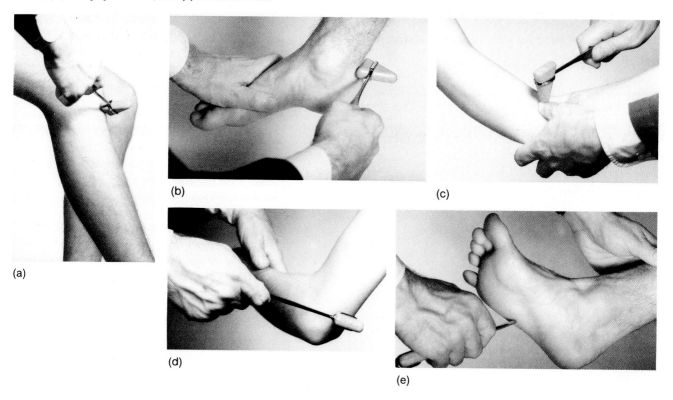

Name _____

Date _____

Section _____

The Reflex Arc and Reflexes

Part A

Complete the following statements:

1. _____ are routes followed by nerve impulses as they pass through the nervous system.

2. A reflex center in the central nervous system is composed of one or more _____ .

3. A(an) _____ is the behavioral unit of the nervous system.

4. A(an) _____ is the simplest behavioral act.

5. A knee-jerk reflex employs only _____ neurons.

6. The effector of the knee-jerk reflex is the _____ .

7. The sensory receptors of the knee-jerk reflex are located in the _____ .

8. The knee-jerk reflex helps the body to maintain _____ .

9. The sensory receptors of a withdrawal reflex are located in the _____ .

10. The effectors of a withdrawal reflex are _____ .

Part B

1. Complete the following chart:

Reflex Tested	Response Observed	Effector Involved
Knee-jerk		
Ankle-jerk		
Biceps-jerk		
Triceps-jerk		
Babinski		

2. List the major events that occur in the knee-jerk reflex from the striking of the patellar ligament to the resulting response. _____
_____
_____
_____
_____

3. What characteristics do the reflexes you demonstrated have in common? _____
_____
_____

Laboratory Exercise 28

The Meninges and Spinal Cord

The meninges consist of layers of membranes located between the bones of the skull and vertebral column and the soft tissues of the nervous system.

The spinal cord is a column of nerve fibers that extends down through the vertebral canal. Together with the brain, it makes up the central nervous system.

Neurons within the spinal cord provide a two-way communication system between the brain and body parts outside the central nervous system. The cord also contains the processing centers for spinal reflexes.

Purpose of the Exercise

To review the characteristics of the meninges and the spinal cord and to observe the major features of these structures.

Learning Objectives

After completing this exercise, you should be able to:

1. Name the layers of the meninges, and describe the structure of each.
2. Identify the major features of the spinal cord.
3. Locate the ascending and descending tracts of the spinal cord.

Materials Needed:

compound microscope
prepared microscope slide of the following:
    spinal cord cross section (ox or human)
spinal cord model

For Demonstration:

preserved spinal cord with meninges intact

Procedure A—Meninges

1. Review the section entitled "Meninges" in chapter 11 of the textbook.
2. Complete Part A of Laboratory Report 28 on page 197.

Demonstration

Observe the preserved section of spinal cord. Note the heavy covering of dura mater, which is firmly attached to the cord on each side by a set of ligaments (denticulate ligaments) originating in the pia mater. The intermediate layer of meninges, the arachnoid mater, is devoid of blood vessels, but in life the space beneath this layer contains cerebrospinal fluid. The pia mater, which is closely attached to the surface of the spinal cord, contains many blood vessels. What are the functions of these layers? _____
_____
_____

Procedure B—Structure of the Spinal Cord

1. Review the section entitled "Spinal Cord" in chapter 11 of the textbook.
2. As a review activity, label figures 28.1, 28.2, and 28.3.
3. Complete Parts B and C of the laboratory report.

Figure 28.1 Label the features of the meninges and spinal cord.

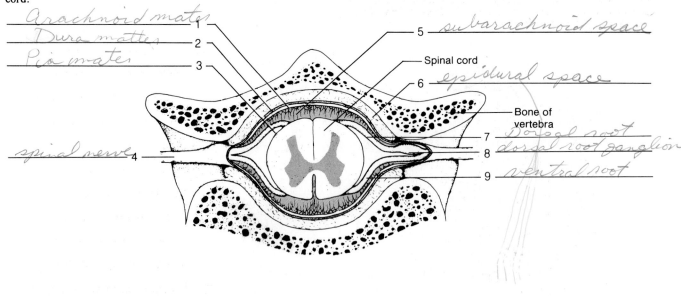

Arachnoid mater 1
Dura mater 2
Pia mater 3

5 subarachnoid space
Spinal cord
6 epidural space
Bone of vertebra
7 Dorsal root
8 dorsal root ganglion
9 ventral root

spinal nerve 4

Figure 28.2 Label this cross section of the spinal cord.

Posterior

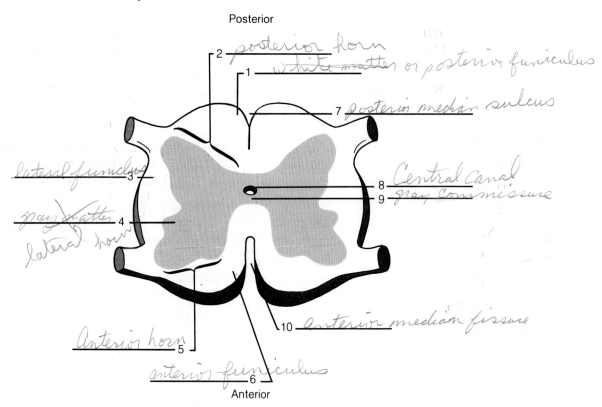

2 posterior horn
1 white matter or posterior funiculus
7 posterior median sulcus
lateral funiculus 3
8 Central Canal
9 gray Commissure
gray matter 4
lateral horn
10 anterior median fissure
Anterior horn 5
anterior funiculus 6
Anterior

Figure 28.3 Label the ascending and descending tracts of the spinal cord by placing the correct numbers in the spaces provided. (Note: These tracts are not visible as individually stained structures on microscope slides.)

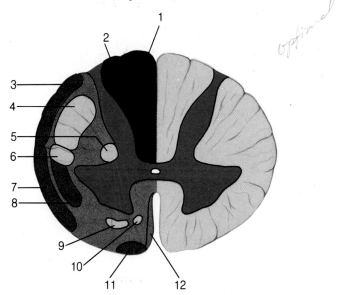

optional

_12_ Anterior corticospinal tract

_9_ Anterior reticulospinal tract

_7_ Anterior spinocerebellar tract

_11_ Anterior spinothalamic tract

_2_ Fasciculus cuneatus

_1_ Fasciculus gracilis

_4_ Lateral corticospinal tract

_5_ Lateral reticulospinal tract

_8_ Lateral spinothalamic tract

_10_ Medial reticulospinal tract

_3_ Posterior spinocerebellar tract

_6_ Rubrospinal tract

4. Obtain a prepared microscope slide of a spinal cord cross section. Use the low power of the microscope to locate the following features:

posterior median sulcus

anterior median fissure

central canal

gray matter

    gray commissure

    posterior horn

    lateral horn

    anterior horn

white matter

    posterior funiculus

    lateral funiculus

    anterior funiculus

5. Observe the model of the spinal cord, and locate the features listed in step 4. Also identify the following:

dorsal roots of spinal nerves

dorsal root ganglia

ventral roots of spinal nerves

6. Use a pencil to trace the path of a sensory nerve impulse from a skin receptor to the cerebral sensory cortex in figure 28.4; similarly, trace a motor nerve impulse from the cerebral motor cortex to a skeletal muscle in figure 28.5.

7. Complete Part D of the laboratory report.

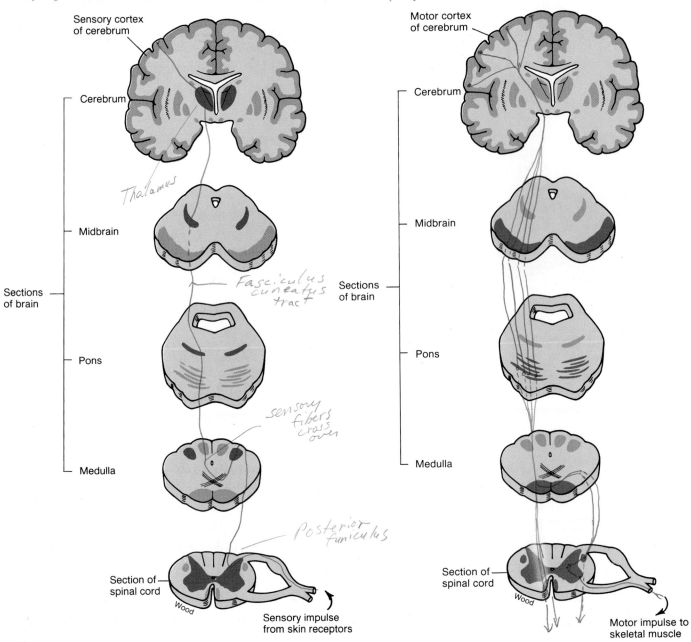

Figure 28.4 With a pencil, trace the pathway of a nerve impulse from the skin receptor to the cerebral cortex.

Sensory cortex of cerebrum

Cerebrum

Sections of brain

Thalamus

Midbrain

Fasciculus cuneatus tract

Pons

sensory fibers cross over

Medulla

Posterior funiculus

Section of spinal cord

Wood

Sensory impulse from skin receptors

Figure 28.5 With a pencil, trace the pathway of a nerve impulse from the motor cortex to a skeletal muscle.

Motor cortex of cerebrum

Cerebrum

Sections of brain

Midbrain

Pons

Medulla

Section of spinal cord

Wood

Motor impulse to skeletal muscle

Name _____

Date _____

Section _____

The Meninges and Spinal Cord

Part A

Match the terms in column A with the descriptions in column B. Place the letter of your choice in the space provided.

Column A		Column B
A. arachnoid mater	_B_ 1.	band of pia mater that attaches dura mater to cord
B. denticulate	_C_ 2.	channel through which venous blood flows
ligament	_D_ 3.	outermost layer of meninges
C. dural sinus	_F_ 4.	follows irregular contours of spinal cord surface
D. dura mater	_G_ 5.	contains cerebrospinal fluid
E. epidural space	_A_ 6.	thin, netlike membrane
F. pia mater	_E_ 7.	separates dura mater from bone of vertebra
G. subarachnoid space		

Part B

Complete the following statements:

1. Each of the thirty-one segments of the spinal cord gives rise to a pair of _spinal nerve_.

2. The bulge in the spinal cord that gives off nerves to the arms is called the _cervical enlar_.

3. The bulge in the spinal cord that gives off nerves to the legs is called the _lumbar enlar_.

4. The _____ is a groove that extends the length of the spinal cord, posteriorly.

5. In a spinal cord cross section, the _gray matter posterior horn_ appears as the upper wings of a butterfly.

6. The cell bodies of motor neurons are found in the _anterior horn_ horns of the spinal cord.

7. The _gray commm_ connects the gray matter on the left and right sides of the spinal cord.

8. The _Central Canal_ of the spinal cord contains cerebrospinal fluid and is continuous with the ventricles of the brain.

9. The white matter of the spinal cord is divided into anterior, lateral, and posterior _funiculi_.

10. The longitudinal bundles of nerve fibers within the spinal cord comprise major nerve pathways called _nerve tract_.

Part C

Match the nerve tracts in column A with the descriptions in column B. Place the letter of your choice in the space provided.

Column A		Column B
A. corticospinal	_A_ 1.	ascending tract whose fibers cross over in the medulla oblongata
B. fasciculus gracilis	_B_ 2.	descending tract whose fibers conduct motor impulses to sweat glands
C. lateral spinothalamic	_A_ ___ 3.	descending tract whose fibers conduct motor impulses to skeletal muscles
D. posterior spinocerebellar	_C_ ___ 4.	ascending tract necessary for coordination of skeletal muscles
E. reticulospinal	___ 5.	ascending tract whose fibers cross over in the spinal cord

Part D *Important*

Identify the features indicated in the spinal cord cross section of figure 28.6.

1. _____

2. _____

3. _____

4. _____

5. _____

6. _____

7. _____

8. _____

Figure 28.6 Micrograph of a spinal cord cross section.

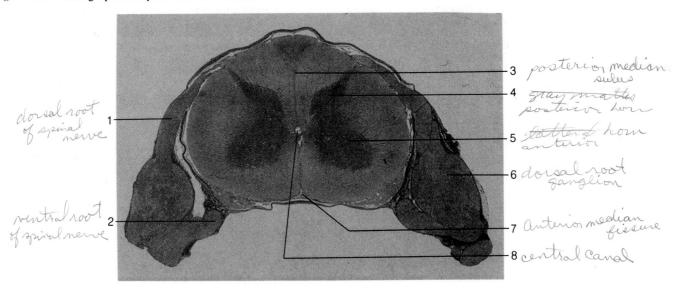

dorsal root of spinal nerve 1

ventral root of spinal nerve 2

3 *posterior median sulcus*

4 *gray matter posterior horn*

5 *gray matter horn anterior*

6 *dorsal root ganglion*

7 *anterior median fissure*

8 *central canal*

The Brain and Cranial Nerves

The brain, which is the largest and most complex part of the nervous system, contains nerve centers associated with sensory functions and is responsible for sensations and perceptions. It issues motor commands to skeletal muscles and carries on higher mental activities. It also functions to coordinate muscular movements and contains centers and nerve pathways necessary for the regulation of internal organs.

Twelve pairs of cranial nerves arise from the underside of the brain and are designated by number and name. Although most of these nerves conduct both sensory and motor impulses, some contain only sensory fibers associated with special sense organs. Others are composed primarily of motor fibers and are involved with the activities of muscles and glands.

Purpose of the Exercise

To review the structural and functional characteristics of the human brain and cranial nerves.

Learning Objectives

After completing this exercise, you should be able to:

1. Identify the major structural parts of the human brain.
2. Locate the major functional regions of the brain.
3. Identify each of the cranial nerves.
4. List the functions of each cranial nerve.

Materials Needed:
  dissectible model of the human brain
  preserved human brain
  anatomic charts of the human brain

Procedure A—Human Brain

1. Review the section entitled "Brain" in chapter 11 of the textbook.
2. As a review activity, label figures 29.1, 29.2, 29.3, and 29.4.
3. Complete Part A of Laboratory Report 29 on page 203.
4. Observe the dissectible model and the preserved specimen of the human brain. Locate each of the following features:

cerebrum
cerebral hemispheres
corpus callosum
convolutions
sulci
   central sulcus
   lateral sulcus
fissures
   longitudinal fissure
   transverse fissure
lobes
   frontal lobe
   parietal lobe
   temporal lobe
   occipital lobe
   insula
cerebral cortex

basal ganglia
   caudate nucleus
   putamen
   globus pallidus
ventricles
   lateral ventricles
   third ventricle
   fourth ventricle
   choroid plexuses
brain stem
   diencephalon
      thalamus
      hypothalamus
      optic chiasma
      mammillary bodies
      pineal gland

Figure 29.1 Label this diagram by placing the correct numbers in the spaces provided.

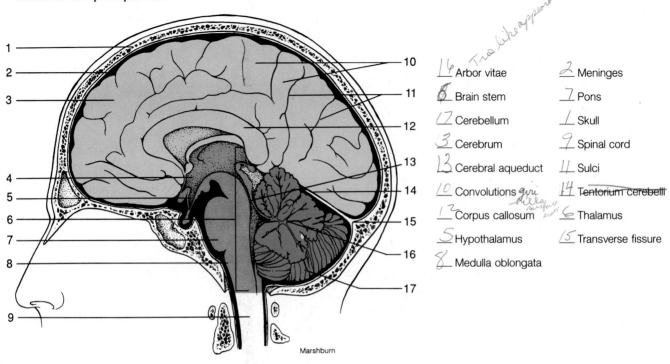

Marshburn

Tree like appearance

14 Arbor vitae   2 Meninges
6 Brain stem   7 Pons
17 Cerebellum   1 Skull
3 Cerebrum   9 Spinal cord
13 Cerebral aqueduct   11 Sulci
10 Convolutions *gyri*   14 Tentorium cerebelli
12 Corpus callosum *hills surface area*   6 Thalamus
5 Hypothalamus   15 Transverse fissure
8 Medulla oblongata

Figure 29.2 Label the lobes of the cerebrum.

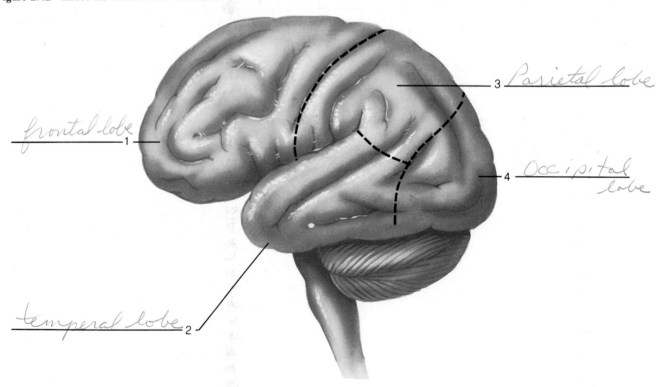

frontal lobe 1
temperal lobe 2
Parietal lobe 3
Occipital lobe 4

Figure 29.3 Label the functional areas of the cerebrum.

primary motor — 1

5 —

6 —

2 —

7 —

3 —

4 —

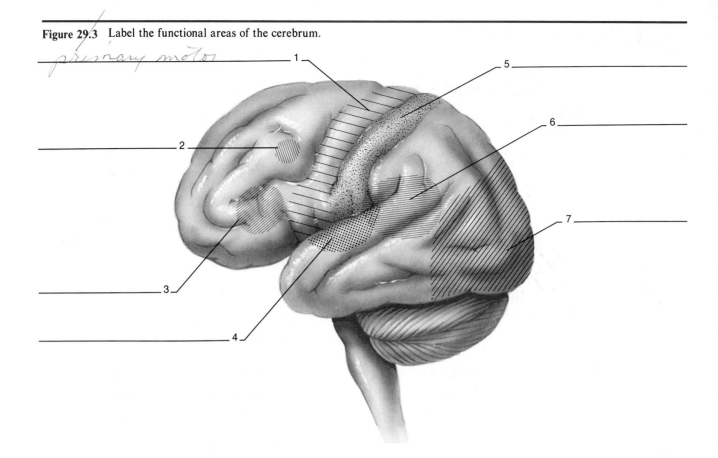

Figure 29.4 Label the features indicated in this illustration of the brain.

7 — *Longitudinal fissure*

8 — *Right cerebral hemisphere*

Caudate nucleus — 1

Putamen — 2

9 — *Corpus callosum*

Globus pallidus — 3

Thalamus
hypothalamic nuclei — 4
brain stem — 5

10 — *4th venticle*

11 — *Cerebellum*

6 —

12 — *spinal cord*

Sims/Schenk

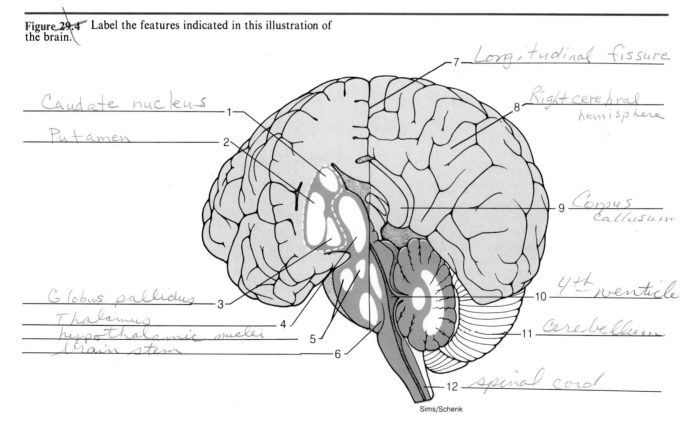

Figure 29.5 Provide the names of the cranial nerves.

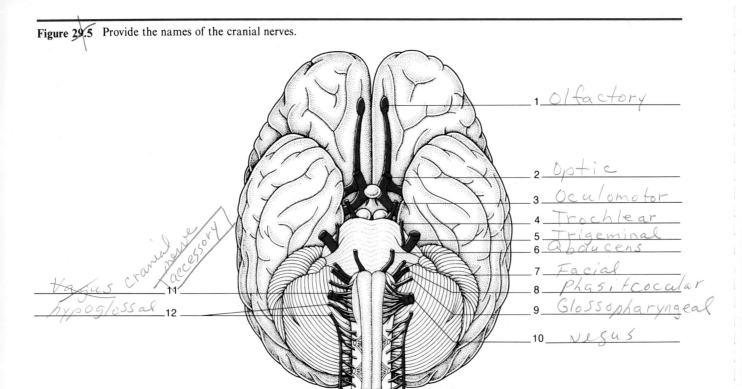

1 Olfactory
2 Optic
3 Oculomotor
4 Trochlear
5 Trigeminal
6 Abducens
7 Facial
8 Phasitcocular
9 Glossopharyngeal
10 Vagus

vagus cranial nerve
accessory
11
hypoglossal 12

midbrain

    cerebral aqueduct

    cerebral peduncles

    corpora quadrigemina

pons

medulla oblongata

cerebellum

    vermis

    cerebellar cortex

    arbor vitae

    cerebellar peduncles

5. Use colored pencils to shade the labeled areas in figure 29.3 that represent the following functional regions of the cerebrum:

primary motor area

frontal eye field

Broca's area

cutaneous sensory area

auditory area

visual area

general interpretative area

6. Complete Part B of the laboratory report.

Procedure B—Cranial Nerves

1. Review the section entitled "Cranial Nerves" in chapter 11 of the textbook.
2. As a review activity, label figure 29.5.
3. Observe the model and preserved specimen of the human brain, and locate as many of the following cranial nerves as possible:

olfactory nerves

optic nerves

oculomotor nerves

trochlear nerves

trigeminal nerves

abducens nerves

facial nerves

vestibulocochlear nerves

glossopharyngeal nerves

vagus nerves

accessory nerves

hypoglossal nerves

4. Complete Part C of the laboratory report.

The Brain and Cranial Nerves

Part A For Quiz

Match the terms in column A with the descriptions in column B. Place the letter of your choice in the space provided.

Column A		Column B
A.	central sulcus	J 1. structure formed by the crossing-over of the optic nerves
B.	cerebral cortex	F 2. conical process to which pituitary gland is attached
C.	convolution	K 3. cone-shaped structure in roof of diencephalon
D.	corpus callosum	D 4. connects cerebral hemispheres
E.	falx cerebelli	C 5. ridge on surface of cerebrum
F.	infundibulum	A 6. separates frontal and parietal lobes
G.	insula	H 7. part of brain stem between diencephalon and pons
H.	midbrain	L 8. rounded bulge on underside of brain stem
I.	medulla oblongata	I 9. enlarged continuation of spinal cord
J.	optic chiasma	E 10. separates cerebellar hemispheres
K.	pineal gland	M 11. separates occipital lobe from cerebellum
L.	pons	G 12. cerebral lobe located deep within lateral sulcus
M.	tentorium cerebelli	B 13. thin layer of gray matter on surface of cerebrum

Part B

Complete the following chart:

Structure	Location	Major Functions
Broca's area	frontal lobe	coordinate speach
Cardiac center	madula logot	heart rate
Cerebellar peduncles		to + from cerebellum
Cerebral peduncles	motor pathway to cerebrum	
Corpora quadrigemina	posterior midbrain	visual
Frontal eye fields		eye lids
Hypothalamus	torus	function Homyostasis
Limbic system		produce + control emotion
Prefrontal areas	anter front lobe	emotion
Respiratory center	midulla olblonge	deapth + rate breathing
Reticular formation		wakeful alertness
Thalamus	ypper dycephian	sensory + motor impul
Vasomotor center	midulla olb	regulate blood plessur

Part C

Indicate which cranial nerve (or nerves) is (or are) most closely associated with each of the following functions:

1. Sense of hearing — *vestibulcocular*
2. Sense of taste — *3/4 Facial 1/3 Glossa*
3. Sense of sight — *Optic*
4. Sense of smell — *Olfactory*
5. Sense of equilibrium — *estibularcocular*
6. Conducting sensory impulses from upper teeth — *trigeminal max branch*
7. Conducting sensory impulses from lower teeth — *trigem mandibul brand*
8. Raising eyelids — *oculobe motor nerve*
9. Focusing lenses of eyes — *oculobe motor nerve*
10. Adjusting amount of light entering eye — *" "*
11. Moving eyes — *" , trochlear, abdusents*
12. Stimulating salivary secretions — *facial, glosal farengeno*
13. Movement of trapezius and sternocleidomastoid muscles — *mastoid muscle (spinal excesory nerve)*
14. Muscular movements associated with speech — *vagus, selery hypogloss*
15. Muscular movements associated with swallowing — *glos farengen, ascel hypoglosal vagus*

Dissection of the Sheep Brain

Mammalian brains have many features in common, and because human brains may not be available, sheep brains often are dissected as an aid to understanding mammalian brain structure. However, as in the cat, the adaptations of the sheep differ from the adaptations of the human so that comparisons of their structural features may not be precise.

Purpose of the Exercise

To observe the major features of the sheep brain and to compare these features with those of the human brain.

Learning Objectives

After completing this exercise, you should be able to:

1. Identify the major features of the sheep brain.
2. Locate the larger cranial nerves of the sheep brain.
3. List several differences and similarities between the sheep brain and the human brain.

Materials Needed:
   dissectible model of human brain
   preserved sheep brain
   dissecting tray
   dissection instruments
   long knife

Procedure

1. Obtain a preserved sheep brain and rinse it thoroughly in water to remove as much of the preserving fluid as possible.

2. Examine the surface of the brain for the presence of meninges. (The outermost layers of these membranes may have been lost during removal of the brain from the cranial cavity.) If meninges are present, locate the following:

   dura mater—the thick, opaque outer layer

   arachnoid mater—the delicate, transparent middle layer that is attached to the undersurface of the dura mater

   pia mater—the thin, vascular layer that adheres to the surface of the brain

3. Remove any remaining dura mater by pulling it gently from the surface of the brain.

4. Position the brain with its ventral surface down in the dissecting tray. Study figure 30.1 and locate the following features on the specimen:

   cerebral hemispheres

   convolutions

   sulci

   longitudinal fissure

   frontal lobe

   parietal lobe

   temporal lobe

   occipital lobe

   cerebellum

   medulla oblongata

5. Gently separate the cerebral hemispheres along the longitudinal fissure, and expose the transverse band of white fibers within the fissure that connects the hemispheres. This band is the *corpus callosum.*

Figure 30.1 Dorsal surface of the sheep brain.

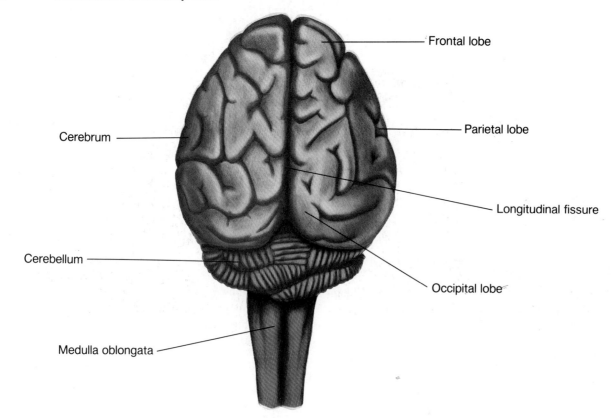

- Frontal lobe
- Cerebrum
- Parietal lobe
- Longitudinal fissure
- Cerebellum
- Occipital lobe
- Medulla oblongata

Figure 30.2 Gently bend the cerebellum and medulla oblongata away from the cerebrum to expose the pineal gland.

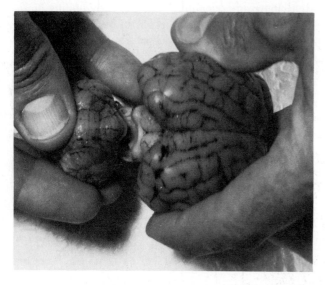

6. Bend the cerebellum and medulla oblongata slightly downward and away from the cerebrum (figure 30.2). This will expose the *pineal gland* in the upper midline and the *corpora quadrigemina*, which consists of four rounded structures associated with the midbrain.

7. Position the brain with its ventral surface upward. Study figures 30.3 and 30.4, and locate the following features on the specimen:

   longitudinal fissure

   olfactory bulbs

   optic nerves

   optic chiasma

   optic tract

   mammillary bodies

   infundibulum

   pituitary gland (this structure may be missing)

   midbrain

   pons

8. Although some of the cranial nerves may be missing or are quite small and difficult to find, locate as many of the following as possible:

   oculomotor nerves

   trochlear nerves

   trigeminal nerves

   abducens nerves

208

Figure 30.3 Lateral surface of the sheep brain.

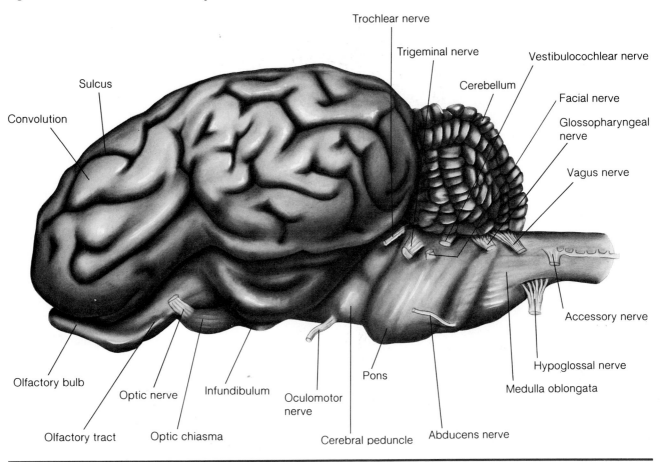

Trochlear nerve

Trigeminal nerve

Vestibulocochlear nerve

Cerebellum

Facial nerve

Glossopharyngeal nerve

Vagus nerve

Sulcus

Convolution

Accessory nerve

Hypoglossal nerve

Medulla oblongata

Olfactory bulb

Optic nerve

Infundibulum

Oculomotor nerve

Pons

Olfactory tract

Optic chiasma

Cerebral peduncle

Abducens nerve

Figure 30.4 Ventral surface of the sheep brain.

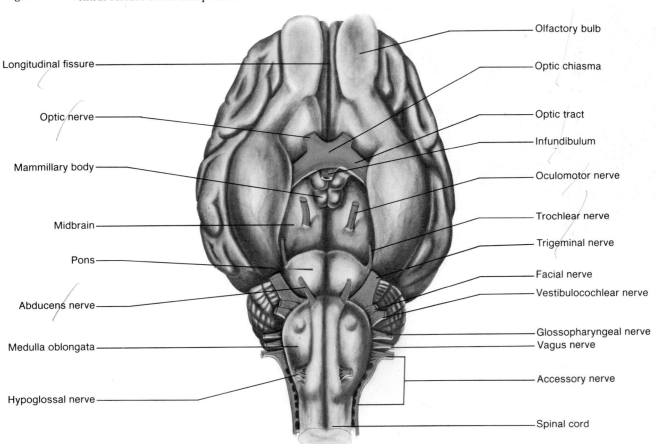

Longitudinal fissure

Optic nerve

Mammillary body

Midbrain

Pons

Abducens nerve

Medulla oblongata

Hypoglossal nerve

Olfactory bulb

Optic chiasma

Optic tract

Infundibulum

Oculomotor nerve

Trochlear nerve

Trigeminal nerve

Facial nerve

Vestibulocochlear nerve

Glossopharyngeal nerve

Vagus nerve

Accessory nerve

Spinal cord

Figure 30.5 Midsagittal section of the sheep brain.

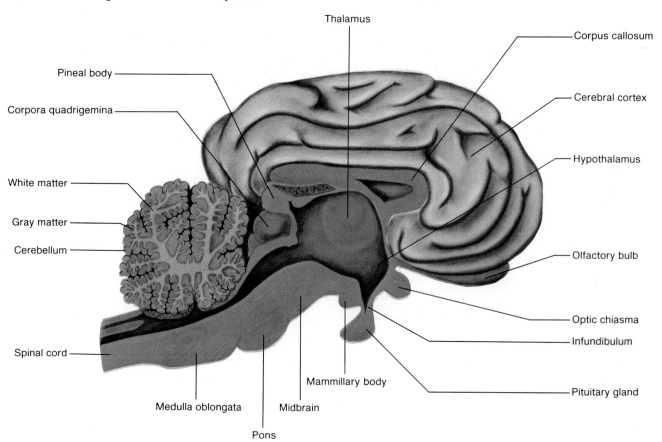

facial nerves

vestibulocochlear nerves

glossopharyngeal nerves

vagus nerves

accessory nerves

hypoglossal nerves

9. Using a long, sharp knife, cut the sheep brain along the midline to produce a midsagittal section. Study figure 30.5, and locate the following features on the specimen:

cerebrum

    cerebral cortex

    white matter

    gray matter

  olfactory bulb

  corpus callosum

  brain stem

diencephalon

    optic chiasma

    infundibulum

    pituitary gland

    mammillary bodies

    thalamus

    hypothalamus

    pineal gland

midbrain

    corpora quadrigemina

    cerebral peduncles

pons

medulla oblongata

10. Dispose of the sheep brain as directed by the laboratory instructor.

11. Complete Parts A and B of Laboratory Report 30 on page 211.

laboratory report 30

Name _____

Date _____

Section _____

Dissection of the Sheep Brain

Part A

Answer the following questions:

1. How do the relative sizes of the sheep and human cerebral hemispheres differ? _____
   _____
   _____

2. How do the convolutions and sulci of the sheep cerebrum compare with the human cerebrum in numbers? _____
   _____
   _____

3. What is the significance of the differences you noted in your answers for questions 1 and 2 above? _____
   _____
   _____

4. What difference did you note in the structures of the sheep cerebellum and the human cerebellum? _____
   _____
   _____

5. How do the sizes of the olfactory bulbs of the sheep brain compare with those of the human brain? _____
   _____
   _____

6. Based on their relative sizes, which of the cranial nerves seem to be most highly developed in the sheep brain? ___
   _____
   _____

7. What is the significance of the observations you noted in your answers for questions 5 and 6 above? _____
   _____
   _____

Part B

Prepare a list of at least five features to illustrate ways in which the brains of sheep and humans are similar.

1. _____

2. _____

3. _____

4. _____

5. _____

Receptors and Somatic Senses

S ensory receptors are sensitive to changes that occur within the body and its surroundings. When they are stimulated, they initiate nerve impulses that travel into the central nervous system. As a result of the brain interpreting such sensory impulses, the person may experience particular sensations.

The sensory receptors found in skin, muscles, joints, and visceral organs are associated with somatic senses. These senses include touch, pressure, temperature, and the senses of muscle movement and body position.

Purpose of the Exercise

To review the characteristics of sensory receptors and somatic senses and to investigate some of the somatic senses associated with the skin.

Learning Objectives

After completing this exercise, you should be able to:

1. Name five general types of receptors.
2. Explain how a sensation results.
3. List the somatic senses.
4. Determine the distribution of touch, heat, and cold receptors in a region of skin.
5. Determine the two-point threshold of a region of skin.

Materials Needed:

marking pen	blunt metal probe
millimeter rule	three beakers (250 ml)
bristle	hot water
forceps	cold water (ice water)

For Demonstration:

prepared microscope slides of Meissner's and pacinian corpuscles
compound microscope

Procedure A—Somatic Receptors

1. Review the sections entitled "Receptors and Sensations" and "Somatic Senses" in chapter 12 of the textbook.
2. Complete Part A of Laboratory Report 31 on page 216.

Demonstration

Observe the Meissner's corpuscle with the microscope set up by the laboratory instructor. This type of receptor is abundant in the outer regions of the dermis, such as in the fingertips, soles, lips, and external genital organs. It is also responsible for the sensation of light touch (see figure 31.1 and chapter 6 of the textbook).

Observe the pacinian corpuscle in the second demonstration microscope. This corpuscle is composed of many layers of connective tissue cells and has a nerve fiber in its central core. Pacinian corpuscles are numerous in the hands, feet, joints, and external genital organs. They are also responsible for the sense of deep pressure (figure 31.2). How are Meissner's and pacinian corpuscles similar? _____

_____

_____

How are they different? _____

_____

_____

_____

Figure 31.1 Meissner's corpuscles, such as this one, are responsible for the sensation of light touch (×400).

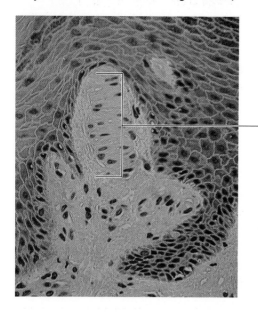

Meissner's corpuscle

Figure 31.2 Pacinian corpuscles, such as this one, are responsible for the sensation of deep pressure (×25).

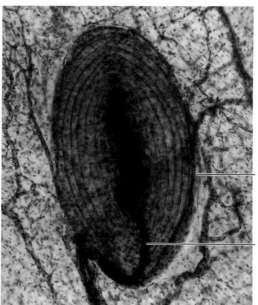

Pacinian corpuscle

Nerve fiber

Procedure B—Sense of Touch

1. Investigate the distribution of touch receptors in your laboratory partner's skin. To do this:
   a. Use a marking pen and a millimeter rule to prepare a square with 2.5 cm on each side on the skin of your partner's inner wrist, near the palm.
   b. Divide the square into smaller squares with 0.5 cm on a side, producing a small grid.
   c. Ask your partner to rest the marked wrist on the tabletop and to keep the eyes closed throughout the remainder of the experiment.
   d. Press the end of a bristle into the skin in some part of the grid, using just enough pressure to cause the bristle to bend.
   e. Ask your partner to report whenever the touch of the bristle is felt. Record the results in Part B of the laboratory report.
   f. Continue this procedure until you have tested twenty-five different locations on the grid.
2. Test two other areas of skin in the same manner, and record the results in Part B of the laboratory report.
3. Answer the questions in Part B of the laboratory report.

Procedure C—Two-Point Threshold

1. Test your partner's ability to recognize the difference between one or two points of skin being stimulated simultaneously. To do this:
   a. Have your partner place a hand with the palm up on the table and close the eyes.
   b. Hold the tips of a forceps tightly together and gently touch the skin of your partner's index finger.
   c. Ask your partner to report if it feels like one or two points are touching the finger.
   d. Allow the tips of the forceps to spread so they are 1 mm apart, press both points against the skin simultaneously, and ask your partner to report as before.
   e. Repeat this procedure, allowing the tips of the forceps to spread more each time until your partner can feel both tips being pressed against the skin. The minimum distance between the tips of the forceps when both can be felt is called the two-point threshold.
   f. Record the two-point threshold for the skin of the index finger in Part C of the laboratory report.
2. Repeat the above procedure to determine the two-point threshold of the palm, the back of the hand, the back of the neck, the lower leg, and the sole. Record the results in Part C of the laboratory report.
3. Answer the questions in Part C of the laboratory report.

Procedure D—Sense of Temperature

1. Investigate the distribution of heat receptors in your partner's skin. To do this:
   a. Mark a square with 2.5 cm sides on your partner's palm.
   b. Prepare a grid by dividing the square into smaller squares, 0.5 cm on a side.
   c. Have your partner rest the marked palm on the table and close the eyes.
   d. Heat a blunt metal probe by placing it in a beaker of hot water for a minute or so. *(Be sure the probe does not get so hot that it burns the skin.)*
   e. Wipe the probe dry and touch it to the skin on some part of the grid.
   f. Ask your partner to report if the probe feels hot. Then record the results in Part D of the laboratory report.
   g. Keep the probe hot, and repeat the procedure until you have tested twenty-five different locations on the grid.

2. Investigate the distribution of cold receptors by repeating the above procedure. Use a blunt metal probe that has been cooled by placing it in ice water for a minute or so. Record the results in Part D of the laboratory report.

3. Answer the questions in Part D of the laboratory report.

Optional Activity

Prepare three beakers of water of different temperatures. One beaker should contain warm water (about 40°C), one should be room temperature (about 22°C), and one should contain cold water (about 10°C). Place the index finger of one hand in the warm water and, at the same time, place the index finger of the other hand in the cold water for about two minutes. Then, simultaneously move both index fingers into the water at room temperature. What temperature do you sense with each finger? How do you explain the resulting sensations? _____
_____
_____
_____

Receptors and Somatic Senses

Part A—Somatic Receptors

Complete the following statements:

1. _____ are receptors that are sensitive to changes in the concentrations of chemicals.

2. Whenever tissues are damaged, _____ receptors are likely to be stimulated.

3. Receptors that are sensitive to temperature changes are called _____ .

4. _____ are sensitive to changes in pressure or to movement of fluid.

5. _____ are sensitive to changes in the intensity of light energy.

6. A feeling that occurs when sensory impulses are interpreted by the brain is called a(an) _____ .

7. The process by which the brain can cause a sensation to seem to come from the receptors being stimulated is called _____ .

8. A sensation may seem to fade away when receptors are continuously stimulated as a result of

   _____ .

9. Meissner's corpuscles are responsible for the sense of _____ .

10. Pacinian corpuscles are responsible for the sense of _____ .

11. Heat receptors are most sensitive to temperatures between _____ .

12. Cold receptors are most sensitive to temperatures between _____ .

13. Pain receptors are widely distributed but are lacking in the _____ .

14. _____ are the only receptors in visceral organs that produce sensations.

15. When visceral pain is felt in some region other than the part being stimulated, the phenomenon is called

    _____ .

Part B—Sense of Touch

1. Record a "+" to indicate where the bristle was felt and a "0" to indicate where it was not felt.

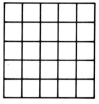

Skin of wrist

2. Show distribution of touch receptors in two other regions of skin.

Region tested _____   Region tested _____

3. Answer the following questions:
    a. How do you describe the pattern of distribution for touch receptors in the regions of the skin you tested?

    _____

    _____

    b. How does the concentration of touch receptors seem to vary from region to region? _____

    _____

Part C—Two-Point Threshold

1. Record the two-point threshold for skin in each of the following regions:

    Index finger ___.5 cm _____
    Palm _____ 1.3 cm _____
    Back of hand ___ 1 cm _____
    Back of neck _____ 1.9 cm _____
    Lower leg _____ 1.2 cm _____
    Sole _____

2. Answer the following questions:
    a. What region of the skin tested has the greatest ability to discriminate two points? _index finger_
    b. What region of skin has the least sensitivity to this test? _neck_
    c. What is the significance of these observations in a and b? _more use of finger_

    _____

    _____

Part D—Sense of Temperature

1. Record a "+" to indicate where heat was felt and a "0" to indicate where it was not felt.

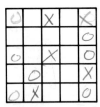

Skin of palm

2. Record a "+" to indicate where cold was felt and a "0" to indicate where it was not felt.

Skin of palm

3. Answer the following questions:

   a. How do temperature receptors appear to be distributed in the skin of the palm? _____

   _____

   b. Compare the distribution and concentration of heat and cold receptors in the skin of the palm. _____

   _____

   _____

Laboratory Exercise 32

Senses of Smell and Taste

The senses of smell and taste are dependent upon chemoreceptors that are stimulated by various chemicals dissolved in liquids. The receptors of smell are found in the olfactory organs, which are located in the upper parts of the nasal cavity and in a portion of the nasal septum. The receptors of taste occur in the taste buds, which are special organs found on the surface of the tongue.

These senses function closely together, because substances that are tasted often are smelled at the same moment and they play important roles in the selection of foods.

Purpose of the Exercise

To review the characteristics of the organs of smell and taste and to investigate the abilities of smell and taste receptors to discriminate various chemical substances.

Learning Objectives

After completing this exercise, you should be able to:

1. Describe the general characteristics of the smell receptors.
2. Describe the general characteristics of the taste receptors.
3. Explain how the senses of smell and taste function together.
4. Determine the time needed for olfactory sensory adaptation to occur.
5. Determine the distribution of taste receptors on the surface of the tongue.

Materials Needed:
For Procedure A—Sense of Smell

Set of substances in stoppered bottles:
cinnamon, sage, vanilla, mustard powder, garlic powder, oil of clove, oil of wintergreen, and perfume

For Procedure B—Sense of Taste

paper cups
cotton swabs
5% sucrose solution
5% NaCl solution
1% acetic acid
0.5% quinine sulfate solution

For Demonstrations:

compound microscope
prepared microscope slides of olfactory epithelium and of taste buds

For Optional Activity:

pieces of apple, potato, carrot, and onion

Procedure A—Sense of Smell

1. Review the section entitled "Sense of Smell" in chapter 12 of the textbook.
2. As a review activity, label figure 32.1.
3. Complete Part A of Laboratory Report 32 on page 223.

Figure 32.1 Label this diagram of the olfactory organ by placing the correct numbers in the spaces provided.

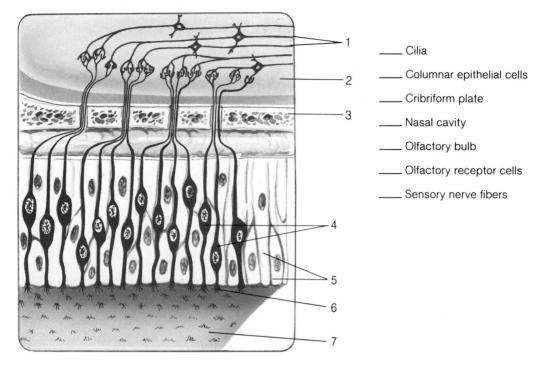

_____ Cilia

_____ Columnar epithelial cells

_____ Cribriform plate

_____ Nasal cavity

_____ Olfactory bulb

_____ Olfactory receptor cells

_____ Sensory nerve fibers

Demonstration

Observe the olfactory epithelium in the microscope set up by the laboratory instructor. The olfactory receptor cells are spindle-shaped, bipolar neurons with spherical nuclei. They also have six to eight cilia at their distal ends. The supporting cells are pseudostratified columnar epithelial cells. However, in this region the tissue lacks goblet cells (see figure 32.2 and chapter 5 of the textbook).

4. Test your laboratory partner's ability to recognize the odors of the bottled substances available in the laboratory. To do this:
   a. Have your partner keep the eyes closed.
   b. Remove the stopper from one of the bottles and hold it about 4 cm under your partner's nostrils for about two seconds.
   c. Ask your partner to identify the odor, then replace the stopper.
   d. Record your partner's response in Part B of the laboratory report.
   e. Repeat steps b–d for each of the bottled substances.
5. Repeat the preceding procedure, using the same set of bottled substances, but present them to your partner in a different sequence. Record the results in Part B of the laboratory report.

6. Wait ten minutes and then determine the time it takes for your partner to experience olfactory sensory adaptation. To do this:
   a. Ask your partner to breathe in through the nostrils and exhale through the mouth.
   b. Remove the stopper from one of the bottles and hold it about 4 cm under your partner's nostrils.
   c. Keep track of the time that passes until your partner is no longer able to detect the odor of the substance.
   d. Record the result in Part B of the laboratory report.
   e. Wait five minutes and repeat this procedure, using a different bottled substance.
   f. Test a third substance in the same manner.
   g. Record the results as before.
7. Complete Part B of the laboratory report.

Procedure B—Sense of Taste

1. Review the section entitled "Sense of Taste" in chapter 12 of the textbook.
2. As a review activity, label figure 32.3.
3. Complete Part C of the laboratory report.

Figure 32.2 Olfactory receptors have cilia at their distal ends (×250).

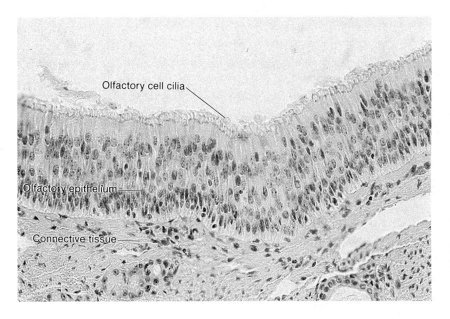

Olfactory cell cilia

Olfactory epithelium

Connective tissue

Figure 32.3 Label this diagram of taste buds by placing the correct numbers in the spaces provided.

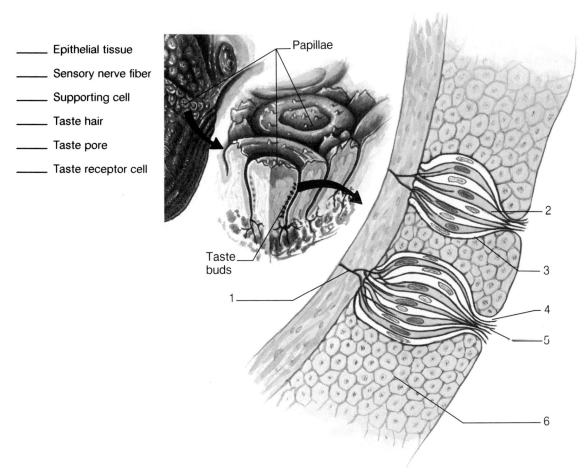

_____ Epithelial tissue

_____ Sensory nerve fiber

_____ Supporting cell

_____ Taste hair

_____ Taste pore

_____ Taste receptor cell

Papillae

Taste buds

1

2

3

4

5

6

Figure 32.4 Taste receptors are found in taste buds such as this one (×250).

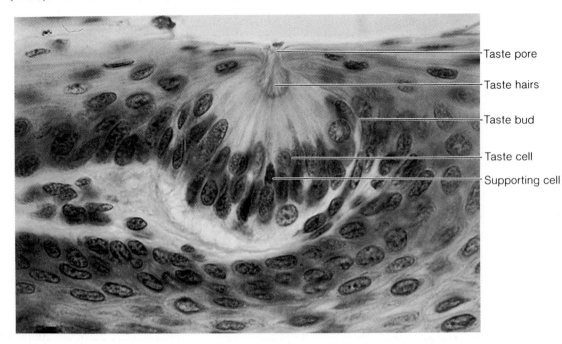

- Taste pore
- Taste hairs
- Taste bud
- Taste cell
- Supporting cell

Demonstration

Observe the oval-shaped taste bud in the microscope set up by the laboratory instructor. Note the surrounding epithelial cells. The taste pore, an opening into the taste bud, may be filled with taste hairs. Within the taste bud there are supporting cells and thinner taste-receptor cells, which often have lightly stained nuclei. (See figure 32.4.)

4. Map the distribution of the receptors for the primary taste sensations on your partner's tongue. To do this:
   a. Ask your partner to rinse the mouth with water and then partially dry the surface of the tongue with a paper towel.
   b. Moisten a clean cotton swab with 5% sucrose solution, and touch several regions of your partner's tongue with the swab.
   c. Each time you touch the tongue, ask your partner to report if a sweet sensation is experienced.
   d. Test the tip, sides, and back of the tongue in this manner.

e. Record your partner's responses in Part D of the laboratory report.
f. Have your partner rinse the mouth and dry the tongue again, and repeat the preceding procedure, using each of the other three test solutions—NaCl, acetic acid, and quinine.
5. Complete Part D of the laboratory report.

Optional Activity

Test your laboratory partner's ability to recognize the tastes of apple, potato, carrot, and onion. To do this:

1. Have your partner close the eyes and hold the nostrils shut.
2. Place a small piece of one of the test substances on your partner's tongue.
3. Ask your partner to identify the substance without chewing or swallowing it.
4. Repeat the procedure for each of the other substances.

How do you explain the results of this experiment? _____

_____

_____

_____

laboratory report 32

Name _____

Date _____

Section _____

Senses of Smell and Taste

Part A

Complete the following statements:

1. Olfactory or smell receptors are _____ neurons.

2. The distal ends of the olfactory neurons are covered by _____ .

3. Before substances can stimulate the olfactory receptors, they must be dissolved in water and must be soluble in _____ .

4. The axons of olfactory receptors pass through small openings in the _____ of the ethmoid bone.

5. Olfactory bulbs lie on either side of the _____ of the ethmoid bone.

6. The sensory impulses pass from the olfactory bulbs through the _____ to the interpreting centers of the brain.

7. The olfactory interpreting centers are located at the base of the _____ lobes of the cerebrum.

8. Olfactory sensations usually fade rapidly as a result of _____ .

9. Olfactory receptors are the only parts of the nervous system that come in contact with the _____ directly.

10. It has been estimated that a person loses about _____ of the olfactory receptors each year.

Part B—Sense of Smell

1. Record the results from the tests of odor recognition in the chart below:

Substance Tested	Odor Reported	
	First Trial	Second Trial

2. Record the results of the olfactory sensory adaptation time in the chart below:

Substance Tested	Adaptation Time

3. Complete the following:
   a. How do you describe your partner's ability to recognize the odors of the substances you tested? _____

   _____

   b. Compare your experimental results with others in the class. Did you find any evidence to indicate that individuals may vary in their abilities to recognize odors? Explain your answer. _____

   _____

   _____

   c. Does the time it takes for sensory adaptation to occur seem to vary with the substances tested? Explain your answer. _____

   _____

Part C

Complete the following statements:

1. Taste cells are modified _____ cells.

2. The opening to a taste bud is called a(an) _____ .

3. The _____ of a taste cell is its sensitive part.

4. Before the taste of a substance can be detected, the substance must be dissolved in _____ .

5. Substances that stimulate taste cells seem to combine with _____ on the surfaces of taste hairs.

6. There are at least _____ kinds of taste cells, although microscopically they all appear to be alike.

7. Sweet receptors are most abundant in the _____ of the tongue.

8. Sour receptors are stimulated mainly by _____ .

9. Salt receptors are stimulated mainly by _____ .

10. Alkaloids usually have a(an) _____ taste.

Part D—Sense of Taste

1. *Taste receptor distribution.* Record a "+" to indicate where a taste sensation seemed to originate and a "0" if no sensation occurred when the spot was stimulated.

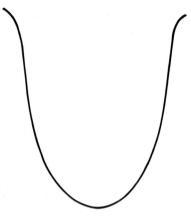

Sucrose (sweet sensation)

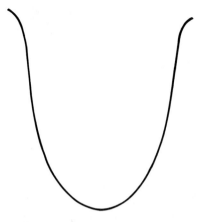

NaCl (salt sensation)

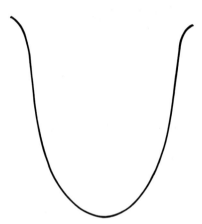

Acetic acid (sour sensation)

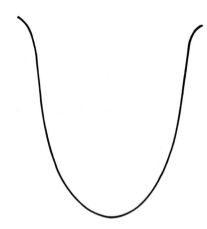

Quinine (bitter sensation)

2. Complete the following:

a. Describe how each type of taste receptor is distributed on the surface of your partner's tongue. _____
_____
_____
_____
_____

b. How do your experimental results compare with the distribution of taste receptors described in chapter 12 of the textbook? _____
_____
_____
_____

Laboratory Exercise 33

The Ear and Hearing

The ear is composed of external, middle, and inner parts. The external structures act to gather sound waves and to direct them inward to the eardrum of the middle ear. The parts of the middle ear, in turn, transmit vibrations from the eardrum to the inner ear, where the hearing receptors are located. As they are stimulated, these receptors initiate nerve impulses to pass over the vestibulocochlear nerve into the auditory cortex of the brain, where the impulses are interpreted and the sensations of hearing are created.

Purpose of the Exercise

To review the structural and functional characteristics of the ear and to conduct some simple hearing tests.

Learning Objectives

After completing this exercise, you should be able to:

1. Identify the major parts of the ear.
2. Describe the functions of the parts of the ear.
3. Trace the pathway of sound vibrations from the eardrum to the hearing receptors.
4. Conduct several simple hearing tests.

Materials Needed:
dissectible ear model
watch
tuning fork (256 cps)
rubber hammer
cotton
meter stick

For Demonstrations:

compound microscope
prepared microscope slide of cochlea (section)
audiometer

Procedure A—Structure and Function of the Ear

1. Review the section entitled "Sense of Hearing" in chapter 12 of the textbook.
2. As a review activity, label figures 33.1, 33.2, and 33.3.
3. Examine the dissectible model of the ear and locate the following features:

external ear

auricle

external auditory meatus

middle ear

tympanic membrane

tympanic cavity

auditory ossicles

malleus

incus

stapes

oval window

tensor tympani

stapedius

auditory tube

inner ear

osseous labyrinth

membranous labyrinth

cochlea

semicircular canals

vestibule

round window

vestibulocochlear nerves

Figure 33.1 Label the major features of the ear.

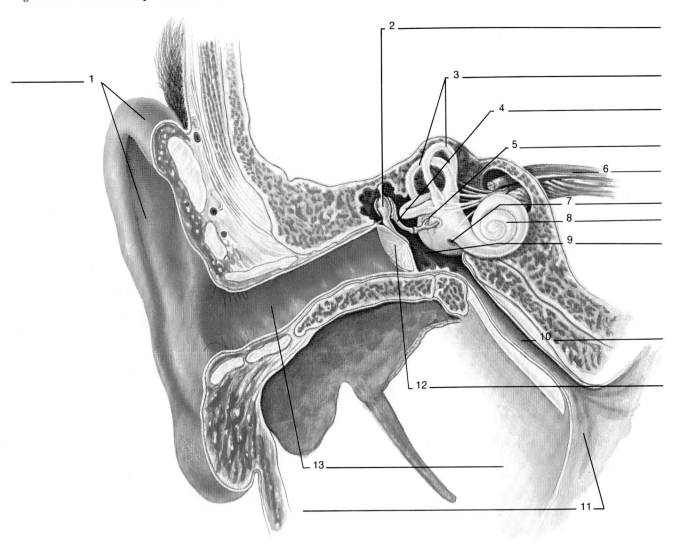

Demonstration

Observe the section of the cochlea in the microscope set up by the laboratory instructor. Locate one of the turns of the cochlea, and using figure 33.3 as a guide, identify the *scala vestibuli, cochlear duct, scala tympani, vestibular membrane, basilar membrane,* and the *organ of Corti* (figure 33.4).

---

4. Complete Parts A and B of Laboratory Report 33 on pages 231–32.

Procedure B—Hearing Tests

Perform the following tests in a quiet room, using your laboratory partner as the test subject.

1. *Auditory acuity test.* To conduct this test:
   a. Have the test subject sit with eyes closed.
   b. Pack one of the subject's ears with cotton.

c. Hold a ticking watch close to the open ear and slowly move it straight out and away from the ear.
d. Have the subject indicate when the sounds of the ticking can no longer be heard.
e. Use a meter stick to measure the distance in centimeters from the ear to the position of the watch.
f. Repeat this procedure to test the acuity of the other ear.
g. Record the test results in Part C of the laboratory report.

2. *Sound localization test.* To conduct this test:
   a. Have the subject sit with eyes closed.
   b. Hold the ticking watch somewhere within the audible range of the subject's ears and ask the subject to point to the watch.

Figure 33.2 Label the features of the inner ear by placing the correct numbers in the spaces provided.

_____ Ampulla   _____ Endolymph   _____ Semicircular canals

_____ Cochlea   _____ Perilymph   _____ Utricle

_____ Cochlear duct   _____ Saccule   _____ Vestibule

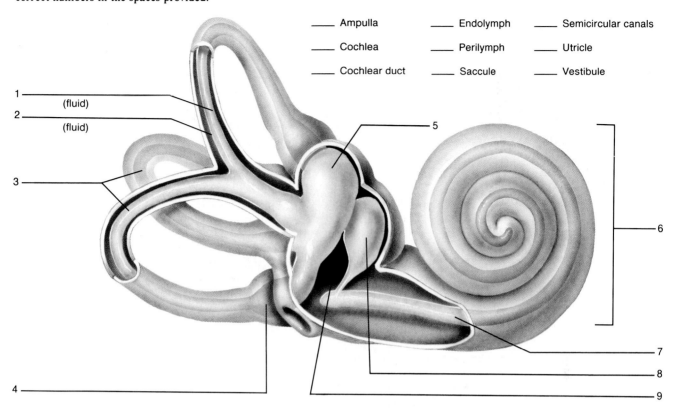

1 _____ (fluid)

2 _____ (fluid)

3

4

5

6

7

8

9

Figure 33.3 Label the major features indicated in this cross section of the cochlea.

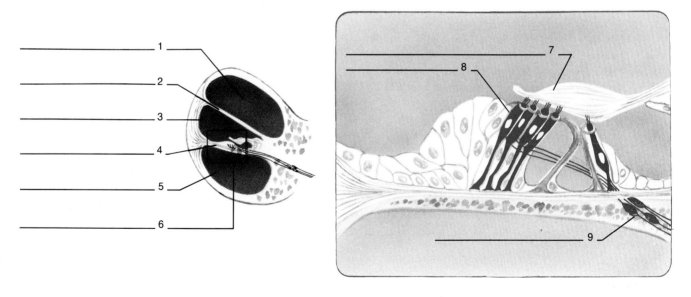

Figure 33.4 A cross section of the cochlea.

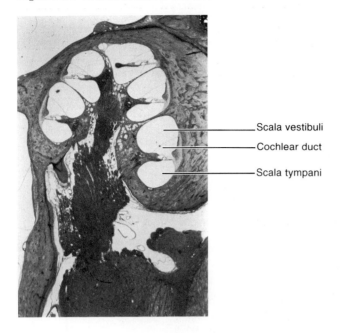

—Scala vestibuli

—Cochlear duct

—Scala tympani

c. Move the watch to another position and repeat the request. In this manner, determine how accurately the subject can locate the watch when it is in each of the following positions: in front of the head, behind the head, above the head, on the right side of the head, and on the left side of the head.

d. Record the test results in Part C of the laboratory report.

3. *Rinne's test.* To conduct this test:

a. Obtain a tuning fork and strike it with a rubber hammer, causing it to vibrate.

b. Place the end of the fork's handle against the subject's mastoid process behind one ear. (Have the prongs of the fork pointed downward and away from the ear and be sure nothing is touching them.)

c. Ask the subject to indicate when the sound is no longer heard.

d. Then, quickly remove the fork from the mastoid process and position it in the air close to the opening of the nearby external auditory meatus.

If hearing is normal, the sound will be heard again; if there is conductive impairment, the sound will not be heard.

4. *Weber's test.* To conduct this test:

a. Strike the tuning fork with the rubber hammer.

b. Place the handle of the fork against the subject's forehead in the midline.

c. Ask the subject to indicate if the sound is louder in one ear than in the other or if it is equally loud in both ears.

If hearing is normal, the sound will be equally loud in both ears. If there is conductive impairment, the sound will appear louder in the affected ear.

d. Have the subject experience the effects of conductive impairment by packing one ear with cotton and repeating Weber's test. Usually the sound appears louder in the plugged (or impaired) ear because extraneous sounds from the room are blocked out.

5. Complete Part C of the laboratory report.

Demonstration

Ask the laboratory instructor to demonstrate the use of the audiometer. This instrument produces sound vibrations of known frequencies that are transmitted to one or both ears of a test subject through earphones. The audiometer can be used to determine the threshold of hearing for different sound frequencies, and in the case of hearing impairment, it can be used to determine the percentage of hearing loss for each frequency.

laboratory report 33

Name _____

Date _____

Section _____

The Ear and Hearing

Part A

Match the terms in column A with the descriptions in column B. Place the letter of your choice in the space provided.

Column A

A. ceruminous gland
B. eustachian tube
C. external auditory meatus
D. malleus
E. membranous labyrinth
F. osseous labyrinth
G. scala tympani
H. scala vestibuli
I. stapedius
J. stapes
K. tectorial membrane
L. tensor tympani
M. tympanic cavity
N. tympanic membrane
O. vestibule

Column B

____ 1. muscle attached to stapes
____ 2. muscle attached to malleus
____ 3. ossicle attached to eardrum
____ 4. air-filled space containing auditory ossicles
____ 5. contacts hairs of hearing receptors
____ 6. leads from oval window to apex of cochlea
____ 7. S-shaped tube leading to eardrum
____ 8. wax-secreting structure
____ 9. cone-shaped semitransparent membrane
____ 10. ossicle attached to oval window
____ 11. connects cochlea and semicircular canals
____ 12. contains endolymph
____ 13. bony canal in temporal bone
____ 14. connects middle ear and pharynx
____ 15. extends from apex of cochlea to round window

Part B

Identify the features indicated in the micrograph of the organ of Corti in figure 33.5.

1. _____

2. _____

3. _____

4. _____

5. _____

Figure 33.5 Identify the features of this organ of Corti.

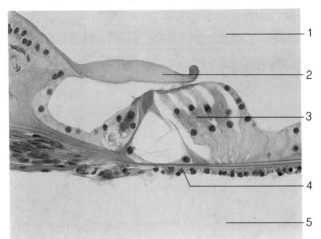

Part C

1. Results of auditory acuity test:

Ear Tested	Audible Distance (cm)
Right	
Left	

2. Results of sound localization test:

Actual Location	Reported Location
Front of head	
Back of head	
Above head	
Right side	
Left side	

3. Prepare a statement to summarize the results of the hearing tests you conducted on your laboratory partner.

_____

_____

_____

_____

_____

_____

_____

Sense of Equilibrium

The sense of equilibrium involves two sets of sensory organs. One set functions to help maintain the stability of the head and body when they are motionless and produces a sense of static equilibrium. The other set is concerned with balancing the head and body when they are moved suddenly and produces a sense of dynamic equilibrium.

The organs associated with the sense of static equilibrium are located within the vestibules of the inner ears, while those associated with the sense of dynamic equilibrium are found within the ampullae of the semicircular canals.

Purpose of the Exercise

To review the structure and function of the organs of equilibrium and to conduct some simple tests of equilibrium.

Learning Objectives

After completing this exercise, you should be able to:

1. Distinguish between static and dynamic equilibrium.
2. Identify the organs of equilibrium and describe their functions.
3. Explain the role of vision in the maintenance of equilibrium.
4. Conduct the Romberg and Bárány's tests of equilibrium.

Materials Needed:
  rotating chair
  bright light

For Demonstration:

  compound microscope
  prepared microscope slide of semicircular canal
    (cross section through ampulla)

Procedure A—Structure and Function of Organs of Equilibrium

1. Review the section entitled "Sense of Equilibrium" in chapter 12 of the textbook.
2. Complete Part A of Laboratory Report 34 on page 235.

Demonstration

Observe the cross section of the semicircular canal through the ampulla in the microscope set up by the laboratory instructor. Note the crista projecting into the lumen of the membranous labyrinth, which in life is filled with endolymph. The space between the membranous and osseous labyrinths normally is filled with perilymph (figure 34.1).

Procedure B—Tests of Equilibrium

Perform the following tests, using a person who is not easily disturbed by dizziness or rotational movement as a test subject. Also have some other students standing close by to help prevent the test subject from falling during the tests. *The tests should be stopped immediately if the test subject begins to feel uncomfortable or nauseated.*

1. *Vision and equilibrium.* To demonstrate the importance of vision in the maintenance of equilibrium:
   a. Have the test subject stand erect on one foot for one minute with the eyes open.
   b. Observe the subject's degree of unsteadiness.
   c. Repeat the procedure with the subject's eyes closed. *Be prepared to prevent the subject from falling.*

Figure 34.1 A micrograph of a crista.

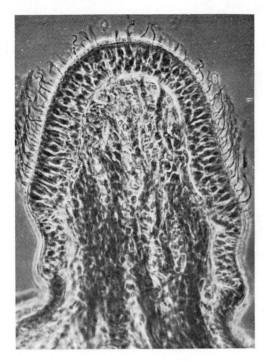

2. *Romberg test.* To conduct this test:
   a. Position the test subject close to a chalkboard with the back toward the board.
   b. Place a bright light in front of the subject so that a shadow of the body is cast on the board.
   c. Have the subject stand erect with feet close together and eyes staring straight ahead for a period of three minutes.
   d. During the test, make marks on the chalkboard along the edge of the shadow of the subject's shoulders to indicate the range of side-to-side swaying.
   e. Measure the maximum sway in centimeters and record the results in Part B of the laboratory report.
   f. Repeat the procedure with the subject's eyes closed.
   g. Position the subject so one side is toward the chalkboard.
   h. Repeat the procedure with the eyes open.
   i. Repeat the procedure with the eyes closed.
      The Romberg test is used to evaluate a person's ability to integrate sensory information from proprioceptors and receptors within the organs of equilibrium and to relay appropriate motor impulses to postural muscles. A person who shows little unsteadiness when standing with feet together and eyes open, but who becomes unsteady when the eyes are closed, has a positive Romberg test.

3. *Bárány's test.* To conduct this test:
   a. Have the test subject sit on a rotating chair with the eyes closed, the head tilted forward about 30°, and the hands gripped firmly to the seat. *Be prepared to prevent the subject and the chair from tipping over.*
   b. Rotate the chair every two seconds for a total of ten rotations.
   c. Abruptly stop the movement of the chair.
   d. Have the subject open the eyes and note the nature of the eye movements and their direction (such eye movements are called nystagmus). Also note the time it takes for the nystagmus to cease.
   e. Record your observations in Part B of the laboratory report.
   f. Allow the subject several minutes of rest, then repeat the procedure with the subject's head tilted nearly 90° onto one shoulder.
   g. After another rest period, repeat the procedure with the subject's head bent forward so that the chin is resting on the chest.
      In this test, when the head is tilted about 30°, the lateral semicircular canals receive maximal stimulation, and the nystagmus is normally from side to side. When the head is tilted at 90°, the superior canals are stimulated and the nystagmus is up and down. When the head is bent forward with the chin on the chest, the posterior canals are stimulated and the nystagmus is rotary.

4. Complete Part B of the laboratory report.

Name _____

Date _____

Section _____

Sense of Equilibrium

Part A

Complete the following statements:

1,2. The organs of static equilibrium are located within two expanded chambers of the membranous labyrinth called the _____ and the _____ .

3. The receptor cells of these organs are found in the wall of the labyrinth in a structure called the _____ .

4. Otoliths are small grains composed of _____ .

5. Sensory impulses travel from the organs of equilibrium to the brain on the _____ nerve.

6. The sensory organ of a semicircular canal lies within a swelling of the canal called the _____ .

7. The sensory organ within a semicircular canal is called a(an) _____ .

8. The _____ of this sensory organ consists of a dome-shaped gelatinous mass.

9. When the head is moved, the fluid inside the membranous portion of a semicircular canal remains stationary because of the _____ of the fluid.

10. Specialized receptors, called _____ , provide the brain with information concerning the position of joints and other body parts.

Part B—Tests of Equilibrium

1. Vision and equilibrium test results:
   a. When the eyes are open, what sensory organs provide information needed to maintain equilibrium? _____
   _____
   _____

   b. When the eyes are closed, what sensory organs provide such information? _____
   _____
   _____

2. Romberg test results:
   a. Record the test results in the following chart.

Conditions	Maximal Movement (cm)
Back toward board, eyes open	
Back toward board, eyes closed	
Side toward board, eyes open	
Side toward board, eyes closed	

   b. Did the test subject's unsteadiness increase when the eyes were closed? _____ What is the significance of this observation? _____ _____ _____

   c. Why would you expect a person with impairment of the organs of equilibrium to become more unsteady when the eyes are closed? _____ _____ _____

3. Bárány's test results:
   a. Record the test results in the following chart.

Position of Head	Description of Eye Movements	Time for Movement to Cease
Tilted 30°		
Tilted 90°		
Tilted forward, chin on chest		

   b. Write a paragraph to summarize the results of this test. _____ _____ _____ _____

The Eye

The eye contains photoreceptors, which are modified neurons located on its inner wall. Other parts of the eye provide protective functions or make it possible to move the eyeball. Still other structures serve to focus light entering the eye so that a sharp image is projected onto the receptor cells. Nerve impulses generated when the receptors are stimulated travel over the optic nerves to the brain, which in turn interprets the impulses and creates the sensation of sight.

Purpose of the Exercise

To review the structure and function of the eye and to dissect a mammalian eye.

Learning Objectives

After completing this exercise, you should be able to:

1. Identify the major parts of an eye.
2. Describe the functions of the parts of an eye.
3. List the structures through which light passes as it travels from the cornea to the retina.
4. Dissect a mammalian eye, and identify its major features.

Materials Needed:
dissectible eye model
compound microscope
prepared microscope slide of a mammalian eye
(median section)
mammalian eye (fresh or preserved)
dissecting tray
dissecting instruments—forceps, sharp scissors, and
dissecting needle

For Optional Activity:

ophthalmoscope

Procedure A—Structure and Function of the Eye

1. Review the sections entitled "Visual Accessory Organs," "Structure of the Eye," and "Visual Receptors" in chapter 12 of the textbook.
2. As a review activity, label figures 35.1, 35.2, and 35.3.
3. Complete Part A of Laboratory Report 35 on page 243.
4. Examine the dissectible model of the eye, and locate the following features:

eyelid

conjunctiva

orbicularis oculi

levator palpebrae superioris

lacrimal apparatus

    lacrimal gland

    canaliculi

    lacrimal sac

    nasolacrimal duct

extrinsic muscles

    superior rectus

    inferior rectus

    medial rectus

    lateral rectus

    superior oblique

    inferior oblique

cornea

sclera

optic nerve

choroid coat

Figure 35.1 Label the parts of the lacrimal apparatus.

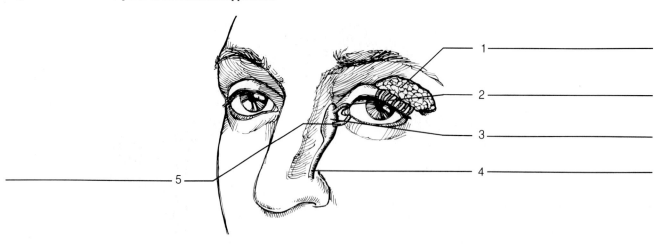

Figure 35.2 Label the extrinsic muscles of the eye.

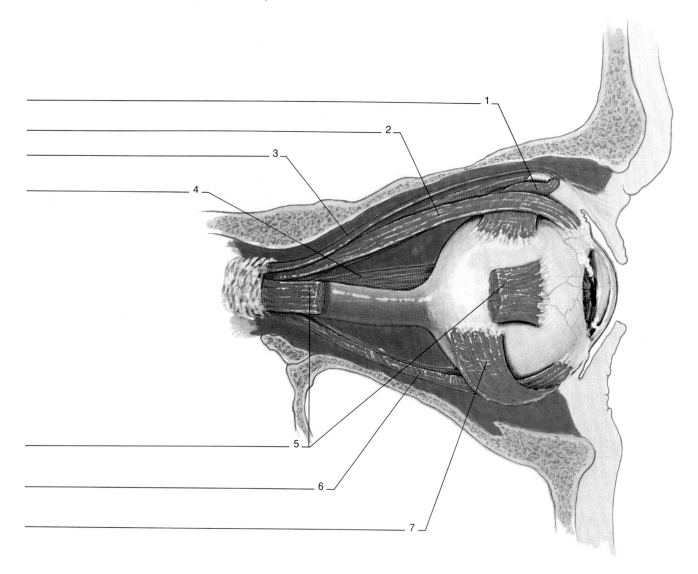

Figure 35.3 Label the parts shown in this section of the eye.

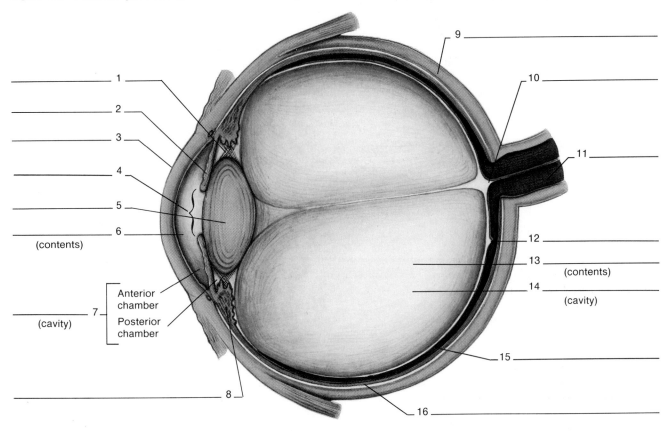

1 _____
2 _____
3 _____
4 _____
5 _____
6 _____
(contents)

7 _____
(cavity)

Anterior chamber
Posterior chamber

8 _____

9 _____
10 _____
11 _____
12 _____
13 _____
(contents)
14 _____
(cavity)

15 _____
16 _____

ciliary body

   ciliary processes

   ciliary muscles

lens

suspensory ligaments

iris

anterior cavity

anterior chamber

posterior chamber

aqueous humor

pupil

retina

   macula lutea

   fovea centralis

   optic disk

posterior cavity

vitreous humor

5. Obtain a microscope slide of a mammalian eye section, and locate as many of the preceding features listed as possible.
6. Observe the posterior portion of the eye wall using high power of the microscope, and locate the sclera, choroid coat, and retina.
7. Examine the retina using high-power magnification, and note its layered structure (figure 35.4). Locate the following:

nerve fibers leading to the optic nerve (innermost layer of the retina)

layer of ganglion cells

layer of bipolar neurons

nuclei of rods and cones

receptor ends of rods and cones

pigmented epithelium (outermost layer of the retina)

Figure 35.4 The cells of the retina are arranged in distinct layers (×200).

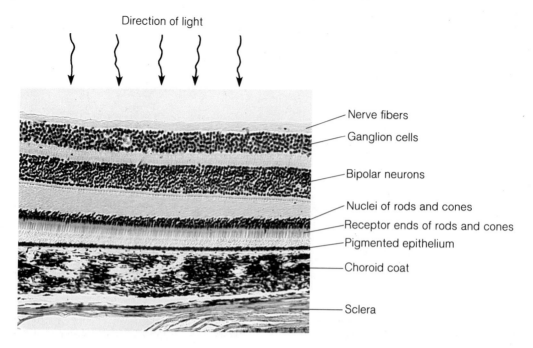

Direction of light

- Nerve fibers
- Ganglion cells
- Bipolar neurons
- Nuclei of rods and cones
- Receptor ends of rods and cones
- Pigmented epithelium
- Choroid coat
- Sclera

Optional Activity

Use an ophthalmoscope to examine the interior of your laboratory partner's eye. This instrument consists of a set of lenses held in a rotating disk, a light source, and some mirrors that reflect the light into the test subject's eye. The examination should be conducted in a dimly lighted room. Have your partner seated and staring straight ahead at eye level. Move the rotating disk of the ophthalmoscope so that the "O" appears in the lens selection window. Hold the instrument in your right hand with the end of your index finger on the rotating disk (figure 35.5). Direct the light at a slight angle from a distance of about 15 cm into the subject's right eye. The light beam should pass along the inner edge of the pupil. Look through the instrument, and you should see a reddish circular area—the interior of the eye. Rotate the disk of lenses to higher values until sharp focus is achieved.

Move the ophthalmoscope to within about 5 cm of the eye being examined, and again rotate the lenses to sharpen the focus (figure 35.6). Locate the optic disk and the blood vessels that pass through it. Also locate the yellowish macula lutea by having your partner stare directly into the light of the instrument (figure 35.7).

Examine the subject's iris by viewing it from the side and by using a lens with a +15 or +20 value.

Procedure B—Eye Dissection

1. Obtain a mammalian eye, place it in a dissecting tray, and dissect it as follows:
   a. Trim away the fat and other connective tissues, but leave the stubs of the *extrinsic muscles* and of the *optic nerve*. This nerve projects outward from the posterior region of the eyeball.

Figure 35.5 An ophthalmoscope is used to examine the interior of the eye.

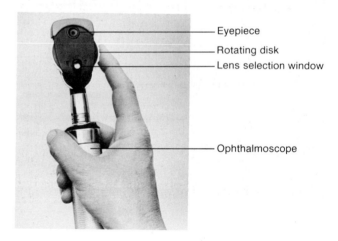

- Eyepiece
- Rotating disk
- Lens selection window
- Ophthalmoscope

b. Note the *conjunctiva*, which lines the eyelid and is reflected over the anterior surface of the eye. Lift some of this thin membrane away from the eye with forceps and examine it.
c. Locate and observe the *cornea, sclera,* and *iris.* Also note the *pupil* and its shape.
d. Use sharp scissors to make a coronal section of the eye. To do this, cut through the wall about 1 cm from the margin of the cornea and continue all the way around the eyeball. Try not to damage the internal structures of the eye (figure 35.8).

Figure 35.6 (*a*) Rotate the disk of lenses until sharp focus is achieved; (*b*) move the ophthalmoscope to within 5 cm of the eye to examine the optic disk.

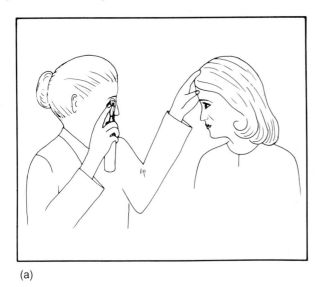

(a)

(b)

Figure 35.7 The interior of the eye as seen using an ophthalmoscope.

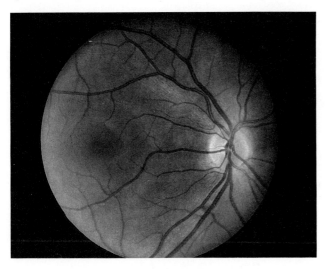

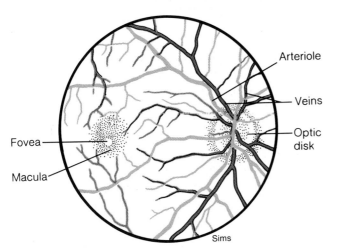

Fovea

Macula

Arteriole

Veins

Optic disk

Sims

Figure 35.8 Prepare a coronal section of the eye.

Margin of cornea

e. Gently separate the eyeball into anterior and posterior portions. Usually the jellylike vitreous humor will remain in the posterior portion, and the lens may adhere to it. Place the parts in the dissecting tray with their contents facing upward.

f. Examine the anterior portion of the eye and locate the *ciliary body,* which appears as a dark, circular structure. Also note the *iris* and the *lens* if it remained in the anterior portion. The lens normally is attached to the ciliary body by many *suspensory ligaments,* which appear as delicate, transparent threads.

g. Use a dissecting needle to gently remove the lens, and examine it. If the lens is transparent, hold it up and look through it at something in the distance.

h. Examine the posterior portion of the eye. Note the *vitreous humor*. This jellylike mass helps to hold the lens in place anteriorly and helps to hold the *retina* against the choroid coat.

i. Carefully remove the vitreous humor and examine the retina. This layer will appear as a thin, nearly colorless membrane that detaches easily from the choroid coat.

j. Locate the *optic disk*—the point where the retina is attached to the posterior wall of the eyeball and where the optic nerve originates. Because there are no receptor cells in the optic disk, this region is also called the "blind spot."

k. Note the iridescent area of the choroid coat beneath the retina. This colored surface is called the *tapetum lucidum*. It serves to reflect light back through the retina, an action that is thought to aid the night vision of some animals. The tapetum lucidum is lacking in the human eye.

l. Discard the tissues of the eye as directed by the laboratory instructor.

2. Complete Part B of the laboratory report.

laboratory report 35

Name _____

Date _____

Section _____

The Eye

Part A

Match the terms in column A with the descriptions in column B. Place the letter of your choice in the space provided.

Column A

A. aqueous humor
B. choroid coat
C. ciliary muscle
D. conjunctiva
E. cornea
F. iris
G. lacrimal gland
H. lysozyme
I. nasolacrimal duct
J. optic disk
K. retina
L. sclera
M. strabismus
N. suspensory ligament
O. vitreous humor

Column B

____ 1. vascular tunic of the eye
____ 2. white of the eye
____ 3. transparent anterior portion of outer tunic
____ 4. inner lining of eyelid
____ 5. secretes tears
____ 6. empties into nasal cavity
____ 7. fills posterior cavity of eye
____ 8. area where optic nerve originates
____ 9. controls light intensity
____ 10. fills anterior chamber of eye
____ 11. contains visual receptors
____ 12. connects lens to ciliary body
____ 13. causes lens to change shape
____ 14. lack of eye alignment
____ 15. antibacterial agent in tears

Complete the following:

16. List the structures and fluids through which light passes as it travels from the cornea to the retina. _____

_____

_____

_____

17. List three ways in which rods and cones differ. _____

_____

_____

_____

_____

Part B

Complete the following:

1. Which layer of the eye was the toughest to cut? _____

2. What kind of tissue do you think is responsible for this quality of toughness? _____

3. How do you compare the shape of the pupil in the dissected eye with your pupil? _____
   _____
   _____

4. Where do you find aqueous humor in the dissected eye? _____
   _____
   _____

5. What is the function of the dark pigment in the choroid coat? _____
   _____
   _____

6. Describe the lens of the dissected eye. _____
   _____
   _____
   _____

7. Describe the vitreous humor of the dissected eye. _____
   _____
   _____
   _____

Laboratory Exercise 36

Visual Tests and Demonstrations

Normal vision (emmetropia) results when light rays from objects in the external environment are refracted by the cornea and the lens of the eye and focused onto the photoreceptors of the retina. Irregular curvatures in the surface of the cornea or lens, inability to change the shape of the lens, or defects in the shape of the eyeball can result in defective vision.

Purpose of the Exercise

To conduct tests for visual acuity, astigmatism, and the ability to accommodate, and to demonstrate the blind spot and certain reflexes of the eye.

Learning Objectives

After completing this exercise, you should be able to:

1. Describe three conditions that can lead to defective vision.
2. Conduct the tests used to evaluate visual acuity, astigmatism, and the ability to accommodate for close vision.
3. Demonstrate the blind spot, photopupillary reflex, accommodation pupillary reflex, and convergence reflex.

Materials Needed:

Snellen eye chart
3″ × 5″ card
astigmatism chart
meter stick
pen flashlight

For Optional Activity:

Ishihara's color plates or other color blindness test

Procedure A—Visual Tests

Perform the following visual tests, using your laboratory partner as a test subject. If your partner usually wears glasses, test each eye with and without the glasses.

1. *Visual acuity test.* Visual acuity (sharpness of vision) can be measured by using a Snellen eye chart (see figure 36.1). This chart consists of several sets of letters in different sizes printed on a white card. The letters near the top of the chart are relatively large, and those in each lower set become smaller. At one end of each set of letters, there is an acuity value in the form of a fraction. One of the sets near the bottom of the chart, for example, is marked 20/20. The normal eye can see these letters clearly from the standard distance of 20 feet and thus is said to have 20/20 vision. The letter at the top of the chart is marked 20/200. The normal eye can read letters of this size from a distance of 200 feet. Thus, an eye that is only able to read the top letter of the chart from a distance of 20 feet is said to have 20/200 vision.

   To conduct the visual acuity test:
   a. Hang the Snellen eye chart on a well-illuminated wall at eye level.
   b. Have your partner stand 20 feet in front of the chart, cover the left eye with a 3″ × 5″ card, and read the smallest set of letters possible.
   c. Record the visual acuity value for that set of letters in Part A of Laboratory Report 36 on pages 249–50
   d. Repeat the procedure, using the left eye.
2. *Astigmatism test.* Astigmatism is a condition that results from a defect in the curvature of the cornea or lens. As a consequence, some portions of the image projected on the retina are sharply focused and other portions are blurred. Astigmatism can be

Figure 36.1 The Snellen eye chart looks like this but is somewhat larger.

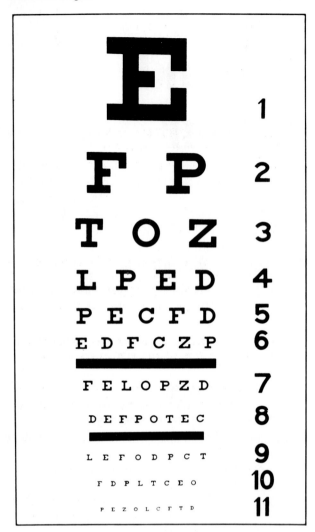

Figure 36.2 Astigmatism is evaluated using a chart such as this one.

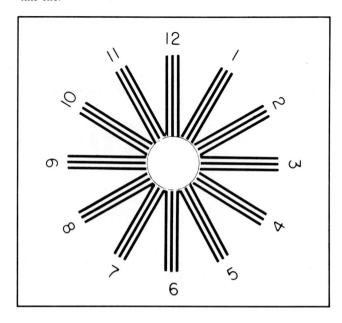

3. *Accommodation test.* Accommodation is the changing of the shape of the lens that occurs when the normal eye is focused for close vision. It involves a reflex in which muscles of the ciliary body are stimulated to contract, releasing tension on the suspensory ligaments that are fastened to the lens capsule. This allows the capsule to rebound elastically, causing the surface of the lens to become more convex. The ability to accommodate is likely to decrease with age because the tissues involved tend to lose their elasticity.

   To evaluate the ability to accommodate:
   a. Hold the end of a meter stick against your partner's chin so that the stick extends outward at a right angle to the plane of the face.
   b. Have your partner close the left eye.
   c. Hold a 3″ × 5″ card with a word typed in the center at the distal end of the meter stick.
   d. Slide the card along the stick toward your partner's open eye, and locate the point *closest to the eye* where your partner can still see the letters of the word sharply focused. This distance is called the *near point of accommodation,* and it tends to increase with age (see chart 36.1).
   e. Repeat the procedure with the right eye closed.
   f. Record the results in Part A of the laboratory report.
4. Complete Part A of the laboratory report.

evaluated by using an astigmatism chart (see figure 36.2). This chart consists of sets of black lines radiating from a central spot like the spokes of a wheel. To a normal eye, these lines appear sharply focused and equally dark; however, if the eye has an astigmatism, some sets of lines appear sharply focused and dark while others are blurred and less dark.

   To conduct the astigmatism test:
   a. Hang the astigmatism chart on a well-illuminated wall at eye level.
   b. Have your partner stand 20 feet in front of the chart, cover the left eye with a 3″ × 5″ card, focus on the spot in the center of the radiating lines, and report which lines, if any, appear more sharply focused and darker.
   c. Repeat the procedure, using the left eye.
   d. Record the results in Part A of the laboratory report.

Chart 36.1 Near point of accommodation	
Age (years)	Average near point (cm)
10	7
20	10
30	13
40	20
50	45
60	90

Optional Activity

Some individuals have defective color vision caused by decreased sensitivity in one or more sets of retinal cones, and they are said to be color-blind. Several tests have been devised to evaluate color vision and to detect such visual problems. If such a test is available in the laboratory, conduct a color blindness test on your laboratory partner. Be sure to read the instructions included with the test before conducting it. What can you conclude from the results of this test? _____

_____

_____

_____

Procedure B—Visual Demonstrations

Perform the following demonstrations with the help of your laboratory partner.

1. *Blind spot demonstration.* There are no photoreceptors in the optic disk, which is located where the nerve fibers of the retina leave the eye and enter the optic nerve. Consequently, this region of the retina is commonly called the blind spot.

    To demonstrate the blind spot:
    a. Close your left eye, hold figure 36.3 about 35 cm away from your face, and stare at the "+" sign in the figure with your right eye.
    b. Move the figure closer to your face as you continue to stare at the "+," until the dot on the figure suddenly disappears. This happens when the image of the dot is focused on the optic disk.

Figure 36.3   Blind spot demonstration.

2. *Photopupillary reflex.* The smooth muscles of the iris function to control the size of the pupil. For example, when the intensity of light entering the eye increases, a photopupillary reflex is triggered, and the circular muscles of the iris are stimulated to contract. As a result, the size of the pupil decreases, and less light enters the eye.

    To demonstrate this reflex:
    a. Ask your partner to sit with the hands tightly covering the eyes for two minutes.
    b. Position a pen flashlight close to one eye with the light shining on the hand that covers the eye.
    c. Ask your partner to remove the hand quickly.
    d. Observe the pupil and note any change in its size.
    e. Have your partner remove the other hand, but keep that uncovered eye shielded from extra light.
    f. Observe both pupils, and note any difference in their sizes.

3. *Accommodation reflex.* To demonstrate the accommodation reflex:
    a. Have your partner stare for several seconds at some dimly illuminated object in the room that is more than 20 feet away.
    b. Observe the size of the pupil of one eye, hold a pencil about 25 cm in front of your partner's face, and have your partner stare at it.
    c. Note any change in the size of the pupil.

4. *Convergence reflex.* To demonstrate the convergence reflex:
    a. Repeat the procedure outlined for the accommodation reflex.
    b. Note any change in the position of the eyeballs as your partner changes focus from the distant object to the pencil.

5. Complete Part B of the laboratory report.

Name ————————————————

Date ————————————————

Section ————————————————

Visual Tests and Demonstrations

Part A

1. Visual acuity test results:

Eye Tested	Acuity Values
Right eye	
Right eye with glasses	
Left eye	
Left eye with glasses	

2. Astigmatism test results:

Eye Tested	Darker Lines
Right eye	
Right eye with glasses	
Left eye	
Left eye with glasses	

3. Accommodation test results:

Eye Tested	Near Point
Right eye	
Right eye with glasses	
Left eye	
Left eye with glasses	

4. Complete the following:

    a. What is meant by 20/70 vision? _____

    _____

    _____

    b. What is meant by 20/10 vision? _____

    _____

    _____

    c. What visual problem is created by astigmatism? _____

    _____

    _____

    d. Why does the near point of accommodation often increase with age? _____

    _____

    _____

    e. Write a short statement to summarize the results of the visual tests you conducted on your partner. _____

    _____

    _____

    _____

    _____

Part B

Complete the following:

1. Explain why an eye has a blind spot. _____

    _____

    _____

    _____

2. Describe the photopupillary reflex. _____

    _____

    _____

    _____

    _____

3. What difference did you note in the size of the pupils when one eye was exposed to bright light and the other eye was shielded from the light? _____

    _____

    _____

    _____

4. Describe the accommodation reflex. _____

    _____

    _____

    _____

5. Describe the convergence reflex. _____

    _____

    _____

    _____

Laboratory Exercise 37

Endocrine System

Endocrine glands act together with parts of the nervous system in helping to control body activities. They secrete hormones that are transported in body fluids and affect cells possessing appropriate receptor molecules. In this way, hormones influence the rate of metabolic reactions, the transport of substances through cell membranes, and the regulation of water and electrolyte balances. By controlling cellular activities, endocrine glands play important roles in the maintenance of homeostasis.

Purpose of the Exercise

To review the structure and function of major endocrine glands and to examine microscopically the tissues of these glands.

Learning Objectives

After completing this exercise, you should be able to:

1. Name and locate the major endocrine glands.
2. Name the hormones secreted by each of the major glands.
3. Describe the principal functions of each hormone.
4. Recognize tissue sections from the pituitary gland, thyroid gland, parathyroid glands, adrenal glands, and pancreas.

Materials Needed:
human manikin
compound microscope
prepared microscope slides of the following:
pituitary gland
thyroid gland
parathyroid gland
adrenal gland
pancreas

For Optional Activity:
water bath equipped with temperature control mechanism (set at 37°C)
laboratory thermometer

Procedure

1. Review the sections entitled "Pituitary Gland," "Thyroid Gland," "Parathyroid Glands," "Adrenal Glands," "Pancreas," and "Other Endocrine Glands" in chapter 13 of the textbook.
2. As a review activity, label figures 37.1, 37.2, 37.3, 37.4, 37.5, and 37.6.
3. Complete Part A of Laboratory Report 37 on pages 259–60.
4. Examine the human manikin and locate the following:

hypothalamus

infundibulum

pituitary gland

anterior lobe

posterior lobe

thyroid gland

parathyroid glands

adrenal glands

adrenal medulla

adrenal cortex

pancreas

pineal gland

thymus gland

ovaries

testes

Figure 37.1 Label the major endocrine glands.

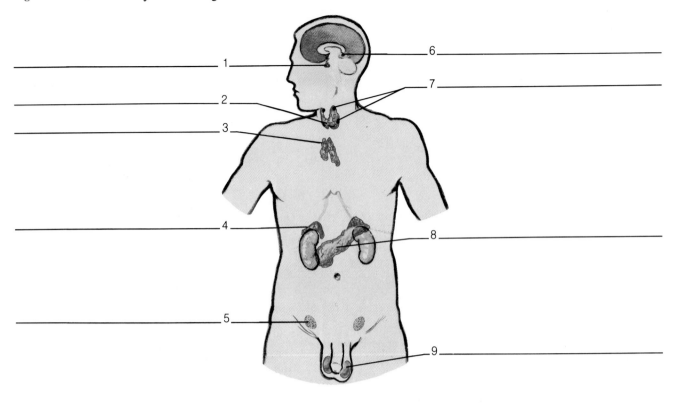

Figure 37.2 Label the features associated with the pituitary gland.

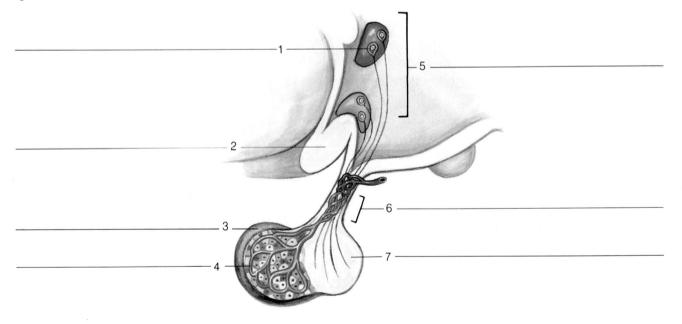

Figure 37.3 Label the features associated with the thyroid gland.

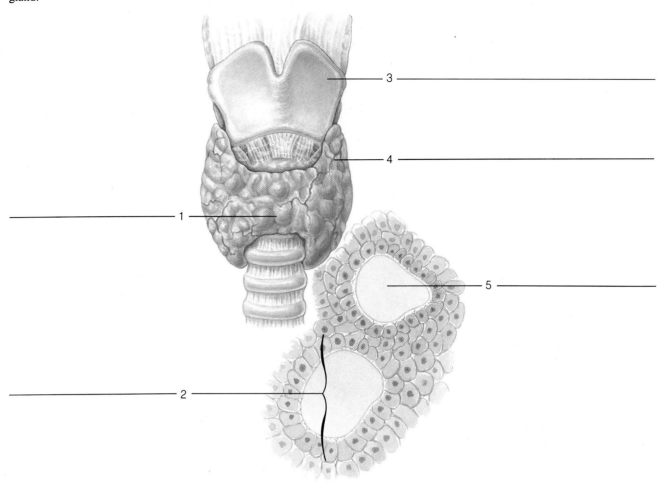

Figure 37.4 Label the features associated with the parathyroid glands.

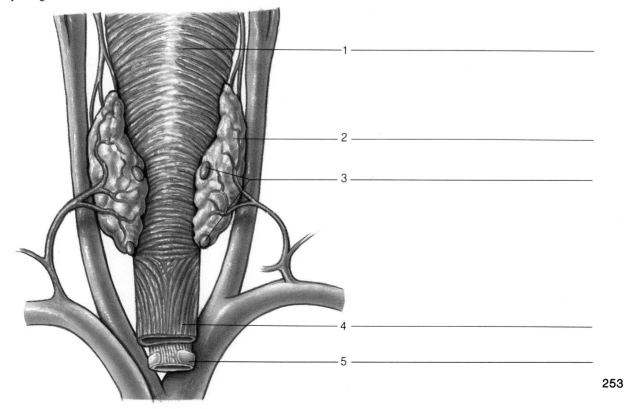

Figure 37.5 Label the features associated with the adrenal gland.

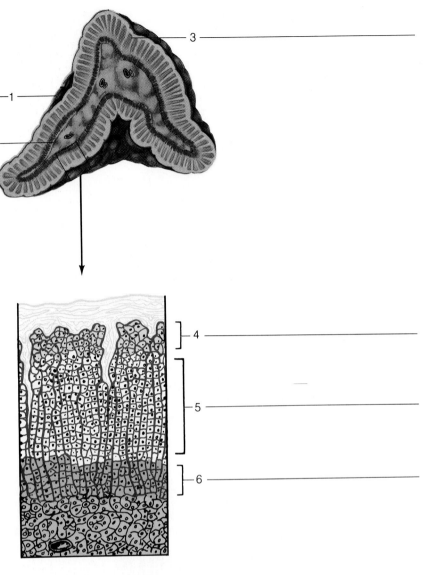

3 _____

1 _____

2 _____

4 _____

5 _____

6 _____

___

Optional Activity

The secretions of endocrine glands are usually controlled by negative feedback systems. As a result, the hormonal concentrations of body fluids remain relatively stable, although they will fluctuate slightly within a normal range. (See the section entitled "Control of Hormonal Secretions" in chapter 13 of the textbook.)

Similarly, the mechanism used to maintain the temperature of a laboratory water bath involves negative feedback. In this case, a temperature-sensitive thermostat in the water allows a water heater to operate whenever the water temperature drops below the thermostat's set point. Then, when the water temperature reaches the set point, the thermostat causes the water heater to turn off (a negative effect), and the water bath begins to cool again.

Use a laboratory thermometer to monitor the temperature of the water bath in the laboratory. Measure the temperature at regular intervals, until you have recorded ten readings. What was the lowest temperature you recorded? _____ The highest temperature? _____ What was the average temperature of the water bath? _____ How is the water bath temperature control mechanism similar to a hormonal control mechanism in the body? _____ _____ _____ _____ _____

___

5. Examine the microscopic tissue sections of the following glands, and identify the features described:

Pituitary gland. To examine the pituitary tissue:
a. Observe the tissues using low-power magnification. (See figure 37.7.)

Figure 37.6 Label the features associated with the pancreas.

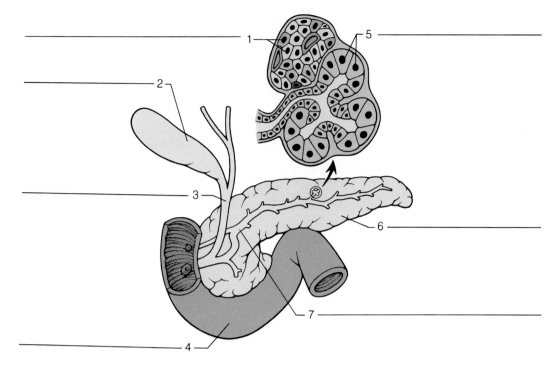

Figure 37.7 Micrograph of the pituitary gland.

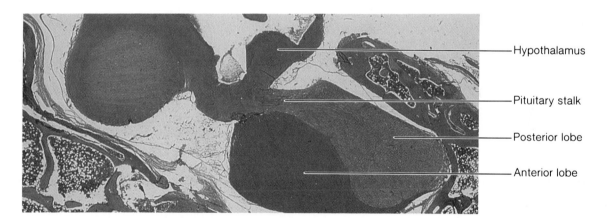

Hypothalamus

Pituitary stalk

Posterior lobe

Anterior lobe

b. Locate the pituitary stalk, the anterior lobe (the largest part of the gland), and the posterior lobe.

c. Observe an area of the anterior lobe with high-power magnification. Locate a cluster of relatively large cells, and identify some acidophil cells, which contain pink-stained granules, and some basophil cells, which contain blue-stained granules.

d. Observe an area of the posterior lobe with high-power magnification. Note the numerous unmyelinated nerve fibers present in this lobe. Also locate some pituicytes, a type of neuroglial cell, scattered among the nerve fibers.

e. Prepare labeled sketches of representative portions of the anterior and posterior lobes in Part B of the laboratory report.

Figure 37.8 Micrograph of the thyroid gland (×100).

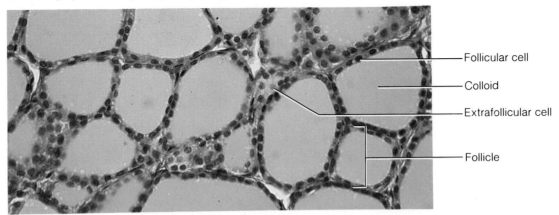

Follicular cell

Colloid

Extrafollicular cell

Follicle

Figure 37.9 Micrograph of the parathyroid gland (×50).

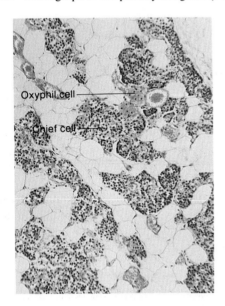

Oxyphil cell

Chief cell

Figure 37.10 Micrograph of the adrenal gland of a mouse.

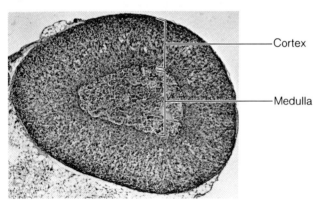

Cortex

Medulla

Thyroid gland. To examine the thyroid tissue:
a. Use low-power magnification to observe the tissue. (See figure 37.8.) Note the numerous follicles, each of which consists of a layer of cells surrounding a colloid-filled cavity.
b. Observe the tissue, using high-power magnification. Note that the cells forming the wall of a follicle are simple cuboidal epithelial cells.
c. Prepare a labeled sketch of a representative portion of the thyroid gland in Part B of the laboratory report.

Parathyroid gland. To examine the parathyroid tissue:
a. Use low-power magnification to observe the tissue. (See figure 37.9.) Note that the gland consists of numerous tightly packed secretory cells.
b. Switch to high-power magnification and locate two types of cells—a smaller form (chief cells) that are arranged in cordlike patterns and a larger form (oxyphil cells) that have distinct cell boundaries and are present in clusters.
c. Prepare a labeled sketch of a representative portion of the parathyroid gland in Part B of the laboratory report.

Adrenal gland. To examine the adrenal tissue:
a. Use low-power magnification to observe the tissue. (See figure 37.10.) Note the thin capsule that covers the gland. Just beneath the capsule there is a relatively thick adrenal cortex. The central portion of the gland is the adrenal medulla. The cells of the cortex are in three poorly defined layers. Those of the outer layer (zona glomerulosa) are arranged irregularly;

Figure 37.11 Micrograph of the pancreas (×100).

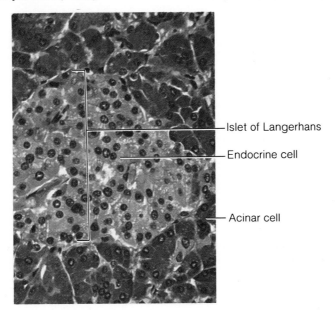

— Islet of Langerhans

— Endocrine cell

— Acinar cell

those of the middle layer (zona fasciculata) are in long cords; and those of the inner layer (zona reticularis) are arranged in an interconnected network of cords. The cells of the medulla are relatively large and irregularly shaped, and they often occur in clusters.

b. Observe each of the layers of the cortex and the cells of the medulla, using high-power magnification.

c. Prepare labeled sketches of representative portions of the cortex and medulla in Part B of the laboratory report.

Pancreas. To examine the pancreas tissue:

a. Use low-power magnification to observe the tissue. (See figure 37.11.) Note that the gland consists largely of deeply stained cells arranged in clusters around secretory ducts. These are acinar cells, and they secrete digestive enzymes. There are circular masses of lightly stained cells scattered throughout the gland. These clumps of cells constitute the islets of Langerhans, and they represent the endocrine portion of the pancreas.

b. Examine an islet, using high-power magnification.

c. Prepare a labeled sketch of a representative portion of the pancreas in Part B of the laboratory report.

laboratory report 37

Name _____

Date _____

Section _____

Part A

Complete the following:

1. Name six hormones secreted by the anterior lobe of the pituitary gland. _____
   _____
   _____
   _____

2. Name two hormones secreted by the posterior lobe of the pituitary gland. _____
   _____

3. Name the pituitary hormone responsible for the following actions:
   a. stimulates ovarian follicle to secrete estrogen _____
   b. causes kidneys to conserve water _____
   c. stimulates body cells to increase in size _____
   d. stimulates secretion from interstitial cells of testes _____
   e. stimulates secretion from thyroid gland _____
   f. causes contraction of uterine wall muscles _____
   g. stimulates secretion from adrenal cortex _____
   h. stimulates milk production _____

4. Name two thyroid hormones that affect metabolic rate. _____
   _____

5. Name a thyroid hormone that acts to lower blood calcium. _____

6. Name a hormone that acts to raise blood calcium. _____

7. Name three target organs of parathyroid hormone. _____
   _____

8. Name two hormones secreted by the adrenal medulla. _____
   _____

9. List five different effects produced by these medullary hormones. _____
   _____
   _____
   _____

10. Name the most important mineralocorticoid secreted by the adrenal cortex. _____

11. List three actions of this mineralocorticoid. _____
_____
_____

12. Name the most important glucocorticoid secreted by the adrenal cortex. _____

13. List three actions of this glucocorticoid. _____
_____
_____

14. Name two hormones secreted by the islets of Langerhans. _____
_____

15. Briefly explain how the actions of these pancreatic hormones complement one another. _____
_____
_____
_____
_____

Part B

Prepare labeled sketches to illustrate representative portions of the following endocrine glands:

Pituitary gland (＿＿X)
(anterior lobe)

Pituitary gland (＿＿X)
(posterior lobe)

Thyroid gland (＿＿X)

Parathyroid gland (＿＿X)

Adrenal gland (_____X)
(medulla)

Adrenal gland (_____X)
(cortex)

Pancreas (_____X)

Laboratory Exercise 38

Organs of the Digestive System

The digestive system includes the organs associated with the alimentary canal and several accessory structures. The alimentary canal, which is a muscular tube, passes through the body from the opening of the mouth to the anus. It includes the mouth, pharynx, esophagus, stomach, small intestine, and large intestine. The canal is adapted to move substances throughout its length and is specialized in various regions to carry on storage, digestion, and absorption of food materials and to eliminate the residues. The accessory organs, which include the salivary glands, liver, gallbladder, and pancreas, secrete products into the alimentary canal that aid digestive functions.

Purpose of the Exercise

To review the structure and function of the digestive organs and to examine the tissues of these organs microscopically.

Learning Objectives

After completing this exercise, you should be able to:

1. Locate the major digestive organs.
2. Describe the functions of these organs.
3. Recognize tissue sections of these organs.
4. Identify the major features of each tissue section.

Materials Needed:
   human manikin
   teeth, sectioned
   paper cup
   compound microscope
   prepared microscope slides of the following:
      parotid gland
      esophagus
      stomach (fundus)
      small intestine (jejunum)
      large intestine

Procedure A—Mouth and Salivary Glands

1. Review the sections entitled "Mouth" and "Salivary Glands" in chapter 14 of the textbook.
2. As a review activity, label figures 38.1, 38.2, and 38.3.
3. Examine your laboratory partner's mouth, and locate the following structures:

   oral cavity

   vestibule

   tongue

      frenulum

      papillae

   lingual tonsils

   palate

      hard palate

      soft palate

      uvula

   palatine tonsils

   pharyngeal tonsils

   gums

   teeth

      incisors

      cuspids

      bicuspids

      molars

263

Figure 38.1 Label the major features of the mouth.

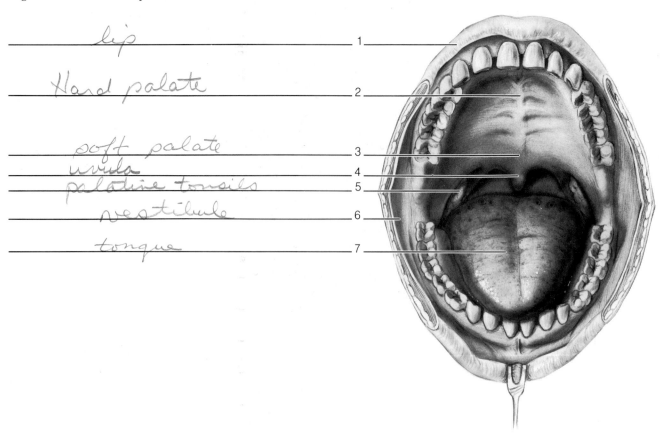

lip	1
Hard palate	2
soft palate	3
uvula	4
palatine tonsils	5
vestibule	6
tongue	7

Figure 38.2 Label the features associated with the major salivary glands.

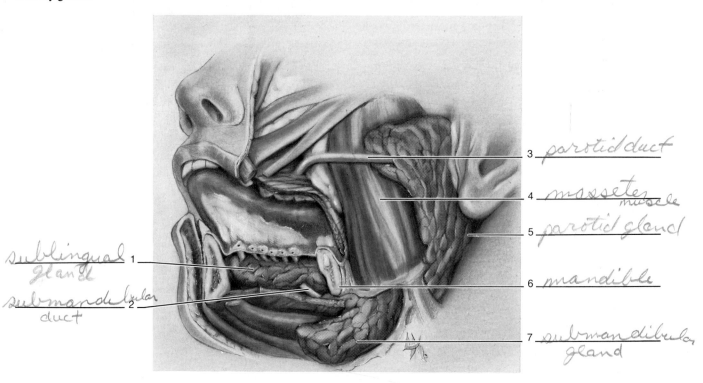

sublingual gland 1
submandibular duct 2

3 *parotid duct*
4 *masseter muscle*
5 *parotid gland*
6 *mandible*
7 *submandibular gland*

Figure 38.3 Label the features of this tooth.

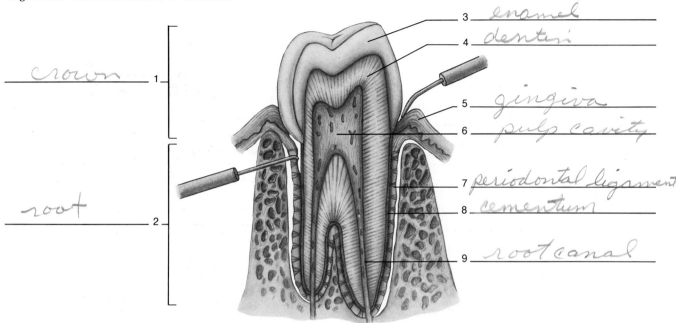

crown _____ 1

root _____ 2

3 _enamel_

4 _dentin_

5 _gingiva_

6 _pulp cavity_

7 _periodontal ligament_

8 _cementum_

9 _root canal_

4. Examine a sectioned tooth, and locate the following features:

crown

enamel

dentin

neck

root

pulp cavity

cementum

root canal

5. Observe the head of the manikin, and locate the following:

parotid salivary gland

Stensen's duct

submandibular salivary gland

Wharton's duct

sublingual salivary gland

6. Examine a microscopic section of parotid gland, using low- and high-power magnification. Note the numerous secretory cells arranged in clusters around small ducts. Also note a larger secretory duct surrounded by lightly stained cuboidal epithelial cells. (See figure 38.4.)

7. Complete Part A of Laboratory Report 38 on page 271.

Figure 38.4 Micrograph of the parotid salivary gland.

Secretory cells

Secretory duct

Procedure B—Pharynx and Esophagus

1. Review the section entitled "Pharynx and Esophagus" in chapter 14 of the textbook.
2. As a review activity, label figure 38.5.
3. Observe the manikin, and locate the following features:

pharynx	laryngopharynx
nasopharynx	constrictor muscles
opening to eustachian tube	epiglottis
	esophagus
oropharynx	esophageal hiatus

Figure 38.5 Label the features associated with the pharynx.

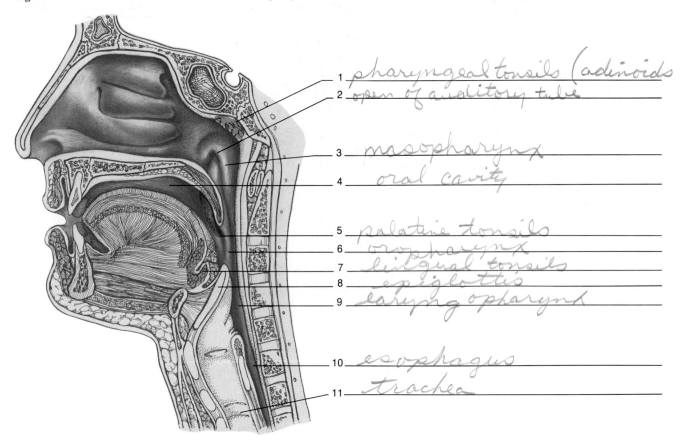

1. pharyngeal tonsils (adenoids
2. open of auditory tube
3. nasopharynx
4. oral cavity
5. palatine tonsils
6. oropharynx
7. lingual tonsils
8. epiglottis
9. laryngopharynx
10. esophagus
11. trachea

4. Have your partner take a swallow from a cup of water, and carefully watch the movements in the anterior region of the neck. What steps in the swallowing process did you observe? _____

_____

_____

_____

5. Examine a microscopic section of esophagus wall, using low-power magnification. Note that the inner lining is composed of stratified squamous epithelium and that there are layers of muscle tissue in the wall. Locate some mucous glands in the submucosa. They appear as clusters of lightly stained cells.

6. Complete Part B of the laboratory report.

Procedure C—The Stomach

1. Review the section entitled "Stomach" in chapter 14 of the textbook.
2. As a review activity, label figures 38.6 and 38.7.

3. Observe the manikin, and locate the following features of the stomach:

rugae	pyloric region
cardiac region	pyloric canal
fundic region	pyloric sphincter
body region	

4. Examine a microscopic section of stomach wall, using low-power magnification. Note how the inner lining of simple columnar epithelium dips inward to form gastric pits. The gastric glands are tubular structures that open into the gastric pits. Near the deep ends of these glands, you should be able to locate some intensely stained (bluish) chief cells and some lightly stained (pinkish) parietal cells. (See figure 38.8.) What are the functions of these cells? _____

_____

_____

_____

5. Complete Part C of the laboratory report.

Figure 38.6 Label the major regions of the stomach.

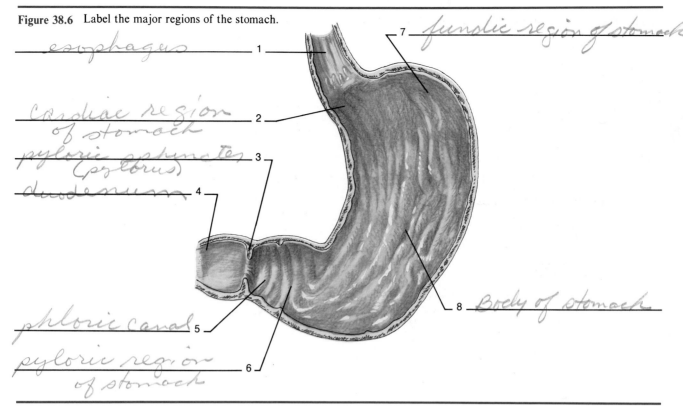

1 — esophagus

2 — cardiac region of stomach

3 — pyloric sphincter (pylorus)

4 — duodenum

5 — phloric canal

6 — pyloric region of stomach

7 — fundic region of stomach

8 — Body of stomach

Figure 38.7 Label this diagram by placing the correct numbers in the spaces provided.

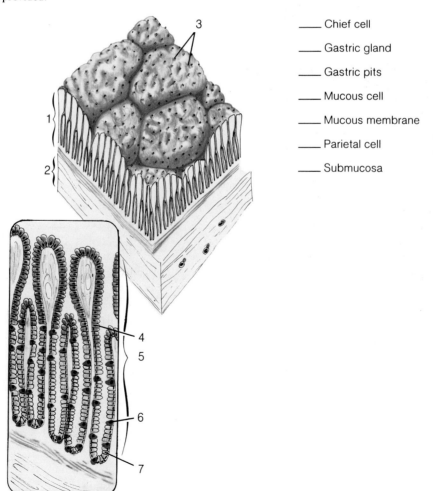

_____ Chief cell

_____ Gastric gland

_____ Gastric pits

_____ Mucous cell

_____ Mucous membrane

_____ Parietal cell

_____ Submucosa

Figure 38.8 Micrograph of the stomach wall (×100).

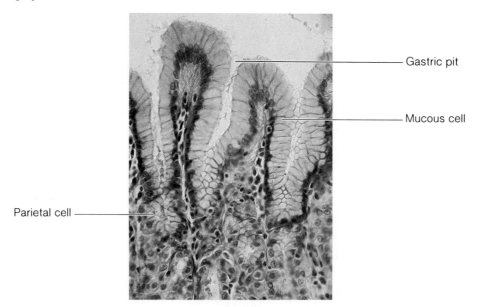

Gastric pit

Mucous cell

Parietal cell

Procedure D—Pancreas and Liver

1. Review the sections entitled "Pancreas" and "Liver" in chapter 14 of the textbook.
2. As a review activity, label figure 38.9.
3. Observe the manikin, and locate the following structures:

   pancreas

   pancreatic duct

   liver

      right lobe

      quadrate lobe

      caudate lobe

      left lobe

   falciform ligament

   coronary ligament

   gallbladder

   hepatic ducts

   common hepatic duct

   cystic duct

   common bile duct

   sphincter of Oddi

4. Complete Part D of the laboratory report.

Procedure E—Small and Large Intestines

1. Review the sections entitled "Small Intestine" and "Large Intestine" in chapter 14 of the textbook.
2. As a review activity, label figure 38.10.
3. Observe the manikin, and locate each of the following features:

   small intestine

      duodenum

      jejunum

      ileum

   mesentery

   ileocecal valve

   large intestine

      cecum

      vermiform appendix

      ascending colon

      transverse colon

      descending colon

      sigmoid colon

   rectum

   anal canal

   anal columns

   anal sphincter

      internal sphincter

      external sphincter

   anus

Figure 38.9 Label the features associated with the liver and pancreas.

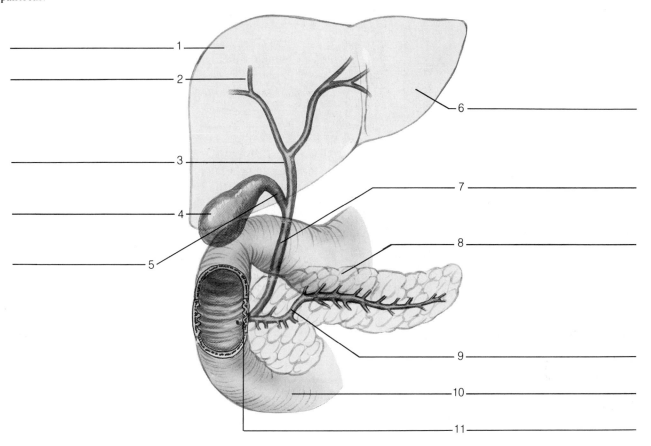

Figure 38.10 Label this diagram by placing the correct numbers in the spaces provided.

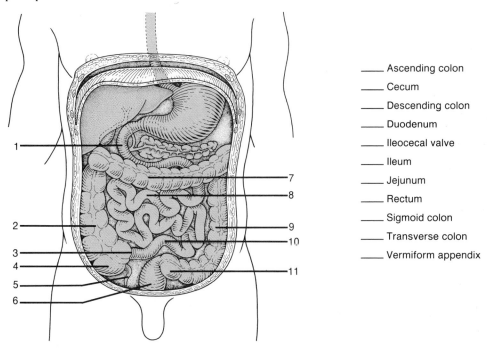

_____ Ascending colon

_____ Cecum

_____ Descending colon

_____ Duodenum

_____ Ileocecal valve

_____ Ileum

_____ Jejunum

_____ Rectum

_____ Sigmoid colon

_____ Transverse colon

_____ Vermiform appendix

Figure 38.11 Micrograph of the duodenal wall (×100).

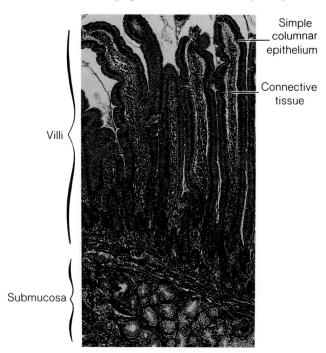

Villi {

Submucosa {

Simple columnar epithelium

Connective tissue

4. Examine a microscopic section of small intestine wall, using low-power magnification. Note the villi that extend into the lumen of the tube. Study a single villus, using high-power magnification. Note the core of connective tissue and the covering of simple columnar epithelium that contains some lightly stained goblet cells. (See figure 38.11.) What is the function of these villi? _____
_____
_____
_____
_____

5. Examine a microscopic section of large intestine wall. Note the lack of villi. Also note the tubular mucous glands that open on the surface of the inner lining and the numerous lightly stained goblet cells. (See figure 38.12.) What is the function of the mucus secreted by these glands? _____
_____
_____
_____

6. Complete Part E of the laboratory report.

Figure 38.12 Micrograph of the large intestinal wall (×40).

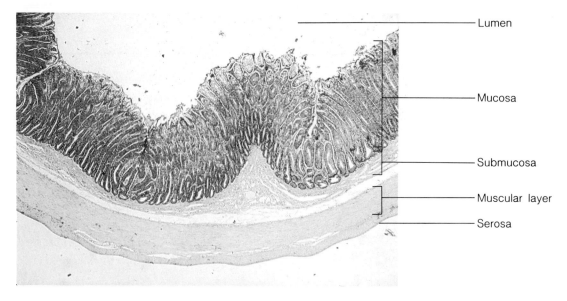

Lumen

Mucosa

Submucosa

Muscular layer

Serosa

laboratory report 38

Name _____

Date _____

Section _____

Organs of the Digestive System

Part A

Match the terms in column A with the descriptions in column B. Place the letter of your choice in the space provided.

Column A		Column B
A.	adenoids	____ 1. bonelike substance beneath tooth enamel
B.	amylase	____ 2. smallest of major salivary glands
C.	crown	____ 3. tooth specialized for grinding
D.	dentine	____ 4. chamber between tongue and palate
E.	frenulum	____ 5. projections on tongue surface
F.	incisor	____ 6. cone-shaped projection of soft palate
G.	molar	____ 7. secretes digestive enzymes
H.	oral cavity	____ 8. attaches tooth to jaw
I.	palate	____ 9. chisel-shaped tooth
J.	papillae	____ 10. roof of oral cavity
K.	periodontal ligament	____ 11. space between cheeks and lips
		____ 12. anchors tongue to floor of mouth
L.	serous cell	____ 13. lymphatic tissue in wall of pharynx
M.	sublingual gland	____ 14. portion of tooth projecting beyond gum
N.	uvula	____ 15. splits starch into disaccharides
O.	vestibule	

Part B

Complete the following:

1. The part of the pharynx above the soft palate is called the _____ .

2. The middle part of the pharynx is called the _____ .

3. The lower portion of the pharynx is called the _____ .

4. The eustachian tube opens through the wall of the _____ .

5. During swallowing, circular muscles called _____ pull the walls of the pharynx inward.

6. List six major steps in the swallowing reflex. _____
_____
_____
_____
_____
_____

7. The esophagus passes through the _____ as it descends into the thorax.

8. The esophagus penetrates the diaphragm through an opening called the _____ .

9. _____ is the main secretion of the esophagus.

10. The esophagus is about _____ cm long.

Part C

Complete the following:

1. Name the four regions of the stomach. _____
_____
_____

2. Name the valve that prevents regurgitation of food from the intestine back into the stomach. _____

3. Name three types of secretory cells found in gastric glands. _____
_____

4. Name the gastric cells that secrete digestive enzymes. _____
_____

5. Name the gastric cells that secrete hydrochloric acid. _____
_____

6. Name the most important digestive enzymes secreted in the stomach. _____
_____

7. Name the substance needed for efficient absorption of vitamin B_{12}. _____

8. Name a hormone secreted by the stomach. _____

9. Name a hormone that causes gastric motility to decrease. _____

10. Name a reflex that slows the movement of chyme from the stomach into the small intestine. _____

Part D

Match the terms in column A with the descriptions in column B. Place the letter of your choice in the space provided.

Column A			Column B
A.	amylase	____ 1.	activates protein-digesting enzyme
B.	bile salts	____ 2.	causes emulsification of fats
C.	cholecystokinin	____ 3.	carries on phagocytosis
D.	enterokinase	____ 4.	iron-storing substance
E.	ferritin	____ 5.	carbohydrate-digesting enzyme
F.	Kupffer's cell	____ 6.	fat-digesting enzyme
G.	lipase	____ 7.	protein-digesting enzyme
H.	nuclease	____ 8.	stimulates pancreas to secrete digestive enzymes
I.	secretin	____ 9.	stimulates pancreas to secrete bicarbonate ions
J.	trypsin	____ 10.	nucleic acid-digesting enzyme

Part E

Complete the following:

1. Name the three portions of the small intestine. _____

2. Name the fold of peritoneum that suspends portions of the intestine from the posterior abdominal wall. _____

3. Name the lymphatic capillary found in an intestinal villus. _____ _____

4. Name the tubular glands found between the bases of the intestinal villi. _____

5. Name the circular folds that characterize the inner lining of the small intestine. _____

6. Name five digestive enzymes secreted by intestinal glands. _____

7. Name the four portions of the colon. _____

8. Name two muscles that act to close the anal opening. _____

9. Name the valve that controls movement of material between the small and large intestines. _____

10. Name the projection of lymphatic tissue attached to the cecum. _____

Cat Dissection: Digestive System

I n this laboratory exercise, you will dissect the major digestive organs of an embalmed cat. As you observe these organs, compare them with those of the human by observing the parts of the human manikin. However, keep in mind that the cat is a carnivore, and that the organs of its digestive system are adapted to capturing, holding, eating, and digesting the bodies of other animals. Because humans are adapted to eating and digesting a greater variety of foods, comparisons of the cat and the human digestive organs may not be precise.

Purpose of the Exercise

To examine the major digestive organs of the cat and to compare these organs with those of the human.

Learning Objectives

After completing this exercise, you should be able to:

1. Locate and identify the major digestive organs of the embalmed cat.
2. Identify the corresponding organs in the human manikin.
3. Compare the digestive system of the cat with that of the human.

Materials Needed:
　　embalmed cat
　　dissecting tray
　　dissecting instruments
　　bone cutter
　　disposable gloves
　　human manikin
　　hand lens

Procedure

1. Place the embalmed cat in the dissecting tray with its ventral side down.
2. Locate the major salivary glands on one side of the head. To do this:
   a. Clear away any remaining fascia and other connective tissue from the region below the ear and near the joint of the mandible.
   b. Identify the *parotid gland,* a relatively large mass of glandular tissue just below the ear. Note its duct (Stensen's duct) that passes over the surface of the masseter muscle and opens into the mouth.
   c. Look for the *submandibular gland* just below the parotid gland, near the angle of the jaw.
   d. Locate the *sublingual gland* that is adjacent and medial to the submandibular gland (figure 39.1).
3. Examine the oral cavity. To do this:
   a. Use scissors to cut through the soft tissues at the angle of the mouth.
   b. When you reach the bone of the jaw, use a bone cutter to cut through the bone, thus freeing the mandible.
   c. Open the mouth wide, and locate the following features:

   cheek

   lip

   vestibule

   palate

   　hard palate

   　soft palate

   palatine tonsils (small, rounded masses of glandular tissue in the lateral wall of the soft palate)

Figure 39.1 Major salivary glands of the cat.

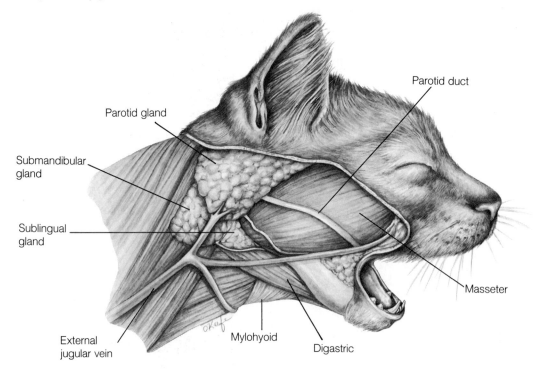

Parotid gland

Submandibular gland

Sublingual gland

External jugular vein

Mylohyoid

Digastric

Parotid duct

Masseter

tongue

frenulum

papillae (examine with a hand lens)

4. Examine the teeth of the maxilla. How many incisors, canines (cuspids), bicuspids, and molars are present? _____

_____

5. Complete Part A of Laboratory Report 39 on page 279.

6. Position the cat with its ventral side up, and open the abdominal cavity. To do this:

   a. Use scissors to make an incision through the body wall along the midline from the symphysis pubis to the sternum.

   b. Make a lateral incision through the body wall along either side of the inferior border of the diaphragm and along the bases of the thighs.

   c. Reflect the flaps created in the body wall as you would open a book, and expose the contents of the abdominal cavity.

   d. Note the *parietal peritoneum* that forms the inner lining of the abdominal wall. Also note the *greater omentum*, a structure composed of a double layer of peritoneum that hangs from the border of the stomach and covers the lower abdominal organs like a fatty apron.

   e. Gently lift away the greater omentum from the underlying organs. Use scissors to cut the greater omentum loose from its attachment, and discard it (figure 39.2 and reference plate 13).

7. Examine the *liver,* which is located just beneath the diaphragm and is attached to the central portion of the diaphragm by the *falciform ligament.* Also, locate the *spleen,* which is posterior to the liver on the left side. The liver has five lobes—a right medial, right lateral (which is subdivided into two parts by a deep cleft), left medial, left lateral, and caudate lobe. The caudate lobe is the smallest lobe and is located in the median line, where it projects into the curvature of the stomach. The caudate lobe is covered by a sheet of mesentery, the *lesser omentum,* that connects the liver to the stomach. (See figure 39.3 and reference plate 14.) Find the greenish *gallbladder* on the inferior surface of the liver on the right side. Also note the *cystic duct* by which the gallbladder is attached to the *common bile duct.* Trace the common bile duct to its connection with the duodenum.

8. Locate the *stomach* in the upper left side of the abdominal cavity. At its anterior end, note the union of the *esophagus,* which descends through the diaphragm. Identify the *cardiac, fundic, body,* and *pyloric regions* of the stomach. Use scissors to make an incision along the convex border of the stomach from the cardiac region to the pylorus. Note the folded inner lining of the stomach, and examine the *pyloric sphincter,* which creates a constriction between the stomach and small intestine.

276

Figure 39.2 Ventral view of abdominal organs of the cat.

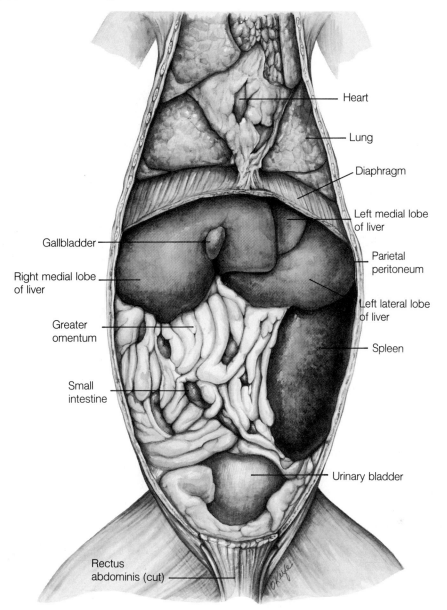

Heart

Lung

Diaphragm

Left medial lobe of liver

Parietal peritoneum

Left lateral lobe of liver

Spleen

Urinary bladder

Gallbladder

Right medial lobe of liver

Greater omentum

Small intestine

Rectus abdominis (cut)

9. Locate the *pancreas*. It appears as a grayish, two-lobed elongated mass of glandular tissue. One lobe lies dorsal to the stomach and extends across to the duodenum. The other lobe is enclosed by the *mesentery*, which supports the duodenum (figure 39.3 and reference plate 14).

10. Trace the *small intestine*, beginning at the pyloric sphincter. The first portion, the *duodenum*, travels posteriorly for several centimeters. Then it loops back around a lobe of the pancreas. The proximal half of the remaining portion of the small intestine is the *jejunum*, and the distal half is the *ileum*.

Note how the mesentery supports the small intestine from the dorsal body wall. The small intestine terminates on the right side where it joins the large intestine.

11. Locate the *large intestine*, and identify the *cecum, ascending colon, transverse colon*, and *descending colon*. Also locate the *rectum*, which extends through the pelvic cavity to the *anus*. Make an incision at the junction between the ileum and cecum, and look for the *ileocecal valve*.

12. Complete Part B of the laboratory report.

Figure 39.3 Ventral view of the cat's abdominal organs with intestines reflected to the right side.

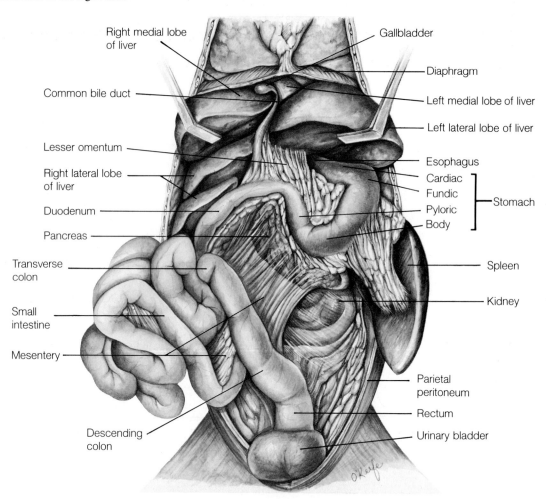

Right medial lobe of liver

Common bile duct

Lesser omentum

Right lateral lobe of liver

Duodenum

Pancreas

Transverse colon

Small intestine

Mesentery

Descending colon

Gallbladder

Diaphragm

Left medial lobe of liver

Left lateral lobe of liver

Esophagus

Cardiac

Fundic

Pyloric

Body

Stomach

Spleen

Kidney

Parietal peritoneum

Rectum

Urinary bladder

O'Keefe

Name _____

Date _____

Section _____

Cat Dissection: Digestive System

Part A

Complete the following:

1. Compare the locations of the major salivary glands of the human with those of the cat. _____

   _____

   _____

   _____

   _____

2. Compare the types and numbers of teeth present in the cat's maxilla with those of the human. _____

   _____

   _____

   _____

3. In what ways do the cat's teeth seem to be adapted to a special diet? _____

   _____

   _____

   _____

   _____

4. What part of the human soft palate is lacking in the cat? _____

5. What do you think is the function of the transverse ridges (rugae) in the hard palate of the cat? _____

   _____

   _____

   _____

6. How do the papillae on the surface of the human tongue compare with those of the cat? _____

   _____

   _____

   _____

Part B

Complete the following:

1. Describe how the peritoneum and mesenteries are associated with the organs in the abdominal cavity. _____

_____

_____

_____

_____

_____

_____

2. Describe the inner lining of the stomach. _____

_____

_____

_____

_____

3. Compare the structure of the human liver with that of the cat. _____

_____

_____

_____

_____

4. Compare the structure and location of the human pancreas with those of the cat. _____

_____

_____

_____

_____

5. What feature of the human cecum is lacking in the cat? _____

_____

Laboratory Exercise 40

Action of a Digestive Enzyme

The digestive enzyme in salivary secretions is called amylase. This enzyme splits starch molecules into sugar (disaccharide) molecules, which is the first step in the digestion of complex carbohydrates.

As in the case of other enzymes, amylase is a protein, whose activity is affected by exposure to certain environmental factors, including excessive heat, radiation, electricity, and certain chemicals.

Purpose of the Exercise

To investigate the action of amylase and the effect of heat on its enzymatic activity.

Learning Objectives

After completing this exercise, you should be able to:

1. Describe the action of amylase.
2. Test a solution for the presence of starch or the presence of sugar.
3. Test the effects of varying temperatures on the activity of amylase.

Materials Needed:

amylase solution	wax marker
beakers (50 and 500 ml)	IKI solution
distilled water	medicine dropper
funnel	ice
ring stand and ring	water bath (37°C)
gauze	porcelain test plate
1% starch solution	Benedict's solution
graduated cylinder (10 ml)	hot plate
test tubes	test tube rack

Procedure A—Amylase Activity

1. Mark three clean test tubes as tubes 1, 2, 3, and prepare the tubes as follows:

   Tube 1: Add 6 ml of distilled water.
   Tube 2: Add 5 ml of starch solution and 1 ml of distilled water.
   Tube 3: Add 5 ml of starch solution and 1 ml of amylase solution.

2. Shake the tubes well to mix the contents, and place them in a warm water bath (37°C) for one hour.
3. At the end of the hour, test the contents of each tube for the presence of starch. To do this:
   a. Place 1 ml of the solution to be tested in a depression of a porcelain test plate.
   b. Next add two drops of IKI solution and note the color of the mixture. If the solution becomes blue-black, starch is present.
   c. Record the results in Part A of Laboratory Report 40 on page 283.
4. Test the contents of each tube for the presence of sugar (monosaccharides in this instance). To do this:
   a. Place 1 ml of the solution to be tested in a clean test tube.
   b. Add 5 ml of Benedict's solution.
   c. Place the test tube in a beaker of boiling water for two minutes, and allow the liquid to cool slowly.
   d. Note the color of the liquid. If the solution becomes green, yellow, or red, sugar is present.
   e. Record the results in Part A of the laboratory report.
5. Complete Part A of the laboratory report.

Procedure B—Effect of Heat

1. Mark four clean test tubes as tubes 4, 5, 6, and 7.
2. Add 1 ml of amylase solution to each of the tubes, and expose each solution to a different test temperature for three minutes as follows:

   Tube 4: Place in beaker of ice water (about 0°C).
   Tube 5: Place in rack at room temperature (about 25°C).
   Tube 6: Place in warm water bath (about 37°C).
   Tube 7: Place in beaker of boiling water (about 100°C).

3. Add 5 ml of starch solution to each tube, shake to mix the contents, and return the tubes to their respective test temperatures for one hour.
4. At the end of the hour, test the contents of each tube for the presence of starch and the presence of sugar by following the directions in Procedure A.
5. Complete Part B of the laboratory report.

Optional Activity

Devise an experiment to test the effect of some other environmental factor on amylase activity. For example, you might test the effect of a strong acid by adding a few drops of concentrated hydrochloric acid to a mixture of starch and amylase solutions. Be sure to include a control in your experimental plan. That is, include a tube containing everything except the factor you are testing. Then you will have something with which to compare your results. Carry out your experiment *only if it has been approved by the laboratory instructor.*

Action of a Digestive Enzyme

Part A—Amylase Activity

1. Test results:

Tube	Starch	Sugar
1		
2		
3		

2. Complete the following:

a. Explain the reason for including tube 1 in this experiment. _____ _____ _____ _____

b. What is the importance of tube 2? _____ _____ _____ _____

c. What do you conclude from the results of this experiment? _____ _____ _____ _____

Part B—Effect of Heat

1. Test results:

Tube	Starch	Sugar
4		
5		
6		
7		

2. Complete the following:

   a. What do you conclude from the results of this experiment? _____

   _____

   _____

   _____

   _____

   b. If digestion failed to occur in one of the tubes in this experiment, how can you tell if the amylase was destroyed by the factor being tested or if the amylase activity was just inhibited by the test treatment? _____

   _____

   _____

   _____

   _____

Organs of the Respiratory System

T he organs of the respiratory system include the nose, nasal cavity, sinuses, pharynx, larynx, trachea, bronchial tree, and lungs. They function mainly to process incoming air and to transport it to and from the atmosphere outside the body and the air sacs of the lungs.

In the air sacs, gas exchanges take place between the air and the blood of nearby capillaries. The blood, in turn, transports gases to and from the air sacs and the body cells. This entire process of transporting and exchanging gases between the atmosphere and the body cells is called respiration.

Purpose of the Exercise

To review the structure and function of the respiratory organs and to examine the tissues of some of these organs microscopically.

Learning Objectives

After completing this exercise, you should be able to:

1. Locate the major organs of the respiratory system.
2. Describe the functions of these organs.
3. Recognize tissue sections of the trachea and lung.
4. Identify the major features of these tissue sections.

Materials Needed:
  human skull, sagittal section
  human manikin
  compound microscope
  prepared microscope slides of the following:
    trachea, cross section
    lung, human (normal)

For Demonstrations:

  animal lung with trachea (fresh or preserved)
  prepared microscope slides of the following:
    lung tissue (smoker)
    lung tissue (emphysema)

Procedure A—Respiratory Organs

1. Review the section entitled "Organs of the Respiratory System" in chapter 16 of the textbook.
2. As a review activity, label figures 41.1, 41.2, 41.3, and 41.4.
3. Examine the sagittal section of the human skull, and locate the following features:

nose

  nostrils

nasal cavity

  nasal septum

  nasal conchae

  superior meatus

  middle meatus

  inferior meatus

sinuses

  maxillary sinus

  frontal sinus

  ethmoid sinus

  sphenoid sinus

4. Observe the human manikin, and locate the features listed in step 3 above. Also locate the following:

pharynx

larynx (palpate your own larynx)

  vocal cords

    false vocal cords

    true vocal cords

  thyroid cartilage

Figure 41.1 Label the major features of the respiratory system.

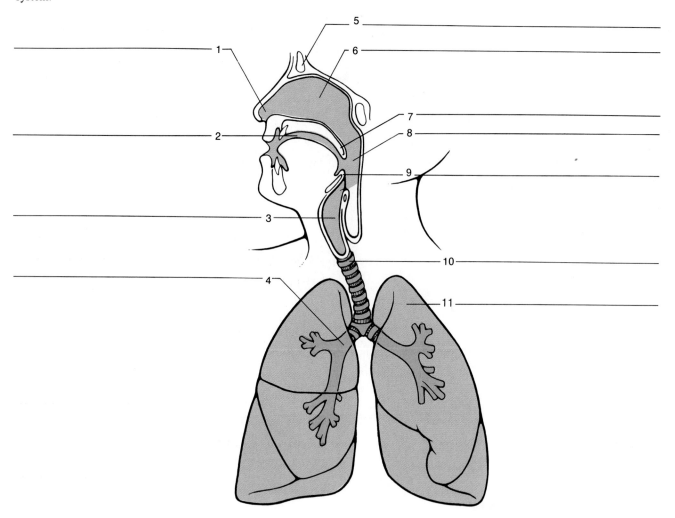

cricoid cartilage

epiglottic cartilage

epiglottis

arytenoid cartilages

corniculate cartilages

cuneiform cartilages

glottis

trachea (palpate your own trachea)

bronchi

   primary bronchi

   secondary bronchi

lung

   hilus

lobes

lobules

visceral pleura

parietal pleura

pleural cavity

5. Complete Part A of Laboratory Report 41 on page 289.

Demonstration

Observe the animal lung and the attached trachea. Identify the larynx, major laryngeal cartilages, trachea, and the cartilaginous rings of the trachea. Open the larynx and locate the vocal folds. Examine the visceral pleura on the surface of a lung, and squeeze a portion of a lung between your fingers. How do you describe the texture of the lung?

Figure 41.2 Label the features of this sagittal section of the head.

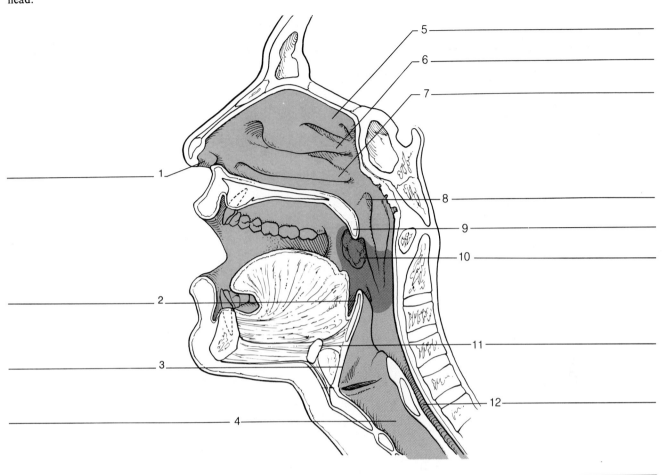

Figure 41.3 Label the major features of the larynx: (a) anterior view and (b) posterior view.

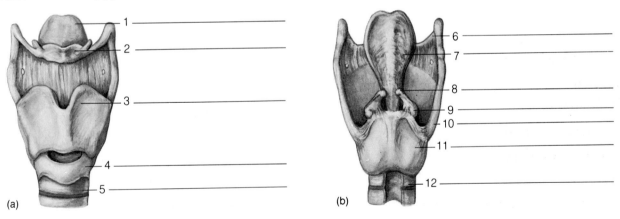

(a)

(b)

Figure 41.4 Label the features of (*a*) the frontal section of the larynx and (*b*) the superior aspect of the vocal cords.

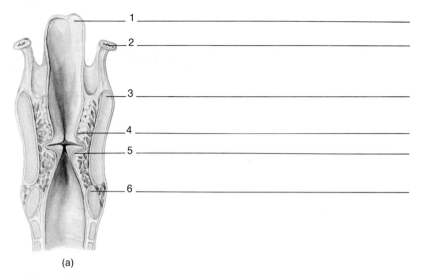

1 _____

2 _____

3 _____

4 _____

5 _____

6 _____

(a)

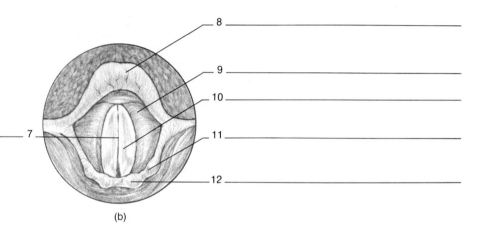

8 _____

9 _____

10 _____

7 _____

11 _____

12 _____

(b)

Procedure B—Respiratory Tissues

1. Obtain a prepared microscope slide of trachea, and use low-power magnification to examine it. Notice the inner lining of pseudostratified columnar epithelium and the deep layer of hyaline cartilage, which represents a portion of a tracheal ring.
2. Use high-power magnification to observe the cilia on the free surface of the epithelial lining. Locate some mucus-secreting goblet cells in the epithelium.
3. Prepare a labeled sketch of a representative portion of the tracheal wall in Part B of the laboratory report.
4. Obtain a prepared microscope slide of human lung. Examine it, using low-power magnification, and note the numerous open spaces of the air sacs. Look for a bronchiole—a tube with a relatively thick wall and a wavy inner lining. Locate the smooth muscle tissue in the wall of this tube.

5. Use high-power magnification to examine the air sacs. Note that their walls are composed of simple squamous epithelium. You may also see sections of blood vessels filled with blood cells.
6. Prepare a labeled sketch of a representative portion of the lung in Part B of the laboratory report.

Demonstration

Examine the prepared microscope slides of the lung tissue of a smoker and a person with emphysema using low-power magnification. How does the smoker's lung tissue compare with that of the normal lung that you examined previously? _____

_____

How does the emphysema patient's lung tissue compare with the normal lung? _____

_____

_____

Name _____

Date _____

Section _____

Organs of the Respiratory System

Part A

Match the terms in column A with the descriptions in column B. Place the letter of your choice in the space provided.

Column A	Column B
A. carina	____ 1. basic unit of the lung
B. cricoid cartilage	____ 2. lowermost portion of larynx
C. epiglottic cartilage	____ 3. serves as resonant chamber
D. vocal cord	____ 4. carries on gas exchange
E. glottis	____ 5. ridge of cartilage between bronchi
F. lobule	____ 6. opening between vocal cords
G. nasal concha	____ 7. fold of mucous membrane
H. pharynx	____ 8. increases surface area of nasal mucous membrane
I. respiratory bronchiole	____ 9. passageway for air and food
J. sinus	____ 10. attached to upper border of the thyroid cartilage

Part B

1. Prepare a labeled sketch of a portion of the tracheal wall.

2. Prepare a labeled sketch of a portion of lung tissue.

Laboratory Exercise 42

Cat Dissection: Respiratory System

In this laboratory exercise, you will dissect the major respiratory organs of an embalmed cat. As you observe these structures in the cat, compare them with those of the human manikin.

Purpose of the Exercise

To examine the major respiratory organs of the cat and to compare these organs with those of the human.

Learning Objectives

After completing this exercise, you should be able to:

1. Locate and identify the major respiratory organs of the embalmed cat.
2. Identify the corresponding organs in the human manikin.
3. Compare the respiratory system of the human with that of the cat.

Materials Needed:
  embalmed cat
  dissecting tray
  dissecting instruments
  disposable gloves
  human manikin

Procedure

1. Place the embalmed cat in a dissecting tray with its ventral side up.
2. Examine the *nostrils* and *nasal septum.*
3. Open the mouth wide to expose the soft palate. Cut through the tissues of the soft palate, and observe the nasopharynx above it. Locate the small openings of the eustachian tubes in the lateral walls of the nasopharynx.

4. Pull the tongue forward, and locate the *epiglottis* at its base. Also identify the *glottis,* which is the opening into the larynx. Locate the *esophagus,* which is dorsal to the larynx. (In the human the term *glottis* refers to the opening between the vocal folds within the larynx.)
5. Open the thoracic cavity, and expose its contents. To do this:
   a. Make a longitudinal incision passing anteriorly from the diaphragm along one side of the sternum. Continue the incision through the neck muscles to the mandible. Try to avoid damaging the internal organs as you cut.
   b. Make a lateral cut on each side along the anterior surface of the diaphragm, and cut the diaphragm loose from the thoracic wall.
   c. Spread the sides of the thoracic wall outward, and use a scalpel to make a longitudinal cut along each side of the inner wall of the rib cage to weaken the ribs. Continue to spread the thoracic wall laterally to break the ribs so that the flaps of the wall will remain open (figure 42.1 and reference plate 15).
6. Dissect the *trachea* in the neck, and expose the *larynx* at the anterior end near the base of the tongue. Note the *tracheal rings,* and locate the lobes of the *thyroid gland* on each side of the trachea, just posterior to the larynx. Also note the *thymus gland,* a rather diffuse mass of glandular tissue extending along the ventral side of the trachea into the thorax. (See reference plate 16.)
7. Examine the *larynx,* and identify the *thyroid cartilage* and *cricoid cartilage* in its wall. Make a longitudinal incision through the ventral wall of the larynx, and locate the *vocal cords* inside.
8. Trace the trachea posteriorly to where it divides into *primary bronchi,* which pass into the lungs.

Figure 42.1 Ventral view of the cat's thoracic cavity.

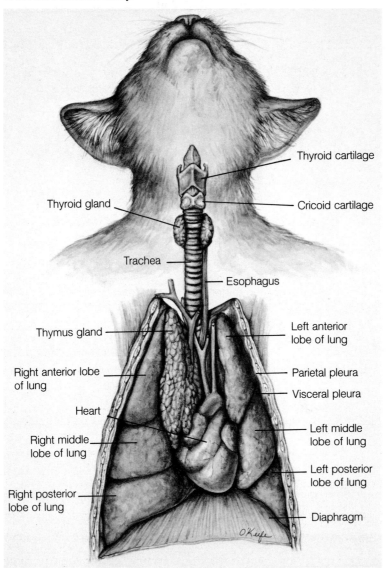

Thyroid cartilage

Cricoid cartilage

Thyroid gland

Trachea

Esophagus

Thymus gland

Left anterior lobe of lung

Right anterior lobe of lung

Parietal pleura

Visceral pleura

Heart

Left middle lobe of lung

Right middle lobe of lung

Left posterior lobe of lung

Right posterior lobe of lung

Diaphragm

O'Keefe

9. Examine the *lungs,* each of which is subdivided into three lobes—an anterior, a middle, and a posterior lobe. Notice the thin membrane, the *visceral pleura,* on the surface of each lung. Also notice the *parietal pleura,* which forms the inner lining of the thoracic wall, and locate the spaces of the *pleural cavities.*

10. Make an incision through a lobe of a lung, and examine its interior. Note the branches of the smaller air passages.

11. Examine the *diaphragm,* and note that its central portion is tendinous.

12. Locate the *phrenic nerve.* This nerve appears as a white thread passing along the side of the heart to the diaphragm.

13. Examine the *mediastinum,* which is composed of the tissues and organs that separate the thoracic cavity into right and left portions.

14. Complete Laboratory Report 42 on pages 293–94.

Name _____

Date _____

Section _____

Cat Dissection: Respiratory System

Complete the following:

1. What is the significance of the eustachian tubes opening into the nasopharynx? _____
_____
_____
_____
_____

2. Distinguish between the glottis and the epiglottis. _____
_____
_____
_____
_____

3. Are the tracheal rings of the cat complete or incomplete? _____ How does this feature compare
with that of the human? _____
_____
_____
_____

4. How does the structure of the primary bronchi compare with that of the trachea? _____
_____
_____
_____

5. Compare the number of lobes in the human lungs with the number of lobes in the cat. _____
_____
_____
_____
_____

6. Describe the attachments of the diaphragm. _____

_____

_____

_____

_____

7. What structures are located within the mediastinum? _____

_____

_____

_____

_____

_____

Laboratory Exercise 43

Breathing and Respiratory Volumes

Breathing involves the movement of air from outside the body into the bronchial tree and alveoli, and the reversal of this air movement. These movements are accompanied by changes in the size of the thoracic cavity that result from muscle contractions and from the elastic recoil of stretched tissues.

The volumes of air that move in and out of the lungs during various phases of breathing are called respiratory air volumes. These volumes can be measured by using an instrument called a spirometer. However, the values obtained vary with a person's age, sex, height, and weight.

Purpose of the Exercise

To review the mechanisms of breathing and to measure certain respiratory air volumes.

Learning Objectives

After completing this exercise, you should be able to:

1. Describe the mechanisms responsible for inspiration and expiration.
2. Define the various respiratory air volumes.
3. Measure these air volumes with a spirometer.

Materials Needed:

spirometer, hand-held
70% alcohol
cotton

For Demonstration:

mechanical model of the respiratory system

Procedure A—Breathing Mechanisms

1. Review the sections entitled "Inspiration" and "Expiration" in chapter 16 of the textbook.
2. Complete Part A of Laboratory Report 43 on page 299.

Demonstration

Observe the mechanical model of the respiratory system. Note that it consists of a heavy glass bell jar with a rubber sheeting tied over its wide open end. Its narrow upper opening is plugged with a rubber stopper, through which a glass "Y" tube is passed. Small rubber balloons are fastened to the arms of the "Y" (figure 43.1). What happens to the balloons when the rubber sheeting is pulled downward?

_____

What happens when the sheeting is pushed upward? _____

_____

_____

How do you explain these changes? _____

_____

_____

_____

Procedure B—Respiratory Air Volumes

1. Review the section entitled "Respiratory Air Volumes" in chapter 16 of the textbook.
2. Complete Part B of the laboratory report.
3. Obtain a hand-held spirometer. Note that the needle can be set at zero by rotating the adjustable dial. The instrument should be held with the dial upward, and air should be blown into the disposable

Figure 43.1 A mechanical model of the respiratory system.

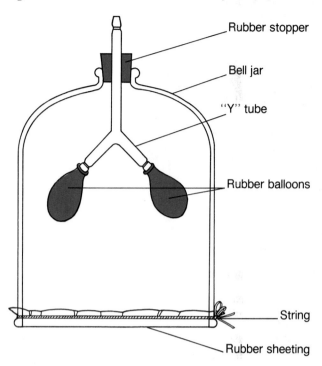

- Rubber stopper
- Bell jar
- "Y" tube
- Rubber balloons
- String
- Rubber sheeting

Figure 43.2 A hand-held spirometer can be used to measure respiratory air volumes.

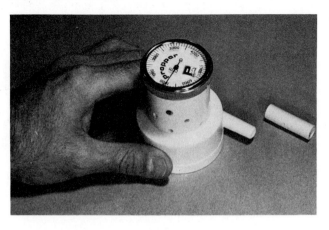

mouthpiece. Movement of the needle indicates the air volume that leaves the lungs (figure 43.2). Before using the instrument, clean it with cotton moistened with 70% alcohol, and place a new mouthpiece over its stem.

4. Tidal volume is the volume of air that enters the lungs during normal, quiet breathing. To measure this volume:
   a. Sit quietly for a few moments.
   b. Position the spirometer dial so that the needle points to zero.
   c. Place the mouthpiece between your lips, and exhale three ordinary expirations into it. *Do not force air out of your lungs; exhale normally.*
   d. Divide the total value indicated by the needle by three, and record this amount as your tidal volume on the chart in Part C of the laboratory report.

5. Expiratory reserve volume is the volume of air in addition to the tidal volume that leaves the lungs during forced expiration. To measure this volume:
   a. Breathe normally for a few moments, and set the needle to zero.
   b. At the end of an ordinary expiration, place the mouthpiece between your lips, and exhale all of the air you can force from your lungs through the spirometer.
   c. Record the results as your expiratory reserve volume in Part C.

6. Vital capacity is the maximum volume of air that can be exhaled after taking the deepest breath possible. To measure this volume:
   a. Breathe normally for a few moments, and set the needle at zero.
   b. Breathe in and out deeply a couple of times, then take the deepest breath possible.
   c. Place the mouthpiece between your lips, and exhale all the air out of your lungs, slowly and forcefully.
   d. Record the value as your vital capacity in Part C, and compare your result with that expected for a person of your age and height listed in charts 43.1 and 43.2.

7. Inspiratory reserve volume (IRV) is the volume of air in addition to the tidal volume that enters the lungs during forced inspiration. Calculate your inspiratory reserve volume by subtracting your tidal volume (TV) and your expiratory reserve volume (ERV) from your vital capacity (VC). That is, IRV = VC − (TV + ERV).

8. Complete Part C of the laboratory report.

Optional Activity

Determine your *minute respiratory volume*. To do this:

1. Sit quietly for awhile, and then to establish your breathing rate, count the number of times you breathe in one minute.
2. Calculate your minute respiratory volume by multiplying your breathing rate by your tidal volume.

$$\overline{\text{(breathing rate)}} \times \overline{\text{(tidal volume)}} = \overline{\text{(minute respiratory volume)}}$$

3. This value indicates the total volume of air that moves into your respiratory passages during each minute of ordinary breathing.

Chart 43.1 Predicted vital capacities for females

Height in centimeters

Age	146	148	150	152	154	156	158	160	162	164	166	168	170	172	174	176	178	180	182	184	186	188	190	192	194
16	2950	2990	3030	3070	3110	3150	3190	3230	3270	3310	3350	3390	3430	3470	3510	3550	3590	3630	3670	3715	3755	3800	3840	3880	3920
17	2935	2975	3015	3055	3095	3135	3175	3215	3255	3295	3335	3375	3415	3455	3495	3535	3575	3615	3655	3695	3740	3780	3820	3860	3900
18	2920	2960	3000	3040	3080	3120	3160	3200	3240	3280	3320	3360	3400	3440	3480	3520	3560	3600	3640	3680	3720	3760	3800	3840	3880
20	2890	2930	2970	3010	3050	3090	3130	3170	3210	3250	3290	3330	3370	3410	3450	3490	3525	3565	3605	3645	3695	3720	3760	3800	3840
22	2860	2900	2940	2980	3020	3060	3095	3135	3175	3215	3255	3290	3330	3370	3410	3450	3490	3530	3570	3610	3650	3685	3725	3765	3800
24	2830	2870	2910	2950	2985	3025	3065	3100	3140	3180	3220	3260	3300	3335	3375	3415	3455	3490	3530	3570	3610	3650	3685	3725	3765
26	2800	2840	2880	2920	2960	3000	3035	3070	3110	3150	3190	3230	3265	3300	3340	3380	3420	3455	3495	3530	3570	3610	3650	3685	3725
28	2775	2810	2850	2890	2930	2965	3000	3040	3070	3115	3155	3190	3230	3270	3305	3345	3380	3420	3460	3495	3535	3570	3610	3650	3685
30	2745	2780	2820	2860	2895	2935	2970	3010	3045	3085	3120	3160	3195	3235	3270	3310	3345	3385	3420	3460	3495	3535	3570	3610	3645
32	2715	2750	2790	2825	2865	2900	2940	2975	3015	3050	3090	3125	3150	3200	3235	3275	3310	3350	3385	3425	3460	3495	3535	3570	3610
34	2685	2725	2760	2795	2835	2870	2910	2945	2980	3020	3055	3090	3130	3165	3200	3240	3275	3310	3350	3385	3425	3460	3495	3535	3570
36	2655	2695	2730	2765	2805	2840	2875	2910	2950	2985	3020	3060	3095	3130	3165	3205	3240	3275	3310	3350	3385	3420	3460	3495	3530
38	2630	2665	2700	2735	2770	2810	2845	2880	2915	2950	2990	3025	3060	3095	3130	3170	3205	3240	3275	3310	3350	3385	3420	3455	3490
40	2600	2635	2670	2705	2740	2775	2810	2850	2885	2920	2955	2990	3025	3060	3095	3135	3170	3205	3240	3275	3310	3345	3380	3420	3455
42	2570	2605	2640	2675	2710	2745	2780	2815	2850	2885	2920	2955	2990	3025	3060	3100	3135	3170	3205	3240	3275	3310	3345	3380	3415
44	2540	2575	2610	2645	2680	2715	2750	2785	2820	2855	2890	2925	2960	2995	3030	3060	3095	3130	3165	3200	3235	3270	3305	3340	3375
46	2510	2545	2580	2615	2650	2685	2715	2750	2785	2820	2855	2890	2925	2960	2995	3030	3060	3095	3130	3165	3200	3235	3270	3305	3340
48	2480	2515	2550	2585	2620	2650	2685	2715	2750	2785	2820	2855	2890	2925	2960	2995	3030	3060	3095	3130	3160	3195	3230	3265	3300
50	2455	2485	2520	2555	2590	2625	2655	2690	2720	2755	2785	2820	2855	2890	2925	2955	2990	3025	3060	3090	3125	3155	3190	3225	3260
52	2425	2455	2490	2525	2555	2590	2625	2655	2690	2720	2755	2790	2820	2855	2890	2925	2955	2990	3020	3055	3090	3125	3155	3190	3220
54	2395	2425	2460	2495	2530	2560	2590	2625	2655	2690	2720	2755	2790	2820	2855	2885	2920	2950	2985	3020	3050	3085	3115	3150	3180
56	2365	2400	2430	2460	2495	2525	2560	2590	2625	2655	2690	2720	2755	2790	2820	2855	2885	2920	2950	2980	3015	3045	3080	3110	3145
58	2335	2370	2400	2430	2460	2495	2525	2560	2590	2625	2655	2690	2720	2750	2785	2815	2850	2880	2920	2945	2975	3010	3040	3075	3105
60	2305	2340	2370	2400	2430	2460	2495	2525	2560	2590	2625	2655	2685	2720	2750	2780	2810	2845	2875	2915	2940	2970	3000	3035	3065
62	2280	2310	2340	2370	2405	2435	2465	2495	2525	2560	2590	2620	2655	2685	2715	2745	2775	2810	2840	2870	2900	2935	2965	2995	3025
64	2250	2280	2310	2340	2370	2400	2430	2465	2495	2525	2555	2585	2620	2650	2680	2710	2740	2770	2805	2835	2865	2895	2925	2955	2990
66	2220	2250	2280	2310	2340	2370	2400	2430	2460	2495	2525	2555	2585	2615	2645	2675	2705	2735	2765	2800	2825	2860	2890	2920	2950
68	2190	2220	2250	2280	2310	2340	2370	2400	2430	2460	2490	2520	2550	2580	2610	2640	2670	2700	2730	2760	2795	2820	2850	2880	2910
70	2160	2190	2220	2250	2280	2310	2340	2370	2400	2425	2455	2485	2515	2545	2575	2605	2635	2665	2695	2725	2755	2780	2810	2840	2870
72	2130	2160	2190	2220	2250	2280	2310	2335	2365	2395	2425	2455	2480	2510	2540	2570	2600	2630	2660	2685	2715	2745	2775	2805	2830
74	2100	2130	2160	2190	2220	2245	2275	2305	2335	2360	2390	2420	2450	2475	2505	2535	2565	2590	2620	2650	2680	2710	2740	2765	2795

From E. DeF. Baldwin and E. W. Richards, Jr., "Pulmonary insufficiency 1, physiologic classification, clinical methods of analysis, standard values in normal subjects," in Medicine 27:243, 1948. Copyright © 1948, The Williams and Wilkins Co., Baltimore. Reprinted from Warren E. Collins, Inc., Braintree, Mass.

Chart 43.2 Predicted vital capacities for males

Height in centimeters

Age	146	148	150	152	154	156	158	160	162	164	166	168	170	172	174	176	178	180	182	184	186	188	190	192	194
16	3765	3820	3870	3920	3975	4025	4075	4130	4180	4230	4285	4335	4385	4440	4490	4540	4590	4645	4695	4745	4800	4850	4900	4955	5005
18	3740	3790	3840	3890	3940	3995	4045	4095	4145	4200	4250	4300	4350	4405	4455	4505	4555	4610	4660	4710	4760	4815	4865	4915	4965
20	3710	3760	3810	3860	3910	3960	4015	4065	4115	4165	4215	4265	4320	4370	4420	4470	4520	4570	4625	4675	4725	4775	4825	4875	4930
22	3680	3730	3780	3830	3880	3930	3980	4030	4080	4135	4185	4235	4285	4335	4385	4435	4485	4535	4585	4635	4685	4735	4790	4840	4890
24	3635	3685	3735	3785	3835	3885	3935	3985	4035	4085	4135	4185	4235	4285	4330	4380	4430	4480	4530	4580	4630	4680	4730	4780	4830
26	3605	3655	3705	3755	3805	3855	3905	3955	4000	4050	4100	4150	4200	4250	4300	4350	4395	4445	4495	4545	4595	4645	4695	4740	4790
28	3575	3625	3675	3725	3775	3820	3870	3920	3970	4020	4070	4115	4165	4215	4265	4310	4360	4410	4460	4510	4555	4605	4655	4705	4755
30	3550	3595	3645	3695	3740	3790	3840	3890	3935	3985	4035	4080	4130	4180	4230	4275	4325	4375	4425	4470	4520	4570	4615	4665	4715
32	3520	3565	3615	3665	3710	3760	3810	3855	3905	3950	4000	4050	4095	4145	4195	4240	4290	4340	4385	4435	4485	4530	4580	4625	4675
34	3475	3525	3570	3620	3665	3715	3760	3810	3855	3905	3950	4000	4045	4095	4140	4190	4225	4285	4330	4380	4425	4475	4520	4570	4615
36	3445	3495	3540	3585	3635	3680	3730	3775	3825	3870	3920	3965	4010	4060	4105	4155	4200	4250	4295	4340	4390	4435	4485	4530	4580
38	3415	3465	3510	3555	3605	3650	3695	3745	3790	3840	3885	3930	3980	4025	4070	4120	4165	4210	4260	4305	4350	4400	4445	4495	4540
40	3385	3435	3480	3525	3575	3620	3665	3710	3760	3805	3850	3900	3945	3990	4035	4085	4130	4175	4220	4270	4315	4360	4410	4455	4500
42	3360	3405	3450	3495	3540	3590	3635	3680	3725	3770	3820	3865	3910	3955	4000	4050	4095	4140	4185	4230	4280	4325	4370	4415	4460
44	3315	3360	3405	3450	3495	3540	3585	3630	3675	3725	3770	3815	3860	3905	3950	3995	4040	4085	4130	4175	4220	4270	4315	4360	4405
46	3285	3330	3375	3420	3465	3510	3555	3600	3645	3690	3735	3780	3825	3870	3915	3960	4005	4050	4095	4140	4185	4230	4275	4320	4365
48	3255	3300	3345	3390	3435	3480	3525	3570	3615	3655	3700	3745	3790	3835	3880	3925	3970	4015	4060	4105	4150	4190	4235	4280	4325
50	3210	3255	3300	3345	3390	3430	3475	3520	3565	3610	3650	3695	3740	3785	3830	3870	3915	3960	4005	4050	4090	4135	4180	4225	4270
52	3185	3225	3270	3315	3355	3400	3445	3490	3530	3575	3620	3660	3705	3750	3795	3835	3880	3925	3970	4010	4055	4100	4140	4185	4230
54	3155	3195	3240	3285	3325	3370	3415	3455	3500	3540	3585	3630	3670	3715	3760	3800	3845	3890	3930	3975	4020	4060	4105	4145	4190
56	3125	3165	3210	3255	3295	3340	3380	3425	3465	3510	3550	3595	3640	3680	3725	3765	3810	3850	3895	3940	3980	4025	4065	4110	4150
58	3080	3125	3165	3210	3250	3290	3335	3375	3420	3460	3500	3545	3585	3630	3670	3715	3755	3800	3840	3880	3925	3965	4010	4050	4095
60	3050	3095	3135	3175	3220	3260	3300	3345	3385	3430	3470	3500	3555	3595	3635	3680	3720	3760	3805	3845	3885	3930	3970	4015	4055
62	3020	3060	3110	3150	3190	3230	3270	3310	3350	3390	3440	3480	3520	3560	3600	3640	3680	3730	3770	3810	3850	3890	3930	3970	4020
64	2990	3030	3080	3120	3160	3200	3240	3280	3320	3360	3400	3440	3490	3530	3570	3610	3650	3690	3730	3770	3810	3850	3900	3940	3980
66	2950	2990	3030	3070	3110	3150	3190	3230	3270	3310	3350	3390	3430	3470	3510	3550	3600	3640	3680	3720	3760	3800	3840	3880	3920
68	2920	2960	3000	3040	3080	3120	3160	3200	3240	3280	3320	3360	3400	3440	3480	3520	3560	3600	3640	3680	3720	3760	3800	3840	3880
70	2890	2930	2970	3010	3050	3090	3130	3170	3210	3250	3290	3330	3370	3410	3450	3480	3520	3560	3600	3640	3680	3720	3760	3800	3840
72	2860	2900	2940	2980	3020	3060	3100	3140	3180	3210	3250	3290	3330	3370	3410	3450	3490	3530	3570	3610	3650	3680	3720	3760	3800
74	2820	2860	2900	2930	2970	3010	3050	3090	3130	3170	3200	3240	3280	3320	3360	3400	3440	3470	3510	3550	3590	3630	3670	3710	3740

From E. DeF. Baldwin and E. W. Richards, Jr., "Pulmonary insufficiency 1, physiologic classification, clinical methods of analysis, standard values in normal subjects," in *Medicine* 27:243, 1948. Copyright © 1948, The Williams and Wilkins Co., Baltimore. Reprinted from Warren E. Collins, Inc., Braintree, Mass.

Name _____

Date _____

Section _____

Breathing and Respiratory Volumes

Part A

Complete the following statements:

1. Breathing also can be called _____ .

2. The weight of air causes a force called _____ .

3. The weight of air at sea level is sufficient to support a column of mercury in a tube _____ mm high.

4. If the pressure inside the lungs decreases, air is pushed into the airways by _____ .

5. Nerve impulses are carried to the diaphragm by the _____ nerve.

6. When the diaphragm contracts, pressure inside the thoracic cavity _____ .

7. The ribs are raised by contraction of the _____ muscles.

8. The attraction of water molecules for one another creates a force called _____ .

9. A mixture of lipoproteins, called _____ , acts to reduce the tendency of alveoli to collapse.

10. The force responsible for normal expiration comes from _____ and _____ .

11. The pressure between the pleural membranes is usually _____ than the atmospheric pressure.

12. Muscles that help to force out more than the normal volume of air by pulling the ribs downward include the _____ .

13. The diaphragm can be forced to move higher than normal by contraction of the _____ muscles.

Part B

Match the air volumes in column A with the definitions in column B. Place the letter of your choice in the space provided.

Column A	Column B
A. expiratory reserve volume	____ 1. volume in addition to tidal volume that leaves the lungs during forced expiration
B. inspiratory reserve volume	
C. residual volume	____ 2. vital capacity plus residual volume
D. tidal volume	____ 3. volume that remains in lungs after most forceful expiration
E. total lung capacity	____ 4. volume that enters or leaves lungs during normal breathing
F. vital capacity	____ 5. volume in addition to tidal volume that enters lungs during forced inspiration
	____ 6. maximum volume a person can exhale after taking the deepest possible breath

Part C

1. Test results for respiratory air volumes:

Respiratory Volume	Expected Value (approximate)	Test Result	Percent of Expected Value (Test result/Expected value × 100)
Tidal volume	500 cc		
Expiratory reserve volume	1100 cc		
Vital capacity	4600 cc		
Inspiratory reserve volume	3000 cc		

2. Complete the following:

   a. How do your test results compare with the expected values? _____

   _____

   _____

   _____

   _____

   b. How does your vital capacity compare with the average value for a person of your age and height? _____

   _____

   _____

   c. What measurement is needed before you can calculate your total lung capacity? _____

   _____

Laboratory Exercise 44

Control of Breathing

Normal breathing is controlled from a poorly defined region of the brain stem called the respiratory center. This center initiates nerve impulses that travel to various muscles, causing rhythmic breathing movements.

Various factors can influence the respiratory center and thus affect the rate and depth of breathing. These factors include stretch of the lung, emotional state, and the presence in the blood of certain chemicals, such as carbon dioxide, hydrogen ions, and oxygen. For example, the breathing rate increases as the blood concentration of carbon dioxide or hydrogen ions increases, or as the concentration of oxygen decreases.

Purpose of the Exercise

To review the mechanisms that control breathing and to investigate some of the factors that affect the rate and depth of breathing.

Learning Objectives

After completing this exercise, you should be able to:

1. Locate the breathing center in the brain.
2. Describe the mechanisms that control normal breathing.
3. List several factors that influence the breathing center.
4. Test the effect of various factors on the rate and depth of breathing.

Materials Needed:
  paper bags, small

For Demonstration:

  flasks   rubber stoppers, two-hole
  glass tubing   limewater

For Optional Activity:

  pneumograph
  physiological recording apparatus

Procedure A—Control of Breathing

1. Review the section entitled "Control of Breathing" in chapter 16 of the textbook.
2. Complete Part A of Laboratory Report 44 on page 305.

Demonstration

When a solution of calcium hydroxide is exposed to carbon dioxide, a chemical reaction occurs and a white precipitate of calcium carbonate is formed as indicated by the following reaction:

$$Ca(OH)_2 + CO_2 \rightarrow CaCO_3 + H_2O$$

Thus, clear limewater can be used to detect the presence of carbon dioxide because the solution becomes cloudy if this gas is bubbled through it.

The laboratory instructor will demonstrate this test for carbon dioxide by drawing some air through limewater in an apparatus like that shown in figure 44.1. Then the instructor will blow an equal volume of expired air through a similar apparatus. Watch for the appearance of a precipitate that causes the limewater to become cloudy. Was there any carbon dioxide in the atmospheric air drawn through the limewater? _____

If so, how did the amount of carbon dioxide in the atmospheric air compare with the amount in the expired air? _____
_____
_____
_____

Procedure B—Factors Affecting Breathing

Perform each of the following tests, using your laboratory partner as a test subject.

1. *Normal breathing.* To determine the subject's normal breathing rate and depth:
   a. Have the subject sit quietly for a few minutes.
   b. After the rest period, ask the subject to count backwards mentally, beginning with 500.

Figure 44.1 Apparatus used to demonstrate the presence of carbon dioxide in air: (*a*) Atmospheric air is drawn through limewater; (*b*) expired air is blown through limewater.

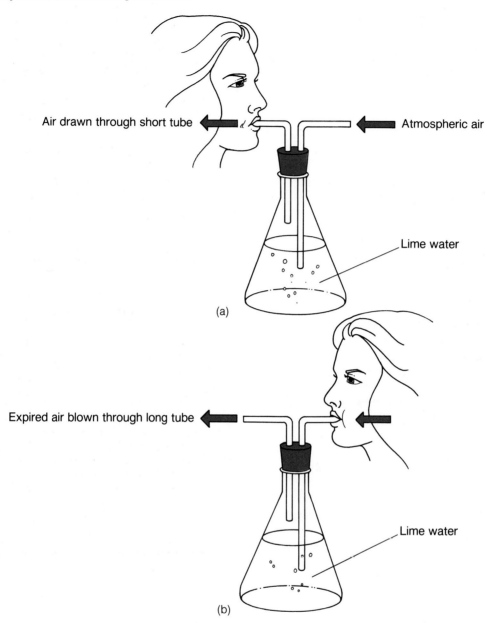

Air drawn through short tube

Atmospheric air

Lime water

(a)

Expired air blown through long tube

Lime water

(b)

c. While the subject is distracted by counting, watch the subject's chest movements, and count the breaths taken in a minute. Use this value as the normal breathing rate (breaths per minute).

d. Note the relative depth of the breathing movements.

e. Record your observations on the chart in Part B of the laboratory report.

2. *Effect of hyperventilation.* To test the effect of hyperventilation on breathing:

a. Seat the subject and *guard to prevent the possibility of the subject falling over.*

b. Have the subject breathe rapidly and deeply for a maximum of two minutes. *If the subject begins to feel dizzy, the hyperventilation should be halted immediately to prevent the subject from fainting.*

c. After the period of hyperventilation, determine the subject's breathing rate and judge the breathing depth.

d. Record the results in Part B.

3. *Effect of rebreathing air.* To test the effect of rebreathing air on breathing:

a. Have the subject sit quietly until the breathing rate returns to normal.

b. Have the subject breathe deeply into a small paper bag that is held tightly over the nose and mouth.

c. After two minutes of rebreathing air, determine the subject's breathing rate and judge the depth of breathing.

d. Record the results in Part B.

4. *Effect of breath holding.* To test the effect of breath holding on breathing:

a. Have the subject sit quietly until the breathing rate returns to normal.

b. Have the subject hold the breath as long as possible.

c. As the subject begins to breathe again, determine the rate of breathing, and judge the depth of breathing.

d. Record the results in Part B.

5. *Effect of exercise.* To test the effect of exercise on breathing:

a. Have the subject sit quietly until breathing rate returns to normal.

b. Have the subject exercise by running in place rapidly for three to five minutes.

c. After the exercise, determine the breathing rate, and judge the depth of breathing.

d. Record the results in Part B.

6. Complete Part B of the laboratory report.

Optional Activity

A *pneumograph* is a device that can be used together with some type of recording apparatus to record breathing movements. The laboratory instructor will demonstrate the use of this equipment to record various movements, such as those that accompany coughing, laughing, yawning, and speaking.

Devise an experiment to test the effect of some factor, such as hyperventilation, rebreathing air, or exercise, on the length of time a person can hold the breath. *After the laboratory instructor has approved your plan,* carry out the experiment, using the pneumograph and recording equipment. What conclusion can you draw from the results of your experiment? _____
_____
_____
_____
_____

Control of Breathing

Part A

Complete the following statements:

1. The respiratory center is located throughout the _____ and _____ of the brain stem.

2. The two major components of the respiratory center are the _____ and _____ .

3. The _____ group establishes the basic rhythm of breathing.

4. The _____ group functions during forceful breathing.

5. The _____ regulates the duration of inspiratory bursts and controls the breathing rate.

6. Chemosensitive areas of the respiratory center are located in the _____ .

7. These chemosensitive areas are stimulated by changes in the blood concentrations of _____ or _____ .

8. As the blood concentration of carbon dioxide increases, the breathing rate _____ .

9. _____ combines with water to form carbonic acid.

10. When carbonic acid dissociates, _____ and hydrogen ions are released.

11. As a result of increased breathing, the blood concentration of carbon dioxide is _____ .

12. The chemoreceptors that sense low blood oxygen concentration are located in the _____ and _____ .

13. As the blood concentration of oxygen decreases, the breathing rate _____ .

14. The inflation reflex involves _____ receptors located in the lungs.

15. The _____ nerve carries sensory impulses from the lungs to the respiratory center.

16. As a result of hyperventilation, breath holding time is _____ .

Part B

1. Record the results of your breathing tests on the chart.

Factor Tested	Breathing Rate (breaths/min)	Breathing Depth (+, ++, +++)
Normal		
Hyperventilation		
Rebreathing air		
Breath holding		
Exercise		

2. Briefly explain the reason for the changes in breathing that occurred in each of the following cases:

   a. Hyperventilation

   _____

   _____

   _____

   _____

   b. Rebreathing air

   _____

   _____

   _____

   _____

   c. Breath holding

   _____

   _____

   _____

   _____

   d. Exercise

   _____

   _____

   _____

   _____

3. Complete the following:

   a. Why is it important to distract a person when you are determining the normal rate of breathing? _____

   _____

   _____

   _____

   b. How can the depth of breathing be measured accurately? _____

   _____

   _____

   _____

   c. Why is it dangerous for a swimmer to hyperventilate in order to hold the breath for a longer period of time?

   _____

   _____

   _____

306

Blood Cells

Blood is a type of connective tissue whose cells are suspended in a liquid intercellular substance. These cells are formed mainly in red bone marrow, and they include red blood cells, white blood cells, and some cellular fragments called platelets.

Red blood cells function to transport gases between the body cells and the lungs, white blood cells serve to defend the body against infections, and platelets play an important role in helping to close breaks in damaged blood vessels.

Purpose of the Exercise

To review the characteristics of blood cells, to examine them microscopically, and to perform a differential white blood cell count.

Learning Objectives

After completing this exercise, you should be able to:

1. Describe the structure and function of red blood cells, white blood cells, and platelets.
2. Identify red blood cells, five types of white blood cells, and platelets on a stained blood slide.
3. Perform a differential white blood cell count.

Materials Needed:
    prepared microscope slides
    human blood (Wright's stain)

For Demonstration:

    microscope slides
    sterile disposable blood lancets
    sterile absorbent cotton
    70% alcohol
    Wright's stain
    distilled water
    compound microscope
    borax (or similar cleaning agent)
    petri dish
    wide cork

For Optional Activity:

    prepared slides of pathological blood, such as eosinophilia, leukocytosis, leukopenia, and lymphocytosis

Warning
Because of the possibility of blood infections being transmitted from one student to another if blood slides are prepared in the classroom, it is suggested that commercially prepared blood slides be used in this exercise. The instructor, however, may wish to demonstrate the procedure for preparing such a slide.

Demonstration

To prepare a stained blood slide:
1. Clean two microscope slides with a cleaning agent (such as borax) and water, and thereafter avoid touching their flat surfaces.
2. Wash the hands thoroughly with soap and water, and dry them with paper towels.
3. Cleanse the end of the middle finger with some sterile cotton moistened with 70% alcohol, and let the finger dry in the air.
4. Remove a sterile disposable blood lancet from its package without touching the sharp end.
5. Puncture the skin on the tip of the middle finger with the lancet, and discard the lancet.
6. Place a drop of blood about 2 cm from the end of a clean microscope slide.
7. Use a second slide to spread the blood across the first slide, as illustrated in figure 45.1.
8. Place the blood slide on a wide cork in the center of a petri dish, and let it dry in the air.
9. Put enough Wright's stain on the slide to cover the smear but not overflow the slide. Count the number of drops of stain that are used.
10. After two to three minutes, add an equal volume of distilled water to the stain, and let the slide stand for four minutes. From time to time, blow gently on the liquid to mix the water and stain.

Figure 45.1 To prepare a blood smear, (*a*) place a drop of blood about 2 cm from the end of a clean slide; (*b*) hold a second slide at about 45° to the first one, allowing the blood to spread along its edge; (*c*) push the second slide over the surface of the first so that it pulls the blood with it; (*d*) observe the completed blood smear.

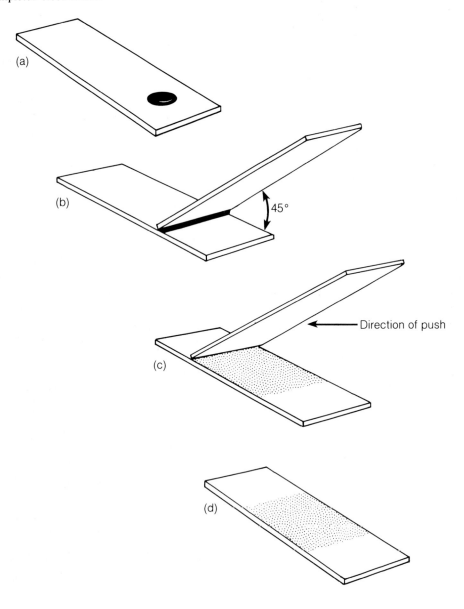

(a)

(b)

45°

Direction of push

(c)

(d)

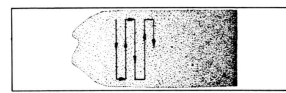

11. Flood the slide with distilled water until the blood smear appears light blue.
12. Tilt the slide to pour off the water, and let the slide dry in the air.

Examine the blood smear with low-power magnification, and locate an area where the blood cells are well distributed. Observe these cells, using high-power magnification, and then with an oil immersion lens if one is available.

Procedure A—Types of Blood Cells

1. Review the sections entitled "Characteristics of Red Blood Cells," "Types of White Blood Cells," and "Blood Platelets" in chapter 17 of the textbook.
2. Complete Part A of Laboratory Report 45 on page 310.
3. Refer to figures 17.10–17.15 in the textbook as an aid for identifying the various types of blood cells, and locate each of the following:

   red blood cell (erythrocyte)

   white blood cell (leukocyte)

      neutrophil

      lymphocyte

      monocyte

      eosinophil

      basophil

   platelet

4. In Part B of the laboratory report, prepare sketches of single blood cells to illustrate each type.

Procedure B—Differential White Blood Cell Count

A differential white blood cell count is performed to determine the percentage of each of the various types of white blood cells present in a blood sample. The test is useful because the relative proportions of white blood cells may change in particular diseases. Neutrophils, for example, usually increase during bacterial infections, while eosinophils may increase during certain parasitic infections and allergic reactions.

1. Make a differential white blood cell count:
   a. Focus on the cells at one end of a prepared blood slide, using high-power magnification or an oil immersion lens.
   b. Move the blood slide back and forth slowly, following a path that avoids passing over the same cells twice (figure 45.2).
   c. Each time you encounter a white blood cell, identify its type, and record it in Part C of the laboratory report.
   d. Continue searching for and identifying white blood cells until you have recorded one hundred cells in the data chart. Because percent means "parts of a hundred," for each type of white blood cell, the total number observed is equal to its percentage in the blood sample.
2. Complete Part C of the laboratory report.

Optional Activity

Obtain a prepared slide of pathological blood that has been stained with Wright's stain. Perform a differential white blood cell count using this slide and compare the results with the values for normal blood listed in "Laboratory Tests of Clinical Importance" in the appendix of the textbook. What differences do you note? _____
_____
_____
_____

Name _____

Date _____

Section _____

Blood Cells

Part A

Complete the following statements:

1. Red blood cells also are called _____ .

2. The shape of a red blood cell can be described as a(an) _____ .

3. The shape of a red blood cell is related to its function of _____ .

4. _____ is the oxygen-carrying substance in a red blood cell.

5. Red blood cells with high oxygen concentrations are bright red in color because of the presence of _____ .

6. Red blood cells cannot reproduce because they lack _____ when they are mature.

7. White blood cells also are called _____ .

8. White blood cells with granular cytoplasm are called _____ .

9. White blood cells lacking granular cytoplasm are called _____ .

10. _____ is another name for a neutrophil.

11. Normally, the most numerous white blood cells are _____ .

12. White blood cells whose cytoplasmic granules stain red in acid stain are called _____ .

13. _____ are normally the least abundant of the white blood cells.

14. _____ are the largest of the white blood cells.

15. _____ are agranulocytes that have relatively large, round nuclei with thin rims of cytoplasm.

16. In red bone marrow, platelets develop from cells called _____ .

17. A platelet is capable of _____ movement.

18. In the presence of damaged blood vessels, platelets release a substance called _____ that causes smooth muscle contraction.

Part B

Sketch a single blood cell of each type in the spaces provided below.

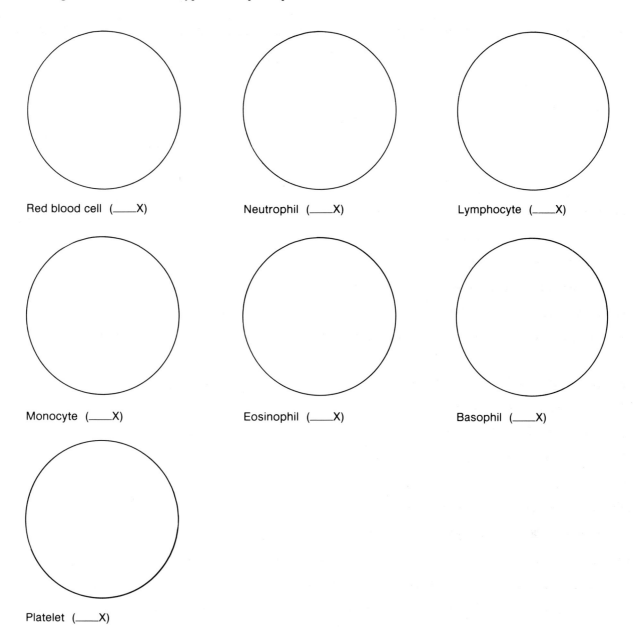

Red blood cell (___X)

Neutrophil (___X)

Lymphocyte (___X)

Monocyte (___X)

Eosinophil (___X)

Basophil (___X)

Platelet (___X)

Part C

1. Differential white blood cell count data chart: As you identify white blood cells, record them on the chart by using a tally system, such as ⊬℟ Ⅱ .

Type of WBC	Number Observed	Total	Percent
Neutrophil			
Lymphocyte			
Monocyte			
Eosinophil			
Basophil			

2. Complete the following:
   a. How do the results of your differential white blood cell count compare with the normal values listed in the appendix of the textbook? _____

   _____

   _____

   _____

   _____

   b. What is the difference between a differential white blood cell count and a total white blood cell count? _____

   _____

   _____

   _____

   _____

Blood Testing—A Demonstration

As an aid for identifying various disease conditions, tests often are performed on blood to determine how its composition compares with normal values. These tests commonly include red blood cell percentage, hemoglobin content, red blood cell count, and total white blood cell count.

Purpose of the Exercise

To observe the blood tests used to determine red blood cell percentage, hemoglobin content, red blood cell count, and total white blood cell count.

Learning Objectives

After completing this exercise, you should be able to:

1. Determine the percentage of red blood cells in a blood sample.
2. Determine the hemoglobin content of a blood sample.
3. Describe how a red blood cell count is performed.
4. Describe how a total white blood cell count is performed.

Materials Needed for Demonstrations:
    sterile disposable blood lancets
    sterile absorbent cotton
    70% alcohol

For Demonstration A:

    heparinized microhematocrit capillary tube
    sealing clay (or Critocaps)
    microhematocrit centrifuge
    microhematocrit reader

For Demonstration B:

    hemoglobinometer
    lens paper
    hemolysis applicator

For Demonstration C:

    hemocytometer
    red blood cell pipette with suction tube and
        mouthpiece
    absorbent paper
    red blood cell diluting fluid

For Demonstration D:

    hemocytometer
    white blood cell pipette with suction tube and
        mouthpiece
    white blood cell diluting fluid

Warning
Because of the possibility of blood infections being transmitted from one student to another during blood testing procedures, it is suggested that the following demonstrations be performed by the instructor.

Demonstration A—Red Blood Cell Percentage

To determine the percentage of red blood cells in a whole blood sample, the cells must be separated from the liquid plasma. This separation can be accomplished rapidly by placing a tube of blood in a centrifuge. The force created by the spinning motion of the centrifuge causes the cells to be packed into the lower end of the tube. Then, the quantities of cells and plasma can be measured, and the percentage of cells (hematocrit or packed cell volume) can be calculated.

1. To determine the percentage of red blood cells in a blood sample:
   a. Lance the end of a finger to obtain a drop of blood. *See the demonstration in Laboratory Exercise 45 (pages 307–9) for directions.*
   b. Touch the drop of blood with the colored end of a heparinized capillary tube. Hold the tube tilted slightly downward so that the blood will move into it easily.
   c. Allow the blood to fill about two-thirds of the length of the tube.
   d. Plug the dry end of the tube by pushing it into sealing clay or adding a plastic Critocap.
   e. Place the sealed tube into one of the numbered grooves of a microhematocrit centrifuge. The tube's sealed end should point outward from the center and should touch the rubber lining on the rim of the centrifuge (figure 46.1).
   f. The centrifuge should be balanced by placing specimen tubes on opposite sides of the moving head, the inside cover should be tightened with a wrench, and the outside cover should be securely fastened.
   g. Run the centrifuge for four minutes.
   h. After the centrifuge has stopped, remove the specimen tube, and note that the red blood cells have been packed into the bottom of the tube. The clear liquid on top of the cells is plasma.
   i. Use a microhematocrit reader to determine the percentage of red blood cells in the tube. If a microhematocrit reader is not available, measure the total length of the blood column (red cells plus plasma) and the length of the red blood cell column alone in millimeters. Divide the red blood cell length by the total blood column length, and multiply the answer by 100 to calculate the percentage of red blood cells.
   j. Record the test result in Part A of Laboratory Report 46 on page 319.
2. Complete Part B of the laboratory report.

Demonstration B—Hemoglobin Content

Although the hemoglobin content of a blood sample can be measured in several ways, a common method makes use of a hemoglobinometer. This instrument is designed to compare the color of light passing through a hemolyzed blood sample with a standard color. The results of the test are expressed in grams of hemoglobin per 100 ml of blood or in percentage of normal values.

1. To measure the hemoglobin content of a blood sample:
   a. Obtain a hemoglobinometer and remove the blood chamber from the slot in its side.

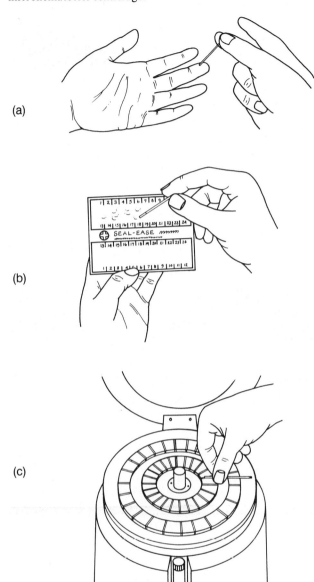

Figure 46.1 Steps in the red blood cell percentage procedure: (*a*) Load a heparinized capillary tube with blood; (*b*) plug the dry end of the tube with sealing clay; (*c*) place the tube in a microhematocrit centrifuge.

(a)

(b)

(c)

   b. Separate the pieces of glass from the metal clip, and clean them with 70% alcohol and lens paper. Note that one of the pieces of glass has two broad U-shaped areas surrounded by depressions. The other piece is flat on both sides.
   c. Obtain a large drop of blood from a finger, as before.
   d. Place the drop of blood on one of the U-shaped areas of the blood chamber glass.

Figure 46.2 Steps in the hemoglobin content procedure: (*a*) Load the blood chamber with blood; (*b*) stir the blood with a hemolysis applicator; (*c*) place the blood chamber in the slot of the hemoglobinometer; (*d*) match the colors in the green area by moving the slide on the side of the instrument.

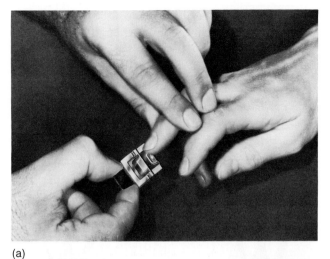

(a)

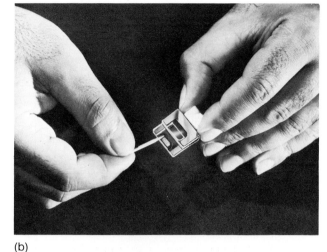

(b)

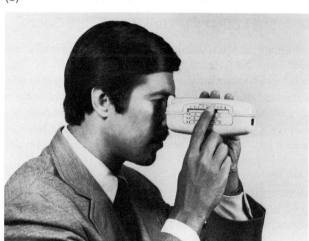

(c)

(d)

e. Stir the blood with the tip of a hemolysis applicator until the blood appears clear rather than cloudy. This usually takes about 45 seconds.

f. Place the flat piece of glass on top of the blood plate and slide both into the metal clip of the blood chamber.

g. Push the blood chamber into the slot on the side of the hemoglobinometer, and make sure that it is in all the way (figure 46.2).

h. Hold the hemoglobinometer in the left hand with the thumb on the light switch on the underside.

i. Look into the eyepiece, and note the green area that is split in half.

j. Slowly move the slide on the side of the instrument back and forth with the right hand until the two halves of the green area look the same.

k. Note the value in the upper scale (grams of hemoglobin per 100 ml of blood), indicated by the mark in the center of the movable slide.

l. Record the test result in Part A of the laboratory report.

2. Complete Part C of the laboratory report.

Figure 46.3 Location of the counting areas of a hemocytometer.

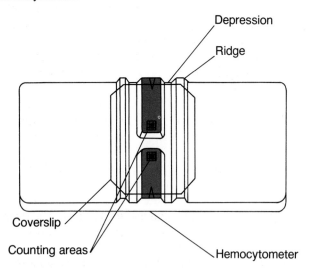

Figure 46.4 The pattern of lines in a counting area of a hemocytometer. The squares marked with "R" are used to count red blood cells, while the squares marked with "W" are used to count white blood cells.

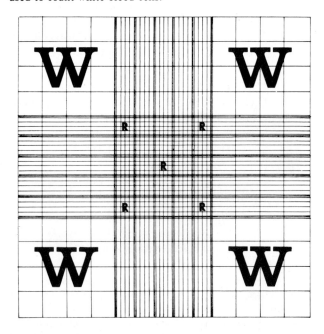

Demonstration C—Red Blood Cell Count

Although modern clinical laboratories commonly use electronic instruments to obtain blood cell counts, a hemocytometer provides a convenient and inexpensive way to count both red and white blood cells. This instrument is a special microscope slide on which there are two counting areas (figure 46.3). Each counting area contains a grid of tiny lines forming nine large squares that are further subdivided into smaller squares (figure 46.4).

The squares on the counting area have known dimensions, and when a coverslip is placed over the area, it is held a known distance above the grid by glass ridges of the hemocytometer. Thus, when a liquid is placed under the coverslip, the volume of liquid covering each part of the grid can be calculated. Also, if the number of blood cells in a tiny volume of blood can be counted, it is possible to calculate the number that must be present in any larger volume.

When red blood cells are counted using a hemocytometer, a small sample of blood is drawn into a special pipette, and the blood is diluted to reduce the number of cells that must be counted. Some of the diluted blood is spread over the counting area, and with the aid of a microscope, all of the cells in the grid areas corresponding to those marked with "R" in figure 46.4 are counted. The total count is multiplied by 10,000 to correct for the dilution and for the fact that only a small volume of blood was observed. The final result provides the number of red blood cells per cubic millimeter in the original blood sample.

1. To perform a red blood cell count:
   a. Examine a hemocytometer, and use a microscope to locate the grid of a counting area. Focus on the large central square of the grid with low-power and then with high-power magnification. Adjust the light intensity so that the lines are clear and sharp.
   b. Wash the hemocytometer with soap and water, and dry it. Place the coverslip over the counting areas and set the instrument aside.
   c. Lance the tip of a finger, as before, to obtain a drop of blood.
   d. Place the mouthpiece of the suction tube attached to a red blood cell pipette between the teeth.
   e. Touch the drop of blood with the tip of the pipette, and use gentle suction to draw blood up to the line marked 0.5. If too much blood enters the pipette, blot the tip of the tube on absorbent paper to lower the level.
   f. Place the tip of the pipette into the red blood cell diluting fluid and draw in enough fluid to reach the line marked 101, which is above the mixing chamber of the pipette.
   g. Plug the tip of the pipette with a thumb and remove the suction tube. Then, hold the pipette horizontally between the thumb and middle finger, and shake the tube for three minutes to mix the blood and diluting fluid thoroughly.
   h. Discard the first four drops from the pipette, and then place a drop of diluted blood at the edge of the coverslip near a counting area of the hemocytometer. If the hemocytometer is properly charged with diluted blood, the space between the counting area and the coverslip will be filled, but no fluid will spill over into the depression on either side.

i. Place the hemocytometer on the microscope stage and focus on the central large square of the counting area with the high-power lens. Adjust the light so that the grid lines and blood cells are clearly visible.

j. Count all the cells in the areas corresponding to those marked with "R" in figure 46.4. To obtain an accurate count, include cells that are touching the lines at the tops and left sides of the squares, but do not count those touching the bottoms and right sides of the squares.

k. Multiply the total count by 10,000, and record the result (cells per cubic millimeter of blood) in Part A of the laboratory report.

l. Clean the pipette and other materials as directed by the laboratory instructor.

2. Complete Part D of the laboratory report.

Demonstration D—Total White Blood Cell Count

A hemocytometer is used to make a total white blood cell count in much the same way that it was used to count red blood cells. However, in the case of white cell counting, a diluting fluid is used that destroys red blood cells. Also, in making the white cell count, all of the cells in the four large squares corresponding to those marked with "W" in figure 46.4 are included. The total count is multiplied by 50 to calculate the total number of white blood cells in a cubic millimeter of the blood sample.

1. To perform a total white blood cell count:
   a. Clean the hemocytometer as before.
   b. Lance the tip of a finger, as before, to obtain a drop of blood.
   c. With the mouthpiece of the suction tube between the teeth, touch the tip of a white blood cell pipette to the blood, and draw in enough to reach the 0.5 line.
   d. Place the tip of the pipette in the white blood cell diluting fluid and draw in enough fluid to reach the 11 line above the mixing chamber of the pipette.
   e. Shake the tube for three minutes, as before.
   f. Discard the first four drops, and charge the hemocytometer with diluted blood as before.
   g. Use low-power magnification to locate the areas of the grid corresponding to those marked with "W" in figure 46.4.
   h. Count all of the cells in the four large squares (remember that the red blood cells were destroyed by the white blood cell diluting fluid), following the same counting rules as before.
   i. Multiply the total by 50, and record the result in Part A of the laboratory report.
   j. Clean the pipette and other materials as directed by the laboratory instructor.

2. Complete Part E of the laboratory report.

laboratory report 46

Name _____

Date _____

Section _____

Blood Testing—A Demonstration

Part A

Blood test data:

Blood Test	Test Results	Normal Values*
Red blood cell percentage		
Hemoglobin content		
Red blood cell count		
White blood cell count		

*See "Laboratory Tests of Clinical Importance" in the appendix of the textbook.

Part B

Complete the following:

1. How does the red blood cell percentage from the demonstration compare with the normal value? _____

_____

_____

_____

2. What conditions might produce a decreased red blood cell percentage? _____

_____

_____

_____

3. What conditions might produce an increased red blood cell percentage? _____

_____

_____

_____

Part C

Complete the following:

1. How does the hemoglobin content from the demonstration compare with the normal value? _____

_____

_____

2. What conditions might produce a decreased hemoglobin content? _____

_____

_____

3. What conditions might produce an increased hemoglobin content? _____

_____

_____

Part D

Complete the following:

1. How does the red blood cell count from the demonstration compare with the normal value? _____

_____

_____

2. What conditions might produce a decreased red blood cell count? _____

_____

_____

3. What conditions might produce an increased red blood cell count? _____

_____

_____

Part E

Complete the following:

1. How does the white blood cell count from the demonstration compare with the normal value? _____

_____

_____

2. What conditions might produce a decreased white blood cell count? _____

_____

_____

3. What conditions might produce an increased white blood cell count? _____

_____

_____

Blood Typing

Blood typing involves identifying substances, called agglutinogens, that are present in red blood cell membranes. Although there are many different agglutinogens associated with human red blood cells, only a few of them are of clinical importance. These include the agglutinogens of the ABO group and those of the Rh group.

To determine which agglutinogens are present, a blood sample is mixed with blood typing sera, which contain known types of agglutinins (antibodies). If a particular agglutinin contacts a corresponding agglutinogen, a reaction occurs and the red blood cells clump together (agglutination). Thus, if blood cells are mixed with serum containing agglutinins that react with agglutinogen A and the cells clump together, agglutinogen A must be present in those cells.

Purpose of the Exercise

To determine the ABO blood types of a blood sample and to observe an Rh blood typing test.

Learning Objectives

After completing this exercise, you should be able to:

1. Explain the basis of ABO blood typing.
2. Determine the ABO type of a blood sample.
3. Explain the basis of Rh blood typing.
4. Describe how the Rh type of a blood sample is determined.

Materials Needed:
ABO blood typing kit

For Demonstrations:

    microscope slide
    70% alcohol
    sterile absorbent cotton
    sterile blood lancet
    toothpick
    anti-D blood serum
    slide warming box

Warning

Because of the possibility of blood infections being transmitted from one student to another if blood testing is performed in the classroom, it is suggested that commercially prepared blood typing kits, containing virus-free human blood, be used for ABO blood typing. The instructor may wish to demonstrate Rh blood typing.

Demonstration A—ABO Blood Typing

1. Review the section entitled "ABO Blood Group" in chapter 17 of the textbook.
2. Complete Part A of Laboratory Report 47 on page 323.
3. Perform the ABO blood type test using the blood typing kit. To do this:
   a. Obtain a clean microscope slide and mark across its center with a wax pencil to divide it into right and left halves. Also write "anti-A" near the edge of the left half and "anti-B" near the edge of the right half (figure 47.1).
   b. Place a small drop of blood on each half of the microscope slide. Work quickly so that the blood will not have time to clot.
   c. Add a drop of anti-A serum to the blood on the left half and a drop of anti-B serum to the blood on the right half.
   d. Use separate toothpicks to stir each sample of serum and blood together, and spread each over an area about as large as a quarter dollar.
   e. Examine the samples for clumping of blood cells (agglutination).
   f. See chart 47.1 for aid in interpreting the test results.
4. Complete Part B of the laboratory report.

Figure 47.1 Slide prepared for ABO blood typing.

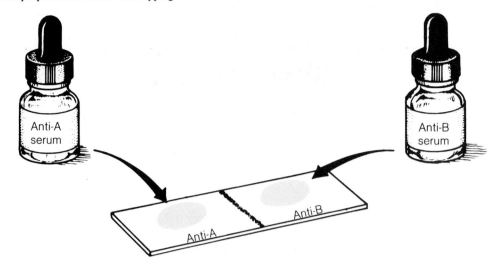

Figure 47.2 Slide warming box used for Rh blood typing.

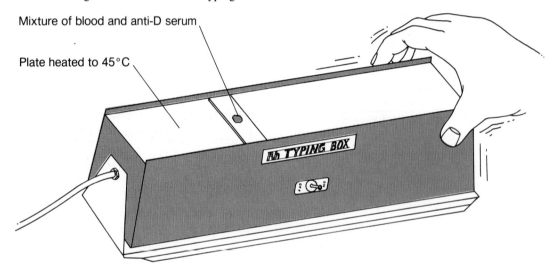

Chart 47.1 Possible reactions of ABO blood typing sera		
Reactions		**Blood type**
Anti-A serum	*Anti-B serum*	
Clumping	No clumping	Type A
No clumping	Clumping	Type B
Clumping	Clumping	Type AB
No clumping	No clumping	Type O

Demonstration B—Rh Blood Typing

1. Review the section entitled "Rh Blood Group" in chapter 17 of the textbook.
2. Complete Part C of the laboratory report.
3. Perform the test used to determine the Rh blood type of a blood sample. To do this:

a. Lance the tip of a finger (see the demonstration in Procedure A of Laboratory Exercise 45 for directions). Place a small drop of blood in the center of a clean microscope slide.

b. Add a drop of anti-D serum to the blood, and mix them together with a clean toothpick.

c. Place the slide on the plate of a warming box that has been prewarmed to 45°C (figure 47.2).

d. Rock the box back and forth slowly to keep the mixture moving, and watch for clumping of the blood cells. When clumping occurs in anti-D serum, the clumps usually are smaller than those that appear in anti-A or anti-B sera, so they may be less obvious. However, if clumping occurs, the blood is called Rh-positive; if no clumping occurs *within two minutes,* the blood is called Rh-negative.

4. Complete Part D of the laboratory report.

laboratory report 47

Name _____

Date _____

Section _____

Blood Typing

Part A

Complete the following statements:

1. The agglutinogens of the ABO blood group are located in the _____ .

2. The blood of every person contains one of (how many possible?) _____ combinations of agglutinogens.

3. Type A blood contains agglutinogen _____ .

4. Type B blood contains agglutinogen _____ .

5. Type A blood contains agglutinin _____ .

6. Type B blood contains agglutinin _____ .

7. Persons with ABO blood type _____ are sometimes called universal recipients.

8. Persons with ABO blood type _____ are sometimes called universal donors.

9. Although agglutinogens are present at the time of birth, agglutinins usually do not appear in the blood until the age of _____ .

10. Agglutinins usually reach a maximum concentration at age _____ .

Part B

Complete the following:

1. What was the ABO type of the blood tested? _____

2. What ABO agglutinogens are present in the red blood cells of this type of blood? _____

3. What ABO agglutinins are present in the plasma of this type of blood? _____

4. If a person with this blood type needed a blood transfusion, what ABO type(s) of blood could be received safely? _____

5. If a person with this blood type were serving as a blood donor, what ABO blood type(s) could receive the blood safely? _____

323

Part C

Complete the following statements:

1. The Rh blood group was named after the _____ .

2. Of the agglutinogens in the Rh group, the most important is _____ .

3. If red blood cells lack Rh agglutinogens, the blood is called _____ .

4. Rh agglutinins form only in persons with _____ type blood.

5. If an Rh-negative person, who is sensitive to Rh-positive blood, receives a transfusion of Rh-positive blood, the donor's cells are likely to _____ .

Part D

Complete the following:

1. What was the Rh type of the blood tested in the demonstration? _____

2. What Rh agglutinogen is present in the red blood cells of this type of blood? _____

3. What Rh agglutinin is present normally in the plasma of this type of blood? _____

4. If a person with this blood type needed a blood transfusion, what type of blood could be received safely?

_____

5. If a person with this blood type were serving as a blood donor, what type of blood could receive the blood safely?

_____

Laboratory Exercise 48

Structure of the Heart

The heart is a muscular pump located within the mediastinum and resting upon the diaphragm. It is enclosed by the lungs, backbone, and sternum and is attached to several large blood vessels at its base. Its distal end extends downward to the left and terminates as a bluntly pointed apex.

The heart and the proximal ends of the attached blood vessels are enclosed by a double-layered pericardium. The inner layer of this membrane consists of a thin covering that is closely applied to the surface of the heart, while the outer layer forms a tough, protective sac surrounding the heart.

Purpose of the Exercise

To review the structural characteristics of the human heart and to examine the major features of a mammalian heart.

Learning Objectives

After completing this exercise, you should be able to:

1. Identify the major structural features of the human heart.
2. Compare the features of the human heart with those of another mammal.

Materials Needed:
  dissectible human heart model
  preserved sheep or other mammalian heart
  dissecting tray
  dissecting instruments

Procedure A—The Human Heart

1. Review the section entitled "Structure of the Heart" in chapter 18 of the textbook.
2. As a review activity, label figures 48.1, 48.2, and 48.3.
3. Complete Part A of Laboratory Report 48 on page 331.

4. Examine the human heart model, and locate the following features:

  heart
    base
    apex
  pericardium
    visceral layer
    parietal layer
  pericardial cavity
  epicardium
  myocardium
  endocardium
  atria
    right atrium
    left atrium
  auricles
  ventricles
    right ventricle
    left ventricle
  atrioventricular orifices
  atrioventricular valves
  atrioventricular sulcus
  interventricular sulci
    anterior sulcus
    posterior sulcus
  superior vena cava
  inferior vena cava
  coronary sinus

Figure 48.1 Label this anterior view of the human heart.

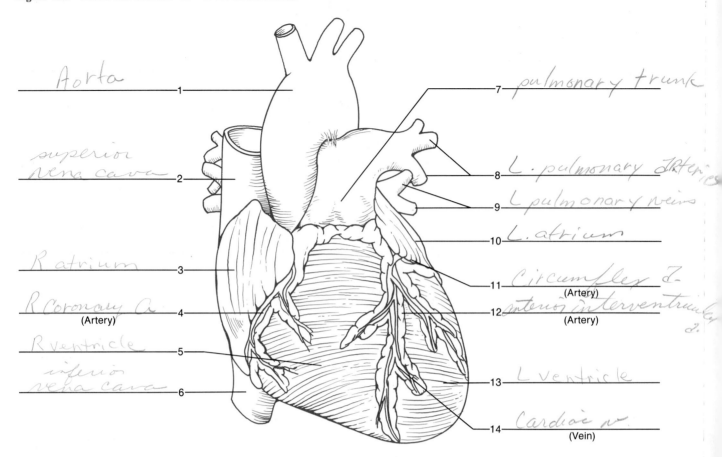

Aorta — 1

superior vena cava — 2

R atrium — 3

R Coronary A — 4
(Artery)

R ventricle — 5

inferior vena cava — 6

7 — pulmonary trunk

8 — L. pulmonary Arteries

9 — L pulmonary veins

10 — L. atrium

11 — Circumflex A.
(Artery)

12 — anterior interventricular A.
(Artery)

13 — L ventricle

14 — Cardiac v.
(Vein)

tricuspid valve

chordae tendineae

papillary muscles

pulmonary trunk

pulmonary arteries

pulmonary semilunar valve

pulmonary veins

bicuspid (mitral) valve

aorta

aortic semilunar valve

left coronary artery

  circumflex artery

  anterior interventricular artery

right coronary artery

  posterior interventricular artery

  marginal artery

cardiac veins

coronary sinus

Optional Activity

Use colored pencils to differentiate the features of the heart in figures 48.1, 48.2, and 48.3. This activity should help you to visualize the relative locations of parts shown from different views in this series of illustrations.

Figure 48.2 Label this posterior view of the human heart.

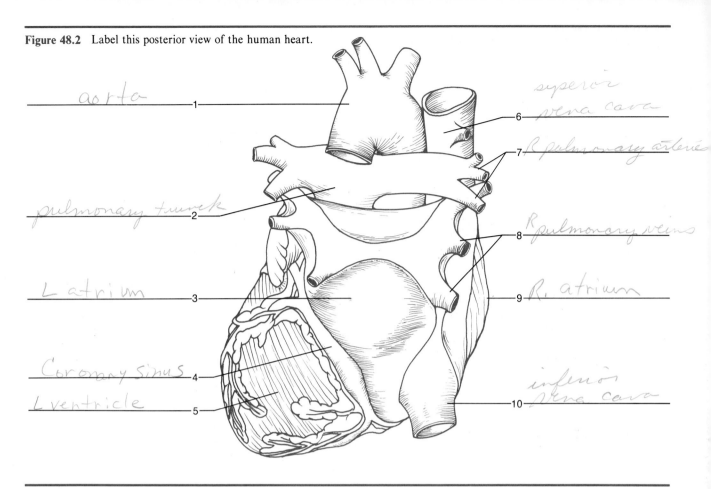

1 _aorta_

2 _pulmonary trunk_

3 _L. atrium_

4 _Coronary sinus_

5 _L ventricle_

6 _superior vena cava_

7 _R pulmonary arteries_

8 _R pulmonary veins_

9 _R. atrium_

10 _inferior vena cava_

Figure 48.3 Label this frontal section of the human heart.

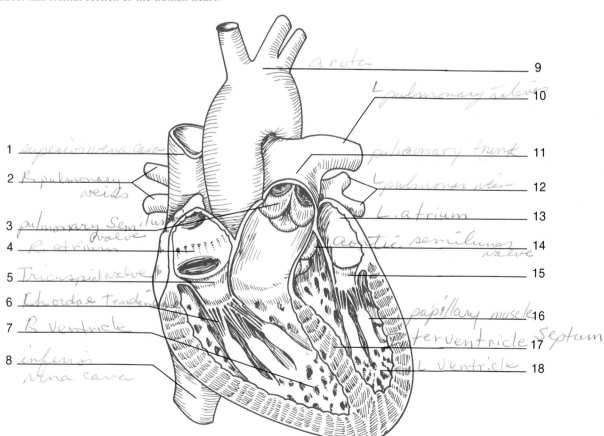

1 _superior vena cava_

2 _R pulmonary veins_

3 _pulmonary Semilunar valve_

4 _R. atrium_

5 _Tricuspid valve_

6 _Chordae Tendineae_

7 _R Ventricle_

8 _inferior vena cava_

9 _aorta_

10 _L pulmonary arteries_

11 _pulmonary trunk_

12 _L pulmonary vein_

13 _L. atrium_

14 _aortic semilunar valve_

15

16 _papillary muscle_

17 _interventricular Septum_

18 _L Ventricle_

Figure 48.4 Ventral side of the sheep heart.

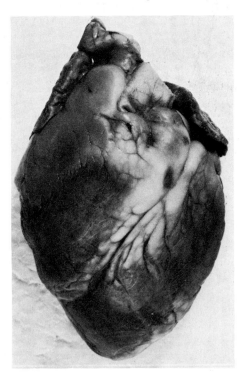

Figure 48.5 Dorsal side of the sheep heart.

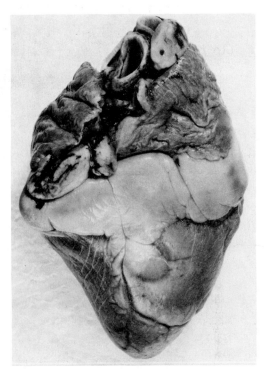

Procedure B—Dissection of a Sheep Heart

1. Obtain a preserved sheep heart. Rinse it in water thoroughly to remove as much of the preservative as possible. Also run water into the large blood vessels to force any blood clots out of the heart chambers.

2. Place the heart in a dissecting tray with its ventral side up (figure 48.4) and proceed as follows:
   a. Although the relatively thick *parietal pericardium* probably is missing, look for traces of this membrane around the origins of the large blood vessels.
   b. Locate the *visceral pericardium* that appears as a thin, transparent layer on the surface of the heart. Use a scalpel to remove a portion of this layer and expose the *myocardium* beneath. Also note the abundance of fat along the paths of various blood vessels. This adipose tissue occurs in the loose connective tissue that underlies the visceral pericardium.
   c. Identify the following:

   right atrium
   right ventricle
   left atrium
   left ventricle
   atrioventricular sulcus
   anterior interventricular sulcus

   d. Carefully remove the fat from the anterior interventricular sulcus, and expose the blood vessels that pass along this groove. They include a branch of the *left coronary artery* (anterior interventricular artery) and a *cardiac vein*.

3. Examine the dorsal surface of the heart (figure 48.5) and proceed as follows:
   a. Identify the *atrioventricular sulcus* and the *posterior interventricular sulcus*.
   b. Locate the stumps of two relatively thin-walled blood vessels that enter the right atrium. Demonstrate this connection by passing a slender probe through them. The upper vessel is the *superior vena cava,* and the lower one is the *inferior vena cava.*

4. Open the right atrium. To do this:
   a. Insert a blade of the scissors into the superior vena cava, and cut downward through the atrial wall (figure 48.6).
   b. Open the chamber, locate the *tricuspid valve,* and examine its leaflets.
   c. Also locate the opening to the *coronary sinus* between the valve and the inferior vena cava.
   d. Run some water through the tricuspid valve to fill the chamber of the right ventricle.
   e. Squeeze the ventricles gently, and watch the leaflets of the valve as the water moves up against them.

Figure 48.6 To open the right atrium, insert a blade of the scissors into the superior vena cava and cut downward.

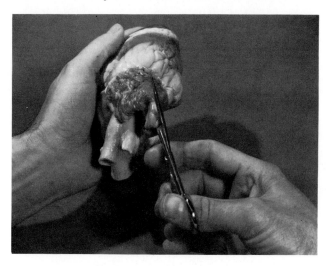

5. Open the right ventricle as follows:
   a. Continue cutting downward through the tricuspid valve and the right ventricular wall until you reach the apex of the heart.
   b. Locate the *chordae tendineae* and the *papillary muscles.*
   c. Find the opening to the *pulmonary trunk,* and use the scissors to cut upward through the wall of the right ventricle. Follow the pulmonary trunk until you have exposed the *pulmonary semilunar valve.*
   d. Examine the valve and its cusps.

6. Open the left side of the heart. To do this:
   a. Insert the blade of the scissors through the wall of the left atrium and cut downward to the apex of the heart.
   b. Open the left atrium, and locate the four openings of the *pulmonary veins.* Pass a slender probe through each opening, and locate the stump of its vessel.
   c. Examine the *bicuspid valve* and its leaflets.
   d. Also examine the left ventricle, and compare the thickness of its wall with that of the right ventricle.

7. Locate the aorta that leads away from the left ventricle and proceed as follows:
   a. Compare the thickness of the aortic wall with that of a pulmonary artery.
   b. Use scissors to cut along the length of the aorta to expose the *aortic semilunar valve* at its base.
   c. Examine the cusps of the valve, and locate the openings of the *coronary arteries* just distal to them.

8. As a review, locate and identify the stumps of each of the major blood vessels associated with the heart.

9. Discard the specimen as directed by the laboratory instructor.

10. Complete Part B of the laboratory report.

Name _____

Date _____

Section _____

Structure of the Heart

Part A

Match the terms in column A with the descriptions in column B. Place the letter of your choice in the space provided.

Column A		Column B
A. aorta	C 1.	upper chamber of the heart
B. atrioventricular sulcus	K 2.	structure from which chordae tendineae originate
C. atrium	D 3.	prevents blood movement from right ventricle to right atrium
D. bicuspid valve	M 4.	double-layered membrane
E. cardiac vein	I 5.	groove separating left and right ventricles
F. coronary artery	D 6.	prevents blood movement from left ventricle to left atrium
G. coronary sinus	N 7.	gives rise to pulmonary arteries
H. endocardium	G 8.	drains blood from myocardium into right atrium
I. interventricular sulcus	H 9.	inner lining of heart chamber
J. myocardium	J 10.	layer composed largely of cardiac muscle tissue
K. papillary muscle	L 11.	potential space containing serous fluid
L. pericardial cavity	E 12.	drains blood from myocardial capillaries
M. pericardium	F 13.	supplies blood to heart muscle
N. pulmonary trunk	A 14.	distributes blood to body parts
O. tricuspid valve	B 15.	groove separating atrial and ventricular portions of heart

Part B

Complete the following:

1. Compare the structure of the tricuspid valve with that of the pulmonary semilunar valve. _____

_____

_____

_____

2. Describe the action of the tricuspid valve when you squeezed the water-filled right ventricle. _____

_____

_____

_____

3. Describe the function of the chordae tendineae and the papillary muscles. _____

_____

_____

_____

4. What is the significance of the difference in thickness between the wall of the aorta and the wall of the pulmonary trunk? _____

_____

_____

_____

5. What is the significance of the difference in thickness of the ventricular walls? _____

_____

_____

_____

6. List in order the major blood vessels, chambers, and valves through which a drop of blood must pass in traveling from a vena cava to the aorta. _____

_____

_____

_____

_____

Laboratory Exercise 49

The Cardiac Cycle

A set of atrial and ventricular contractions constitutes a cardiac cycle. Such a cycle is accompanied by blood pressure changes within the heart chambers, opening and closing of heart valves, and movement of blood in and out of the chambers. These events produce vibrations in the tissues, and thus create the sounds associated with the heartbeat.

A number of electrical changes also occur in the myocardium as it contracts and relaxes. These changes can be detected by using metal electrodes and an instrument called an electrocardiograph. The recording produced by the instrument is an electrocardiogram or ECG.

Purpose of the Exercise

To review the events of a cardiac cycle, to become acquainted with normal heart sounds, and to record an electrocardiogram.

Learning Objectives

After completing this exercise, you should be able to:

1. Describe the major events of a cardiac cycle.
2. Identify the sounds produced during a cardiac cycle.
3. Record an electrocardiogram.
4. Identify the parts of a normal ECG pattern.
5. Describe the phases of a cardiac cycle represented by each part of a normal ECG pattern.

Materials Needed:
For Procedure A—Heart Sounds

> stethoscope
> 70% alcohol
> absorbent cotton

For Procedure B—The Electrocardiogram

> electrocardiograph (or other instrument for recording an ECG)
> electrode cream
> plate electrodes and cables
> lead selector switch

Procedure A—Heart Sounds

1. Review the sections entitled "Cardiac Cycle" and "Heart Sounds" in chapter 18 of the textbook.
2. Complete Part A of Laboratory Report 49 on page 335.
3. Listen to your own heart sounds. To do this:
   a. Obtain a stethoscope, and clean its earpieces by using cotton moistened with alcohol.
   b. Fit the earpieces into your ear canals so that they are directed slightly upward.
   c. Place the diaphragm (or bell) of the stethoscope firmly on the chest over the apex of the heart (figure 49.1) and listen to the sounds. This is a good location to hear the first sound of a cardiac cycle.
   d. Move the diaphragm to the second intercostal space, just to the left of the sternum, and listen to the sounds from this region. You should be able to hear clearly the second sound of the cardiac cycle.
4. Inhale slowly and deeply and exhale slowly while you listen to the heart sounds from each of the locations as before. Note any changes that have occurred in the sounds.
5. Exercise vigorously outside the laboratory for a few minutes, so that other students listening to heart sounds will not be disturbed. After the exercise period, listen to the heart sounds, and note any changes that have occurred in them.
6. Complete Part B of the laboratory report.

Figure 49.1 The first sound of a cardiac cycle can be heard by placing the diaphragm of a stethoscope over the apex of the heart. The second sound can be heard over the second intercostal space, just left of the sternum.

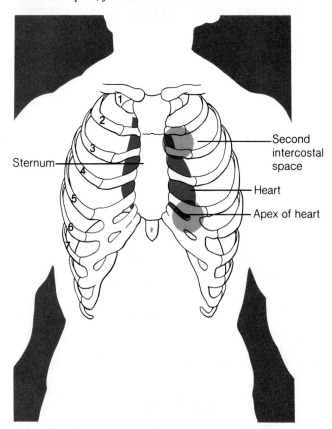

Sternum

Second intercostal space

Heart

Apex of heart

Figure 49.2 To record an ECG, attach electrodes to the wrists and ankles.

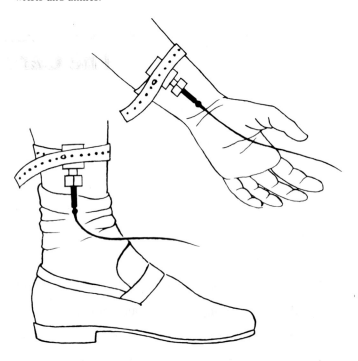

Procedure B—The Electrocardiogram

1. Review the sections entitled "Cardiac Conduction System" and "The Electrocardiogram—A Laboratory Application" in chapter 18 of the textbook.
2. Complete Part C of the laboratory report.
3. The laboratory instructor will demonstrate the proper adjustment and use of the instrument available to record an electrocardiogram.
4. Record your laboratory partner's ECG. To do this:
   a. Have your partner lie on a cot or table close to the electrocardiograph, remaining as relaxed and still as possible.
   b. Apply a small quantity of electrode cream to the skin on the insides of the wrists and ankles. (Any jewelry on the wrists or ankles should be removed.)
   c. Spread some electrode cream over the inner surfaces of four plate electrodes, and attach one to each of the prepared skin areas, using rubber

straps (figure 49.2). Make sure there is good contact between the skin and the metal of the electrodes.
   d. Attach the plate electrodes to the corresponding cables of a lead selector switch. When an ECG recording is made, only two electrodes are used at a time, and the selector switch allows various combinations of electrodes (leads) to be activated.
   e. Turn on the recording instrument, and adjust it as previously demonstrated by the laboratory instructor. The paper speed should be set at 2.5 cm/sec. This is the standard speed for ECG recordings.
   f. Set the lead selector switch to Lead I (right wrist, left wrist electrodes), and record the ECG for one minute.
   g. Set the lead selector switch to Lead II (right wrist, left ankle electrodes), and record the ECG for one minute.
   h. Set the lead selector switch to Lead III (left wrist, left ankle electrodes), and record the ECG for one minute.
   i. Remove the electrodes, and clean the cream from the metal and skin.
5. Complete Part D of the laboratory report.

Name _____

Date _____

Section _____

The Cardiac Cycle

Part A

Complete the following statements:

1. About _____% of the blood from the atria passes into the ventricles before the atrial walls contract.

2. The period during which a heart chamber is contracting is called __*systole*__ . *depolarize*

3. The period during which a heart chamber is relaxing is called __*diastole*__ . *polarize*

4. During ventricular contraction, the tricuspid and bicuspid valves remain __*closed*__ .

5. During ventricular relaxation, the tricuspid and bicuspid valves remain __*open*__ .

6. The semilunar valves open when the pressure in the __*L ventricle*__ exceeds the pressure in the pulmonary trunk and aorta.

7. Heart sounds are due to __*closing of valves*__ in heart tissues created by changes in blood flow.

8. The first sound of a cardiac cycle occurs when the __*bicuspid, Tricuspid*__ are closing.

9. The second sound of a cardiac cycle occurs when the __*semilunars*__ are closing.

10. The sound created when blood leaks back through an incompletely closed valve is called a(an) __*heart murmur*__ .

Part B

Complete the following:

1. What changes did you note in the heart sounds when you inhaled deeply? __*increase*__

_____

_____

2. What changes did you note in the heart sounds following the exercise period? __*increase*__

_____

_____

Part C

Complete the following statements:

1. The fibers of the cardiac conduction system are specialized __*nodal*__ tissue.

2. Normally, the __*sinoatrial*__ node serves as the pacemaker of the heart.

3. The __*atrioventricular*__ node is located in the floor of the right atrium near the interatrial septum.

4. The large fiber on the distal side of the AV node makes up the __*Purkinje fiber*__

5. The fibers that carry cardiac impulses from the interventricular septum into the myocardium are called _____ .

6. An electrocardiogram is a recording of electrical changes occurring in the myocardium during a(an) _cardiac cycle_ . heart attack → see damage

7. Between cardiac cycles, cardiac muscle fibers remain ____relax____ .

8. The P wave is caused by depolarization of the ____atrius____ . (repolarization of atrias)

9. The QRS complex is caused by depolarization of the _ventricle_ .

10. The T wave is caused by repolarization of the _ventricles_ .

Part D

1. Attach a short segment of the ECG recording from each of the three leads you used, and label the waves of each.

Lead I

Lead II

Lead III

2. What differences do you find in the ECG patterns of these leads? _____
_____
_____
_____
_____
_____

3. How much time passed from the beginning of the P wave to the beginning of the QRS complex (P-Q interval) in the ECG from Lead I? _____

4. What is the significance of this P-Q interval? _____
_____
_____
_____
_____

5. How can you determine the heart rate from an electrocardiogram? _____
_____
_____
_____
_____

Laboratory Exercise 50

Factors Affecting the Cardiac Cycle

Although the cardiac cycle is controlled by the SA node serving as the pacemaker, the rate of heart action can be altered by various other factors. These factors include parasympathetic and sympathetic nerve impulses that originate in the cardiac center of the medulla oblongata, changes in body temperature, and concentrations of certain ions.

Purpose of the Exercise

To review the mechanism by which the heartbeat is regulated, to observe the action of a frog heart, and to investigate the effects of various factors on the frog heartbeat.

Learning Objectives

After completing this exercise, you should be able to:

1. Describe the mechanism by which the human cardiac cycle is controlled.
2. List several factors that affect the rate of the heartbeat.
3. Identify the atrial and ventricular contractions, and determine the heart rate from a recording of a frog heartbeat.
4. Test the effects of various factors on the action of a frog heart.

Materials Needed:

physiological recording apparatus
live frog
dissecting tray
dissecting instruments
dissecting pins
frog Ringer's solution in plastic squeeze bottle
thread
small hook
ice
hot plate
calcium chloride, 2% solution
potassium chloride, 5% solution

For Optional Activity:

epinephrine, 1:10,000 solution
acetylcholine, 1:10,000 solution
caffeine, 0.2% solution

Procedure

1. Review the section entitled "Regulation of the Cardiac Cycle" in chapter 18 of the textbook.
2. Complete Part A of Laboratory Report 50 on page 342.
3. Observe the normal action of a frog heart. To do this:
   a. Obtain a live frog, and pith it according to the directions in Procedure C of Laboratory Exercise 19.
   b. Place the frog in a dissecting tray with its ventral side up, and pin its jaw and legs to the tray with dissecting pins.
   c. Use scissors to make a midline incision through the skin from the pelvis to the jaw.
   d. Cut the skin laterally on each side in the pelvic and pectoral regions, and pin the resulting flaps of skin to the tray (figure 50.1).
   e. Remove the exposed pectoral muscles and the sternum, being careful not to injure the underlying organs.
   f. Note the beating heart surrounded by the thin-walled pericardium. Use forceps to pull the pericardium upward, and carefully slit it open with scissors, thus exposing the heart.
   g. Flood the heart with frog Ringer's solution, and keep it moist throughout this exercise.

Figure 50.1 Pin the frog to the dissecting tray and make incisions through the skin as indicated.

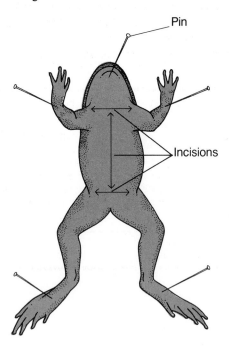

Figure 50.3 Attach the thread from the heart to the recording apparatus so that there is no slack in the thread.

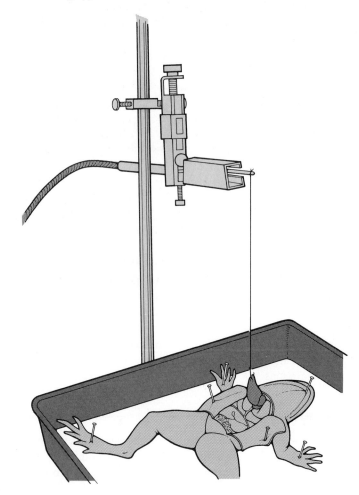

Figure 50.2 Attach a hook and thread to the tip of the ventricle.

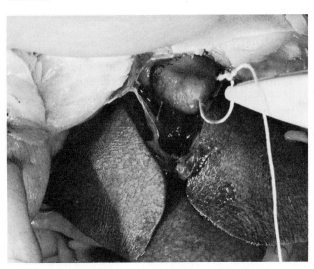

h. Note that the frog heart has only three chambers—two atria and a ventricle. Watch the heart carefully as it beats, and note the sequence of chamber movements during a cardiac cycle.

4. Tie a piece of thread about 45 cm long to a small metal hook, and insert the hook into the tip of the ventricle (figure 50.2). The laboratory instructor will demonstrate how to connect the thread to a physiological recording apparatus so that you can record the frog heart movements. The thread should be adjusted so that there is no slack in it,

but at the same time it should not be so taut that it pulls the heart out of its normal position (figure 50.3).

5. Record the movements of the frog heart for two to three minutes. Identify on the recording the smaller atrial contraction waves and the larger ventricular contraction waves. Also, determine the heart rate (beats per minute) for each minute of recording, and calculate the average rate. Enter the results in Part B of the laboratory report.

6. Test the effect of temperature change on the frog's heart rate. To do this:
   a. Remove as much as possible of the Ringer's solution from around the heart, using a medicine dropper.
   b. Flood the heart with fresh Ringer's solution that has been cooled in an ice water bath to about 10°C.
   c. Record the heart movements, and determine the heart rate as before.
   d. Remove the cool liquid from around the heart, and replace it with room temperature Ringer's solution.

e. After the heart is beating at its normal rate again, flood it with Ringer's solution that has been heated on a hot plate to about 35°C.

f. Record the heart movements, and determine the heart rate as before.

g. Enter the results in Part B of the laboratory report.

7. Complete Part B of the laboratory report.

8. Test the effect of an increased concentration of calcium ions on the frog heart. If the frog heart from the previous experiment is still beating, replace the fluid around it with room temperature Ringer's solution, and wait until its rate is normal. Otherwise, prepare a fresh specimen, and determine its normal rate as before. To perform the test:

a. Flood the frog heart with 2% calcium chloride.

b. Record the heartbeat for about five minutes and note any change in rate.

c. Flood the heart with fresh Ringer's solution until heart rate returns to normal.

9. Test the effect of an increased concentration of potassium ions on the frog heart. To do this:

a. Flood the heart with 5% potassium chloride.

b. Record the heartbeat for about five minutes, and note any change in rate.

10. Complete Part C of the laboratory report.

Optional Activity

Plan an experiment to test the effect of some additional factor on the action of a frog heart. For example, you might test the effect of epinephrine, acetylcholine, caffeine, or some other available substance. If the laboratory instructor approves your plan, perform the experiment and record the heart movements. What do you conclude from the results of your experiment? _____
_____
_____
_____

laboratory report 50

Name _____

Date _____

Section _____

Factors Affecting the Cardiac Cycle

Part A

Complete the following statements:

1. The primary function of the heart is to _pump blood (circulation)_ .

2. The _various receptors_ normally controls the heart rate.

3. Parasympathetic nerve fibers that supply the heart make up part of the _____ nerve.

4. Endings of parasympathetic nerve fibers secrete _____ , which causes the heart rate to decrease.

5. Sympathetic nerve fibers reach the heart by means of _____ nerves.

6. Endings of sympathetic nerve fibers secrete _noraepanephrin_ _adrenalin_ , which causes the heart rate to increase.

7. The cardiac center is located in the _medulla_ of the brain stem.

8. Pressoreceptors located in the walls of the aorta and carotid arteries are sensitive to changes in _blood pressure_ .

9. If pressoreceptors in the walls of the venae cavae are stimulated, the cardiac center sends _____ impulses to the heart.

10. Rising body temperature usually causes the heart rate to _increase_ .

11. Of the ions that affect heart action, the most important are ions of _Ca_ and _K_ .

Part B

1. Describe the actions of the frog heart chambers during a cardiac cycle. _____
   _____
   _____
   _____
   _____

2. Attach a short segment of the normal frog heart recording in the space below. Label the atrial and ventricular waves of one cardiac cycle.

3. Temperature effect results:

Temperature	Heart Rate
10°C	
Room temperature	
35°C	

4. Write a statement that summarizes the effect of temperature on the frog's heart action that was demonstrated by this experiment. _____
   _____
   _____
   _____
   _____
   _____

Part C

Complete the following:

1. Describe the effect of an increased calcium ion concentration on the frog's heart rate. _____

   Hypercalcemia → increase heart rate, can cause prolong contraction

2. Describe the effect of an increased potassium ion concentration on the frog's heart rate. _____

   Hyper Kalemia → decrease heart rate of force of contraction
   very high — cause impulses block → cardiac arrest (stops)

3. In testing the effects of different ions on heart action, why were chlorides used in each case? _____

_−K_
Hypokalemia − abnormal heart rate (arrhythmia)

_−C_
Hypocalcemia − decrease heart rate

Laboratory Exercise 51

Blood Vessels

The blood vessels form a closed system of tubes that carry blood to and from the heart, lungs, and body cells. These tubes include arteries and arterioles that conduct blood away from the heart, capillaries in which exchanges of substances occur between the blood and surrounding tissues, and venules and veins that return blood to the heart.

Purpose of the Exercise

To review the structure and functions of blood vessels and to observe examples of blood vessels microscopically.

Learning Objectives

After completing this exercise, you should be able to:

1. Describe the structure and functions of arteries, capillaries, and veins.
2. Distinguish cross sections of arteries and veins microscopically.
3. Identify the three major layers in the wall of an artery or vein.
4. Identify the types of blood vessels in the web of a frog's foot.

Materials Needed:

compound microscope
prepared microscope slides:
   artery cross section
   vein cross section
live frog
frog Ringer's solution
paper towel
rubber bands

frog board or
   heavy cardboard
ice
dissecting pins
thread
masking tape

For Optional Activity:

ice
hot plate

Procedure

1. Review the section entitled "Blood Vessels" in chapter 18 of the textbook.
2. As a review activity, label figures 51.1 and 51.2.
3. Complete Part A of Laboratory Report 51 on page 348.
4. Obtain a microscope slide of an artery cross section, and examine it using low- and high-power magnification. Identify the three distinct layers of the arterial wall. The inner layer, or *endothelium,* is composed of simple squamous epithelium, and appears as a wavy line due to an abundance of elastic fibers that have recoiled just beneath it. The middle layer consists of numerous concentrically arranged smooth muscle cells with elastic fibers scattered among them. The outer layer contains connective tissue that is rich in collagenous fibers (see figure 51.3).
5. Prepare a labeled sketch of the arterial wall in Part B of the laboratory report.
6. Obtain a slide of a vein cross section, and examine it as you did the artery cross section. Identify the three layers of the wall, and prepare a labeled sketch in Part B of the laboratory report.
7. Complete Part B of the laboratory report.
8. Observe the blood vessels in the webbing of a frog's foot. To do this:
   a. Obtain a live frog, and wrap its body in a moist paper towel, leaving one foot extending outward. Secure the towel with rubber bands.
   b. Place the frog on a piece of heavy cardboard with the foot near the hole in one corner.
   c. Fasten the wrapped body to the board with masking tape.

Figure 51.1 Label the wall of this artery by placing the correct numbers in the spaces provided.

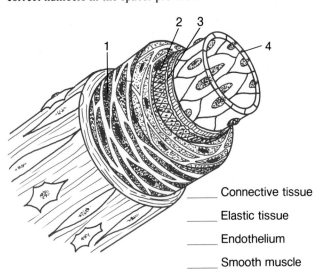

_____ Connective tissue

_____ Elastic tissue

_____ Endothelium

_____ Smooth muscle

Figure 51.2 Label this arteriole by placing the correct numbers in the spaces provided.

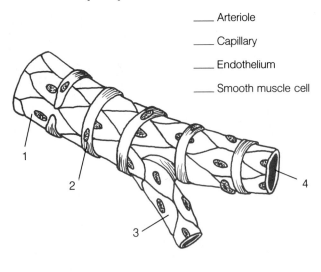

_____ Arteriole

_____ Capillary

_____ Endothelium

_____ Smooth muscle cell

Figure 51.3 Cross section of an artery and a vein.

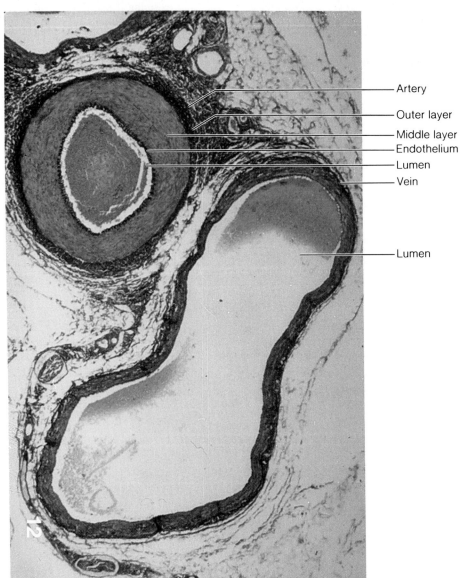

Artery

Outer layer

Middle layer

Endothelium

Lumen

Vein

Lumen

Figure 51.4 Spread the web of the foot over the hole and secure it to the board with pins and thread.

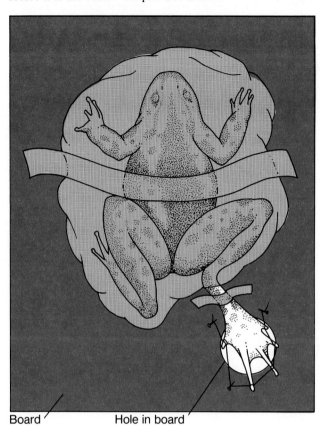

Board   Hole in board

d. Carefully spread the web of the foot over the hole, and secure it to the board with dissecting pins and thread (figure 51.4). Keep the web moist with frog Ringer's solution.
e. Secure the board on the stage of a microscope with heavy rubber bands, and position it so that the web is beneath the objective lens.
f. Focus on the web, using low-power magnification, and locate some blood vessels. Note the movement of the blood cells and the direction of the blood flow. Identify an arteriole, a capillary, and a venule.
g. Examine each of these vessels with high-power magnification.
9. Complete Part C of the laboratory report.

Optional Activity

Investigate the effect of temperature change on the blood vessels of the frog's foot by flooding the web with a small quantity of ice water. Observe the blood vessels with low-power magnification, and note any changes in their diameters or the rate of blood flow. Remove the ice water, and replace it with water heated to about 35°C. Repeat your observations. What do you conclude from this experiment?

_____
_____
_____
_____

Name _____

Date _____

Section _____

Blood Vessels

Part A

Complete the following statements:

1. Tissue called _____ forms the inner linings of blood vessels.

2. The _____ of an artery wall contains many smooth muscle cells.

3. The _____ of an artery wall is composed largely of connective tissue.

4. When the smooth muscle in a blood vessel wall contracts, the vessel is said to be _____ .

5. When the smooth muscle in a blood vessel wall relaxes, the vessel is said to be _____ .

6. A vessel that connects an arteriole directly to a venule, thus bypassing the capillaries, is called a(an) _____ .

7. The smallest blood vessels are called _____ .

8. The pores of capillary walls within muscle tissues are relatively _____ .

9. The resistance to movement of substances between the capillaries and tissues of the brain is called the _____ .

10. Tissues with high rates of metabolism tend to have _____ densities of capillaries.

11. _____ are composed of smooth muscles that encircle the entrances to capillaries and thus can control the distribution of blood within tissues.

12. The process called _____ provides the most important means of transfer through capillary walls.

13. Oxygen and carbon dioxide move easily through cell membranes because they are soluble in _____ .

14. Water moves through capillary walls by passing through _____ .

15. Filtration results when substances are forced through capillary walls by _____ .

16. The presence of plasma proteins in blood increases its _____ pressure.

17. Excess tissue fluid is returned to the venous circulation by means of _____ .

18. The substance called _____ causes an increase in capillary membrane permeability.

19. _____ in certain veins prevents blood from flowing away from the heart.

20. _____ are vessels that serve as blood reservoirs.

Part B

1. Sketch and label a section of an arterial wall.

2. Sketch and label a section of a venous wall.

3. Describe the differences you noted in the structures of the arterial and venous walls. Mention each of the three layers of the wall. _____

_____

_____

_____

_____

_____

Part C

Complete the following:

1. How did you distinguish between arterioles and venules when you observed the vessels in the web of the frog's foot? _____

_____

_____

_____

_____

_____

2. How did you recognize capillaries in the web? _____

_____

_____

_____

_____

3. What differences did you note in the rate of blood flow through the arterioles, capillaries, and venules? _____

_____

_____

_____

_____

4. Did you observe any evidence of precapillary sphincter activity? Explain your answer. _____

_____

_____

_____

_____

Laboratory Exercise 52

Pulse Rate and Blood Pressure

The surge of blood that enters the arteries each time the ventricles of the heart contract causes the elastic walls of these vessels to swell. Then, as the ventricles relax, the walls recoil. This alternate expanding and recoiling of an arterial wall can be felt as a pulse in vessels that run close to the surface of the body.

The force exerted by the blood pressing against the inner walls of arteries also creates blood pressure. This pressure reaches a maximum during ventricular contraction, and then drops to its lowest level while the ventricles are relaxed.

Purpose of the Exercise

To examine the pulse, determine the pulse rate, measure blood pressure, and investigate the effects of body position and exercise on pulse rate and blood pressure.

Learning Objectives

After completing this exercise, you should be able to:

1. Determine the pulse rate.
2. Test the effect of various factors on pulse rate.
3. Measure blood pressure, using a sphygmomanometer.
4. Test the effect of various factors on blood pressure.
5. Calculate pulse pressure and mean arterial pressure from blood pressure readings.

Materials Needed:
  sphygmomanometer
  stethoscope
  70% alcohol
  absorbent cotton

For Demonstration:

  pulse pick-up transducer or plethysmogram
  physiological recording apparatus

Procedure

1. Review the sections entitled "Arterial Blood Pressure" and "Measurement of Arterial Blood Pressure—A Laboratory Application" in chapter 18 of the textbook.
2. Complete Part A of Laboratory Report 52 on page 355.
3. Examine your laboratory partner's radial pulse. To do this:
   a. Have your partner sit quietly, remaining as relaxed as possible.
   b. Locate the pulse by placing your index and middle fingers over the radial artery on the anterior surface of the wrist. Do not use your thumb for sensing the pulse, because you may feel a pulse coming from an artery in the thumb itself.
   c. Note the characteristics of the pulse. That is, could it be described as regular or irregular, strong or weak, hard or soft?
   d. To determine the pulse rate, count the number of pulses that occur in one minute.
4. Repeat the procedure, and determine the pulse rate in each of the following conditions:
   a. immediately after lying down,
   b. five minutes after lying down,
   c. immediately after standing,
   d. five minutes after standing,
   e. immediately after three minutes of strenuous exercise, and
   f. five minutes after exercise.
5. Complete Part B of the laboratory report.

Figure 52.1 Blood pressure commonly is measured using a sphygmomanometer.

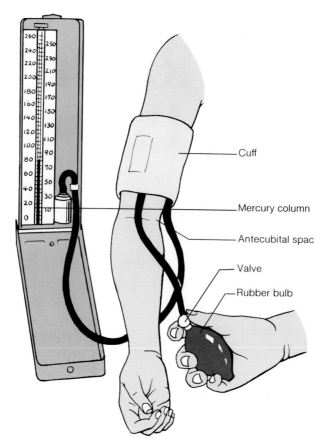

Cuff

Mercury column

Antecubital spac

Valve

Rubber bulb

Demonstration

If the equipment is available, the laboratory instructor will demonstrate how a photoelectric pulse pick-up transducer or plethysmogram can be used together with a physiological recording apparatus to record the pulse. Such a recording allows an investigator to analyze certain characteristics of the pulse more precisely than is possible using a finger to examine the pulse. For example, the pulse rate can be determined very accurately from a recording, and the heights of the pulse waves provide information concerning the blood pressure.

6. Measure your laboratory partner's arterial blood pressure. To do this:
   a. Obtain a sphygmomanometer and a stethoscope.
   b. Clean the earpieces of the stethoscope with cotton moistened in alcohol.
   c. Have your partner sit quietly, remaining as relaxed as possible.
   d. Wrap the cuff of the sphygmomanometer around the upper left arm so that its lower border is about 2.5 cm above the bend inside the elbow (figure 52.1).

   e. Locate the brachial artery by palpating its pulse in the antecubital space in front of the elbow.
   f. Place the diaphragm of the stethoscope over the brachial artery so that you can hear any sounds (Korotkoff sounds) coming from it.
   g. Close the valve on the neck of the rubber bulb connected to the cuff, and pump air from the bulb into the cuff.
   h. When sounds from the brachial artery disappear, watch the pressure indicator or mercury column, and add about 20 mm Hg more pressure to the cuff.
   i. Open the valve of the bulb slowly until the pressure in the cuff drops at a rate of about 2 or 3 mm Hg per second.
   j. Listen for sounds from the brachial artery, and when the first loud tapping sound is heard, record the reading as the systolic pressure.
   k. Continue to listen to the sounds as the pressure drops, and when the sounds suddenly become muffled, record the reading as the diastolic pressure.

l. Release all of the pressure from the cuff.

m. Repeat the procedure until you have three blood pressure measurements, allowing two to three minutes of rest between readings.

n. Average your readings, and enter them in the chart of Part C of the laboratory report.

7. Measure your partner's blood pressure in each of the following conditions:

a. immediately after lying down,

b. five minutes after lying down,

c. immediately after standing,

d. five minutes after standing,

e. immediately after three minutes of strenuous exercise, and

f. five minutes after exercise.

8. Complete Part C of the laboratory report.

laboratory report 52

Name _____

Date _____

Section _____

Pulse Rate and Blood Pressure

Part A

Complete the following statements:

1. The term blood pressure most commonly is used to refer to _____ pressure.

2. The maximum pressure achieved during ventricular contraction is called _____ .

3. The lowest pressure that remains in the arterial system during ventricular relaxation is called

   _____ .

4. The pulse rate is equal to the _____ rate.

5. A pulse that feels full is produced by an elevated _____ .

6. The instrument commonly used to measure systemic arterial blood pressure is called a(an)

   _____ .

7. Blood pressure is expressed in units of _____ .

8. The upper number of the fraction used to record blood pressure indicates the _____ pressure.

9. The difference between the systolic and diastolic pressure is called the _____ .

10. The mean arterial pressure is approximated by adding the _____ pressure to one-third of the pulse pressure.

Part B

1. Enter your observations of pulse characteristics and pulse rates in the chart:

Test Subject	Pulse Characteristics	Pulse Rate
Sitting		
Lying down		
5 minutes later		
Standing		
5 minutes later		
After exercise		
5 minutes later		

2. Write a statement to summarize the effects of body position and exercise on the characteristics and rates of the pulse.

_____

_____

_____

_____

_____

_____

_____

_____

Part C

1. Enter the initial measurements of blood pressure in the chart:

Reading	Blood Pressure
First	
Second	
Third	
Average	

2. Enter your test results in the chart:

Test Subject	Blood Pressure
Lying down	
5 minutes later	
Standing	
5 minutes later	
After exercise	
5 minutes later	

3. Write a statement to summarize the effects of body position and exercise on blood pressure. _____

_____

_____

_____

_____

_____

_____

Major Arteries and Veins

The blood vessels of the cardiovascular system can be divided into two major pathways—the pulmonary circuit and the systemic circuit. Within each circuit arteries transport blood away from the heart. After exchanges of gases, nutrients, and wastes have occurred between the blood and the surrounding tissues, veins return the blood to the heart.

Purpose of the Exercise

To review the major circulatory pathways and to locate the major arteries and veins on anatomical charts and in the manikin.

Learning Objectives

After completing this exercise, you should be able to:

1. Describe the pulmonary and systemic circuits.
2. Locate the major arteries in these circuits on a chart or model.
3. Locate the major veins in these circuits on a chart or model.

Materials Needed:

human manikin
anatomical charts of the circulatory system

Procedure A—The Arterial System

1. Review the section entitled "Arterial System" in chapter 18 of the textbook.
2. As a review activity, label figures 53.1, 53.2, 53.3, 53.4, 53.5, and 53.6.
3. Locate the following arteries on the charts and manikin:

aorta

   ascending aorta

   aortic sinus

arch of aorta

  thoracic aorta

  abdominal aorta

branches of the aorta

  coronary arteries

  brachiocephalic artery

  left common carotid artery

  left subclavian artery

  bronchial artery

  pericardial artery

  esophageal artery

  mediastinal artery

  posterior intercostal artery

  celiac artery

  gastric artery

  splenic artery

  hepatic artery

  phrenic arteries

  superior mesenteric artery

  suprarenal arteries

  renal arteries

  gonadal arteries

    ovarian arteries

    spermatic arteries

  inferior mesenteric artery

  lumbar arteries

  middle sacral artery

  common iliac arteries

Figure 53.1 Label the portions of the aorta and its principal branches.

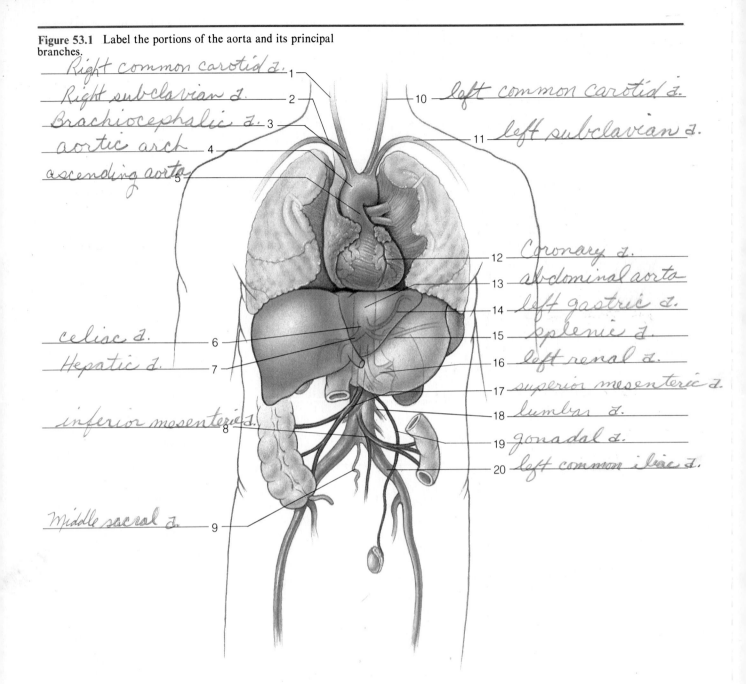

Right common carotid a. — 1
Right subclavian a. — 2
Brachiocephalic a. — 3
aortic arch — 4
ascending aorta — 5

10 — *left common carotid a.*
11 — *left subclavian a.*

12 — *Coronary a.*
13 — *abdominal aorta*
14 — *left gastric a.*
15 — *splenic a.*
16 — *left renal a.*
17 — *superior mesenteric a.*
18 — *lumbar a.*
19 — *gonadal a.*
20 — *left common iliac a.*

celiac a. — 6
Hepatic a. — 7

inferior mesenteric a. — 8

Middle sacral a. — 9

arteries to neck, head, and brain

  vertebral arteries

  basilar artery

  posterior cerebral artery

  thyrocervical arteries

  costocervical arteries

  external carotid arteries

  superior thyroid artery

  lingual artery

facial artery

occipital artery

posterior auricular artery

maxillary artery

superficial temporal artery

internal carotid arteries

ophthalmic artery

posterior communicating artery

anterior choroid artery

Figure 53.2 Label the arteries supplying the neck and head.

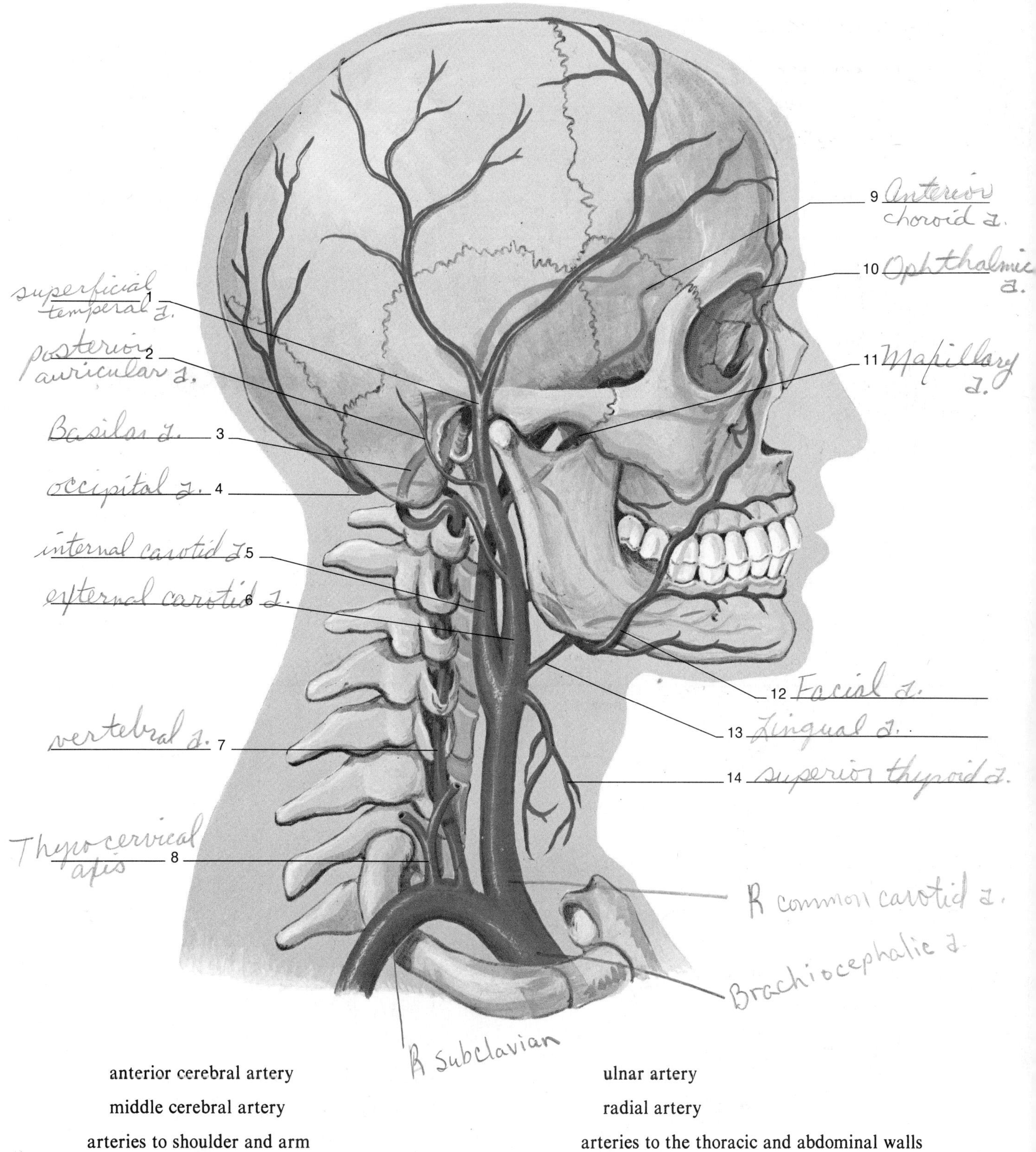

superficial temporal a. 1

posterior auricular a. 2

Basilar a. 3

occipital a. 4

internal carotid a. 5

external carotid a. 6

vertebral a. 7

Thyrocervical axis 8

9 Anterior choroid a.

10 Ophthalmic a.

11 Maxillary a.

12 Facial a.

13 Lingual a.

14 superior thyroid a.

R common carotid a.

Brachiocephalic a.

R Subclavian

anterior cerebral artery

middle cerebral artery

arteries to shoulder and arm

axillary artery

brachial artery

deep brachial artery

ulnar artery

radial artery

arteries to the thoracic and abdominal walls

internal thoracic artery

anterior intercostal artery

posterior intercostal artery

359

Figure 53.3 Label the major arteries of the shoulder and arm.

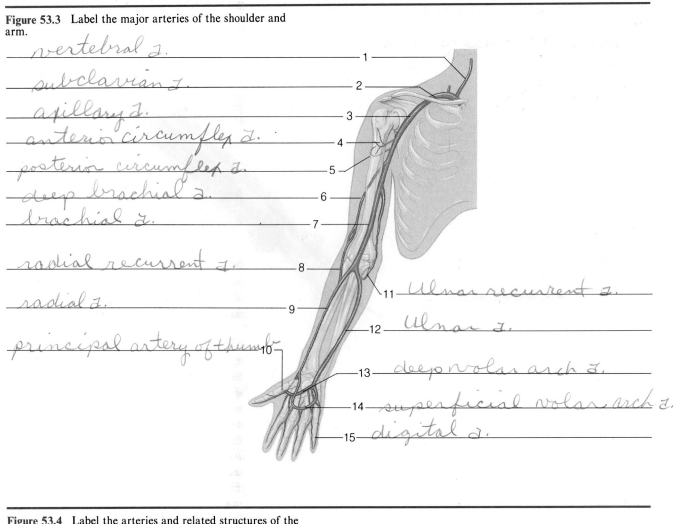

vertebral a.

subclavian a.

axillary a.

anterior circumflex a.

posterior circumflex a.

deep brachial a.

brachial a.

radial recurrent a.

radial a.

principal artery of thumb

Ulnar recurrent a.

Ulnar a.

deep volar arch a.

superficial volar arch a.

digital a.

Figure 53.4 Label the arteries and related structures of the thoracic wall.

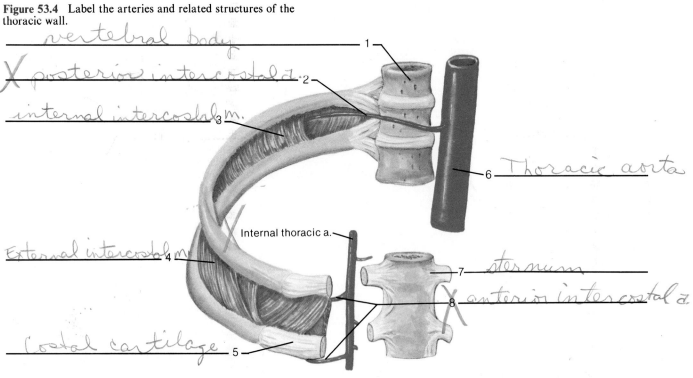

vertebral body

X posterior intercostal a.

internal intercostal m.

External intercostal m.

Costal cartilage

Internal thoracic a.

Thoracic aorta

sternum

X anterior intercostal a.

Figure 53.5 Label the arteries associated with the pelvic cavity.

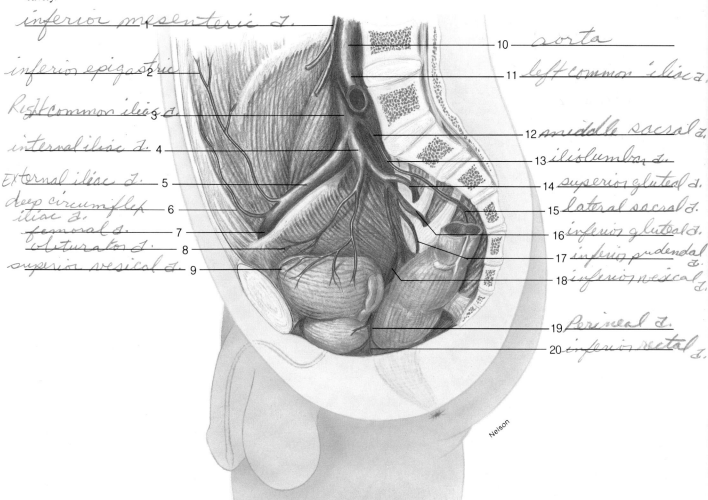

inferior mesenteric a. — 1

inferior epigastric — 2

Right common iliac a. — 3

internal iliac a. — 4

External iliac a. — 5

deep circumflex iliac a. — 6

femoral a. — 7

obturator a. — 8

superior vesical a. — 9

10 — aorta

11 — left common iliac a.

12 — middle sacral a.

13 — iliolumbar a.

14 — superior gluteal a.

15 — lateral sacral a.

16 — inferior gluteal a.

17 — inferior pudendal a.

18 — inferior vesical a.

19 — Perineal a.

20 — inferior rectal a.

Nelson

arteries to pelvis and leg

  common iliac artery

  internal iliac artery

  iliolumbar artery

  superior gluteal artery

  inferior gluteal artery

  internal pudendal artery

  superior vesical artery

  inferior vesical artery

  middle rectal artery

  uterine artery

  external iliac artery

  inferior epigastric artery

  deep circumflex iliac artery

  femoral artery

superficial circumflex iliac artery

superficial epigastric artery

superficial external pudendal artery

deep external pudendal artery

profunda femoris artery

deep genicular artery

popliteal artery

anterior tibial artery

dorsalis pedis artery

posterior tibial artery

peroneal artery

plantar artery

4. Complete Parts A and B of Laboratory Report 53 on page 367.

Figure 53.6 Label the major arteries supplying the pelvis and leg.

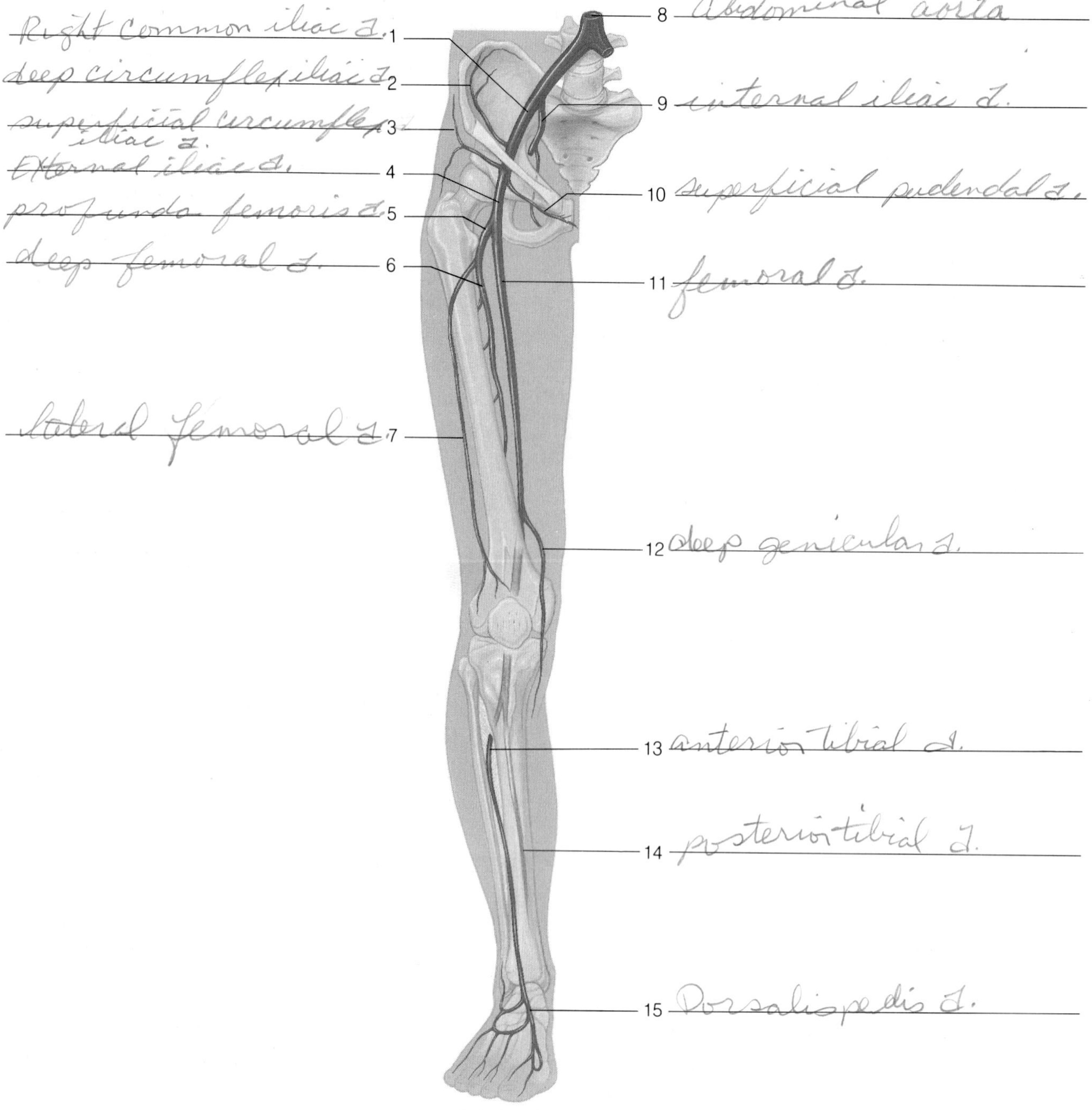

Right common iliac a. — 1

deep circumflex iliac a. — 2

superficial circumflex iliac a. — 3

External iliac a. — 4

profunda femoris a. — 5

deep femoral a. — 6

lateral femoral a. — 7

8 — Abdominal aorta

9 — internal iliac a.

10 — superficial pudendal a.

11 — femoral a.

12 — deep genicular a.

13 — anterior tibial a.

14 — posterior tibial a.

15 — Dorsalis pedis a.

Anterior view

Procedure B—The Venous System

1. Review the section entitled "Venous System" in chapter 18 of the textbook.
2. As a review activity, label figures 53.7, 53.8, 53.9, 53.10, and 53.11.
3. Locate the following veins on the charts and the manikin:

   veins from head, neck, and brain

   external jugular veins

   subclavian veins

   internal jugular veins

   brachiocephalic veins

   superior vena cava

   veins from arm and shoulder

   radial vein

   ulnar vein

   brachial vein

Figure 53.7 Label the major veins of the head and neck.

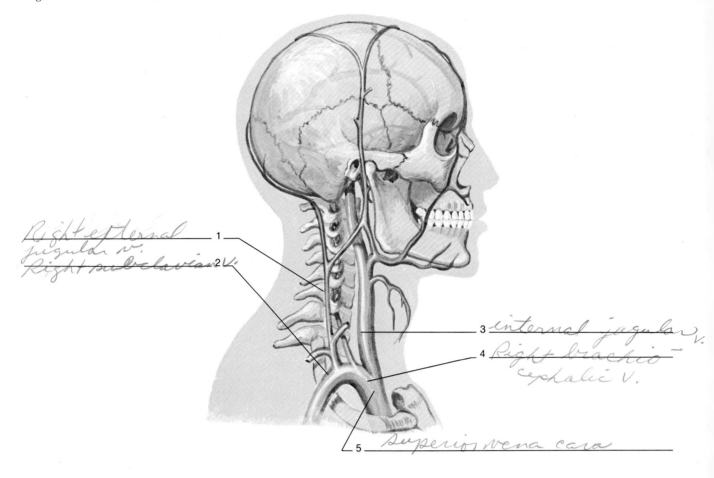

Right external jugular v.
Right subclavian v. 2

1

3 internal jugular v.
4 Right brachio- cephalic v.

5 Superior vena cava

Figure 53.8 Label the veins associated with the thoracic wall.

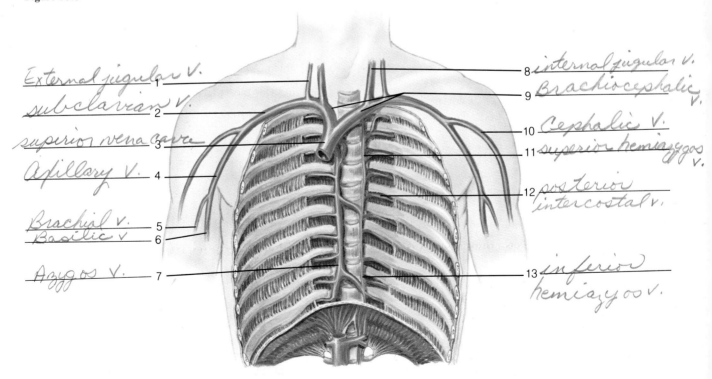

External jugular V. 1
subclavian V. 2
superior vena cava 3
Axillary V. 4

Brachial V. 5
Basilic V. 6

Azygos V. 7

8 internal jugular V.
9 Brachiocephalic V.

10 Cephalic V.
11 superior hemiazygos V.

12 posterior intercostal v.

13 inferior hemiazygos V.

Figure 53.9 Label the veins of the arm and shoulder.

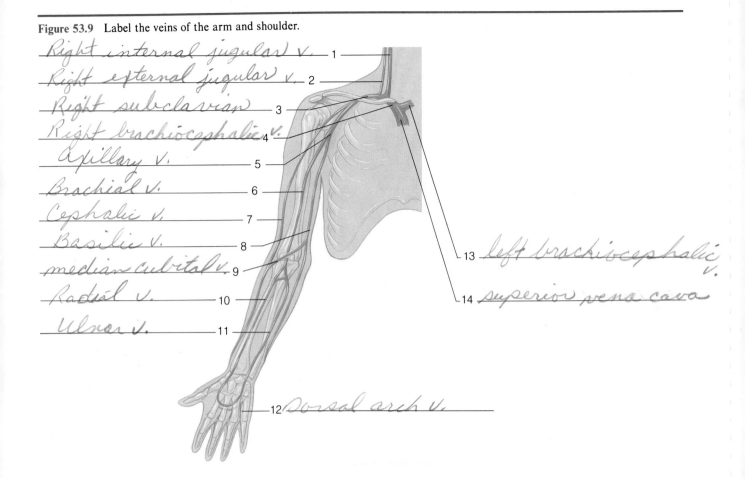

1 — Right internal jugular v.
2 — Right external jugular v.
3 — Right subclavian
4 — Right brachiocephalic v.
5 — Axillary v.
6 — Brachial v.
7 — Cephalic v.
8 — Basilic v.
9 — median cubital v.
10 — Radial v.
11 — Ulnar v.
12 — Dorsal arch v.
13 — left brachiocephalic v.
14 — superior vena cava

basilic vein

brachial vein

axillary vein

cephalic vein

median cubital vein

subclavian vein

veins of the abdominal and thoracic walls

internal thoracic veins

intercostal veins

azygos vein

posterior intercostal veins

superior hemiazygos vein

inferior hemiazygos vein

ascending lumbar veins

veins of the abdominal viscera

portal vein

gastric vein

superior mesenteric vein

splenic vein

inferior mesenteric vein

hepatic vein

lumbar vein

gonadal vein

renal vein

suprarenal vein

phrenic vein

Figure 53.10   Label the veins that drain the abdominal viscera.

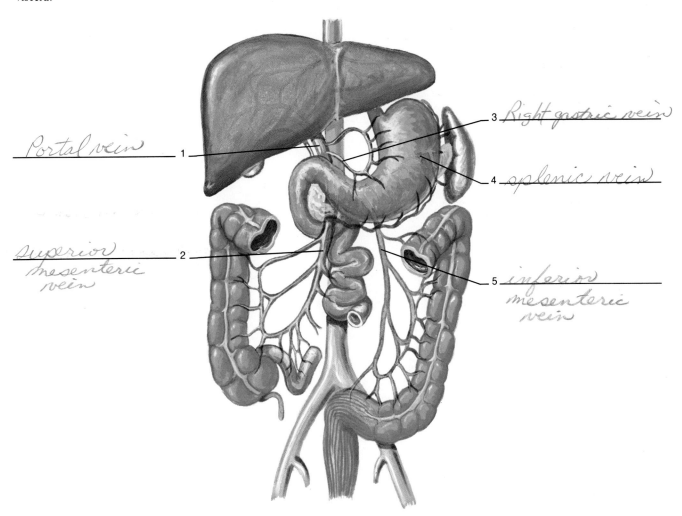

Portal vein — 1

superior mesenteric vein — 2

3 — Right gastric vein

4 — splenic vein

5 — inferior mesenteric vein

veins from leg and pelvis

anterior tibial vein

posterior tibial vein

popliteal vein

femoral vein

great saphenous vein

small saphenous vein

external iliac vein

internal iliac vein

gluteal vein

pudendal vein

vesical vein

rectal vein

uterine vein

vaginal vein

common iliac vein

inferior vena cava

4. Complete Parts C and D of the laboratory report.

Figure 53.11 Label the veins of the pelvis and leg.

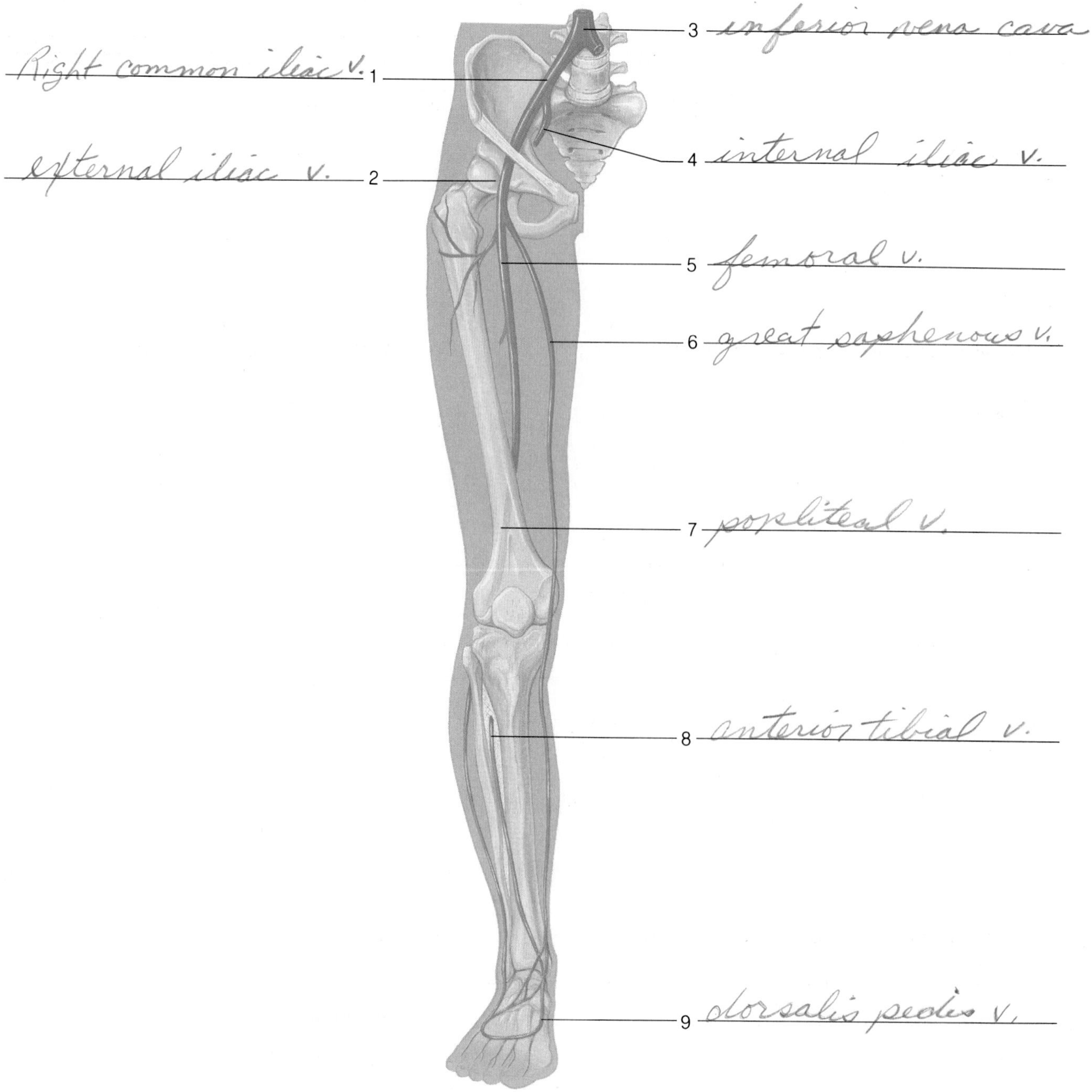

Right common iliac v. 1

external iliac v. 2

3 inferior vena cava

4 internal iliac v.

5 femoral v.

6 great saphenous v.

7 popliteal v.

8 anterior tibial v.

9 dorsalis pedis v.

Anterior view

Name _____

Date _____

Section _____

Major Arteries and Veins

Part A

Match the arteries in column A with the regions supplied in column B. Place the letter of your choice in the space provided.

Column A		Column B
A. anterior tibial	___ 1.	jaw and teeth
B. brachial	___ 2.	larynx and trachea
C. celiac	___ 3.	kidney
D. costocervical	___ 4.	upper digestive tract
E. external carotid	___ 5.	foot and toes
F. inferior mesenteric	___ 6.	gluteal muscles
G. internal carotid	___ 7.	triceps muscle
H. internal iliac	___ 8.	thoracic wall
I. lumbar	___ 9.	posterior abdominal wall
J. phrenic	___ 10.	adrenal gland
K. popliteal	___ 11.	diaphragm
L. renal	___ 12.	lower colon
M. suprarenal	___ 13.	brain
N. thyrocervical	___ 14.	lower arm muscles
O. ulnar	___ 15.	knee joint

Part B

Provide the name of the missing artery in each of the following sequences:

1. brachiocephalic artery, _____ , axillary artery

2. ascending aorta, _____ , thoracic aorta

3. abdominal aorta, _____ , ascending colon

4. abdominal aorta, _____ , descending colon

5. brachiocephalic artery, _____ , external carotid artery

6. subclavian artery, _____ , basilar artery

7. external carotid artery, _____ , lips

8. internal carotid artery, _____ , eye

9. axillary artery, _____ , radial artery

10. common iliac artery, _____ , femoral artery

Part C

Match each vein in column A with the vein it drains into from column B. Place the letter of your choice in the space provided.

Column A		Column B
A. anterior tibial	___ 1.	popliteal
B. basilic	___ 2.	axillary
C. brachiocephalic	___ 3.	inferior vena cava
D. common iliac	___ 4.	subclavian
E. external jugular	___ 5.	brachial
F. femoral	___ 6.	superior vena cava
G. popliteal	___ 7.	femoral
H. radial	___ 8.	external iliac

Part D

Provide the name of the missing vein or veins in each of the following sequences:

1. right axillary vein, _____ , superior vena cava

2. descending colon, _____ , portal vein

3. posterior tibial vein, _____ , femoral vein

4. interior iliac vein, _____ , inferior vena cava

5. medial cubital vein, _____ , axillary vein

6. cephalic vein, _____ , brachiocephalic vein

7. stomach, _____ , portal vein

8. dorsalis pedis vein, _____ , popliteal vein

9. great saphenous vein, _____ , common iliac vein

10. liver, _____ , inferior vena cava

Cat Dissection: Circulatory System

In this laboratory exercise, you will dissect the major organs of the cat circulatory system. As before, while you are examining the organs of the cat, compare them with the corresponding organs of the human manikin.

If the circulatory system of the cat has been injected, the arteries will be filled with red latex and the veins will be filled with blue latex. This will make it easier for you to trace the vessels as you dissect them.

Purpose of the Exercise

To examine the major organs of the cat circulatory system and to compare them with the corresponding organs of the human manikin.

Learning Objectives

After completing this exercise, you should be able to:

1. Locate and identify the major organs of the cat circulatory system.
2. Identify the corresponding organs in the human manikin.
3. Compare the features of the cat circulatory system with those of the human.

Materials Needed:
    embalmed cat
    dissecting tray
    dissecting instruments
    disposable gloves
    human manikin

Procedure A—The Arterial System

1. Place the embalmed cat in the dissecting tray with its ventral side up.
2. Open the thoracic cavity. If this was not done previously, refer to Laboratory Exercise 42, step 5 of the procedure, for directions.
3. Note the location of the heart and the large blood vessels associated with it. Slit the thick *parietal pericardium* that surrounds the heart by cutting with scissors along the midventral line. Note how this membrane is connected to the *visceral pericardium* that is attached to the surface of the heart. Locate the *pericardial cavity,* the potential space between the two layers of the pericardium.
4. Examine the heart (see figures 48.1, 48.2, and 48.3 as well as reference plates 15 and 17), and locate the following:

   right atrium

   left atrium

   right ventricle

   left ventricle

   pulmonary trunk

   aorta

   coronary arteries

5. Use a scalpel to open the heart chambers by making a cut along the coronal plane from its apex to its base. Remove any remaining latex from the chambers. Examine the valves between the chambers, and note the relative thicknesses of the chamber walls. (See reference plate 18.)

Figure 54.1 Arteries of the cat's thorax, neck, and forelimb.

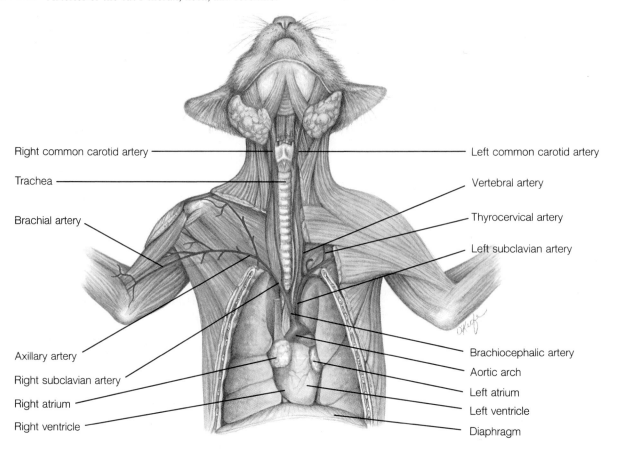

Right common carotid artery

Trachea

Brachial artery

Axillary artery

Right subclavian artery

Right atrium

Right ventricle

Left common carotid artery

Vertebral artery

Thyrocervical artery

Left subclavian artery

Brachiocephalic artery

Aortic arch

Left atrium

Left ventricle

Diaphragm

6. Using figure 54.1 and reference plates 17 and 18 as guides, locate and dissect the following arteries of the thorax and neck:

arch of aorta

brachiocephalic artery (on right only)

right subclavian artery

left subclavian artery

right common carotid artery

left common carotid artery

7. Trace the right subclavian artery into the forelimb, and locate the following arteries:

vertebral artery

thyrocervical artery

subscapular artery

axillary artery

brachial artery

radial artery

ulnar artery

8. Open the abdominal cavity. If this was not done previously, refer to Laboratory Exercise 39, step 6 of the procedure, for directions.

9. Using figure 54.2 and reference plate 19 as guides, locate and dissect the following arteries of the abdomen:

abdominal aorta

celiac artery

superior mesenteric artery

renal arteries

lumbar arteries

gonadal arteries

inferior mesenteric artery

external iliac artery

internal iliac artery

10. Trace the external iliac artery into the right hind limb (see reference plate 20), and locate the following:

femoral artery

deep femoral artery

11. Complete Part A of Laboratory Report 54 on page 375.

Figure 54.2 Arteries of the cat's abdomen and hind limb.

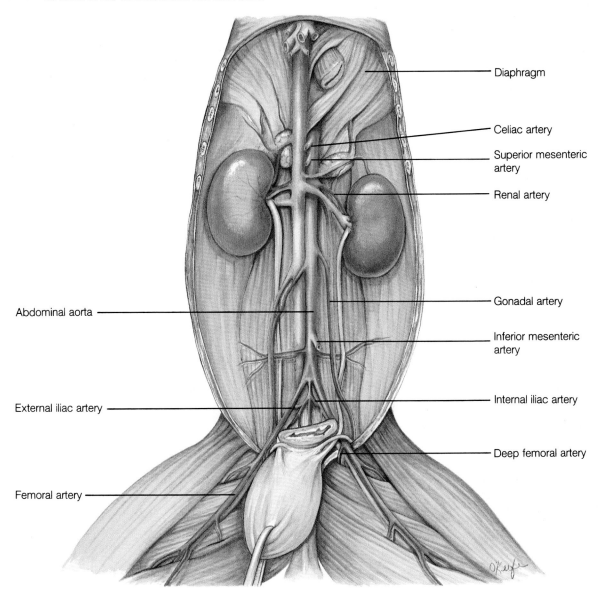

Diaphragm

Celiac artery

Superior mesenteric artery

Renal artery

Gonadal artery

Inferior mesenteric artery

Internal iliac artery

Deep femoral artery

Abdominal aorta

External iliac artery

Femoral artery

Figure 54.3 Veins of the cat's thorax, neck, and forelimb.

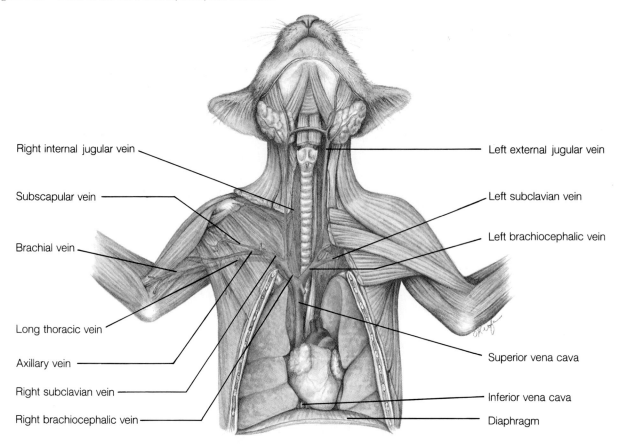

Right internal jugular vein

Subscapular vein

Brachial vein

Long thoracic vein

Axillary vein

Right subclavian vein

Right brachiocephalic vein

Left external jugular vein

Left subclavian vein

Left brachiocephalic vein

Superior vena cava

Inferior vena cava

Diaphragm

Procedure B—The Venous System

1. Examine the heart again, and locate the following veins:

   superior vena cava

   inferior vena cava

   pulmonary veins

2. Using figure 54.3 and reference plates 17 and 18 as guides, locate and dissect the following veins in the thorax and neck:

   right brachiocephalic vein

   left brachiocephalic vein

   right subclavian vein

   left subclavian vein

   internal jugular vein

   external jugular vein

3. Trace the right subclavian vein into the forelimb, and locate the following veins:

   axillary vein

   subscapular vein

   brachial vein

   long thoracic vein

4. Using figure 54.4 and reference plate 19 as guides, locate and dissect the following veins in the abdomen:

   inferior vena cava

   adrenolumbar veins

   renal veins

   lumbar veins

   common iliac veins

   internal iliac vein

   external iliac vein

5. Trace the external iliac vein into the hind limb (see reference plate 20), and locate the following veins:

   femoral vein

   deep femoral vein

6. Complete Part B of the laboratory report.

Figure 54.4 Veins of the cat's abdomen and hind limb.

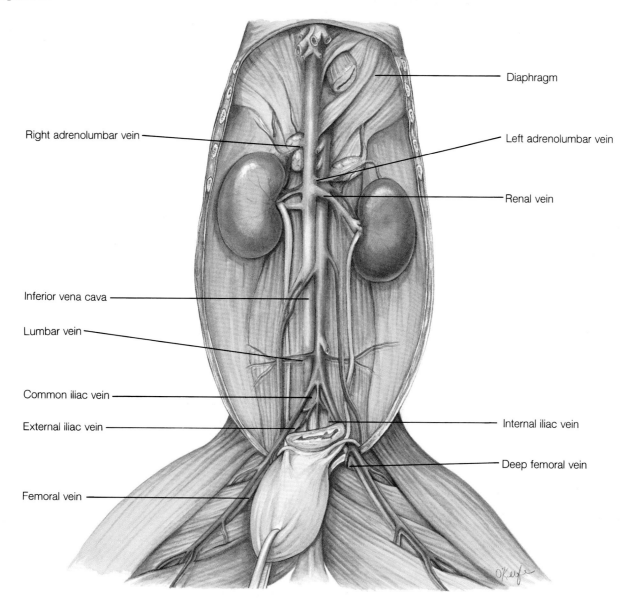

Right adrenolumbar vein

Inferior vena cava

Lumbar vein

Common iliac vein

External iliac vein

Femoral vein

Diaphragm

Left adrenolumbar vein

Renal vein

Internal iliac vein

Deep femoral vein

Cat Dissection: Circulatory System

Part A

Complete the following:

1. Describe the parietal pericardium of the cat heart. _____
   _____
   _____
   _____

2. Describe the relative thicknesses of the walls of the cat heart chambers. _____
   _____
   _____
   _____

3. Explain how the wall thicknesses are related to the functions of the chambers. _____
   _____
   _____
   _____
   _____
   _____

4. Compare the origins of the common carotid arteries of the cat with those of the human. _____
   _____
   _____
   _____
   _____
   _____

5. Compare the iliac arteries of the cat with those of the human. _____
   _____
   _____
   _____
   _____

Part B

Complete the following:

1. Compare the brachiocephalic veins of the cat with those of the human. _____
_____
_____
_____

2. Compare the jugular veins of the cat with those of the human. _____
_____
_____
_____
_____
_____

Lymphatic System

The lymphatic system is closely associated with the cardiovascular system and includes a network of capillaries and vessels that assist in the circulation of body fluids. These lymphatic capillaries and vessels provide pathways through which excess fluid can be transported away from intercellular spaces within tissues and returned to the bloodstream.

The organs of the lymphatic system also help to protect the tissues against infections by filtering particles from lymph and by supporting the activities of lymphocytes that furnish immunity against specific disease-causing agents.

Purpose of the Exercise

To review the structure of the lymphatic system and to observe the microscopic structure of a lymph node.

Learning Objectives

After completing this exercise, you should be able to:

1. Locate and identify the major lymphatic pathways in an anatomical chart or model.
2. Locate and identify the major chains of lymph nodes in an anatomical chart or model.
3. Describe the structure of a lymph node.
4. Identify the major microscopic parts of a lymph node.

Materials Needed:

human manikin   compound microscope
anatomical chart of the   prepared microscope slide:
     lymphatic system   lymph node section

For Demonstration:

embalmed cat with collecting ducts dissected

Procedure A—Lymphatic Pathways

1. Review the section entitled "Lymphatic Pathways" in chapter 19 of the textbook.
2. As a review activity, label figure 55.1.
3. Complete Part A of Laboratory Report 55 on page 380.
4. Observe the human manikin and the anatomical chart of the lymphatic system, and locate the following lymphatic vessels:

lumbar trunk

intestinal trunk

intercostal trunk

bronchomediastinal trunk

subclavian trunk

jugular trunk

thoracic collecting duct

right lymphatic duct

Demonstration

Examine the collecting ducts in the thorax of the embalmed cat. The *thoracic duct* joins the venous system near the junction of the left subclavian and the left external jugular vein (see figure 54.3). Trace this duct posteriorly, and note its beaded appearance, which is due to the presence of valves in the vessel. These valves prevent the backflow of lymph.

Look for the short *right lymphatic duct*, which joins the venous system near the junction of the right subclavian and the right external jugular vein. How do the collecting ducts compare in size? _____

What is the significance of this observation? _____
_____
_____
_____

Figure 55.1 Label the diagram by placing the correct numbers in the spaces provided.

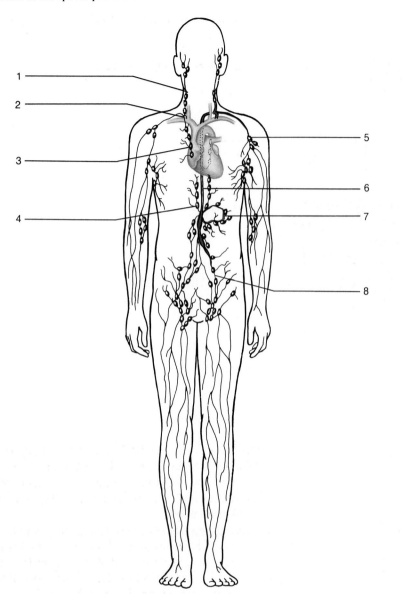

_____ Bronchomediastinal trunk

_____ Intercostal trunk

_____ Intestinal trunk

_____ Jugular trunk

_____ Lumbar trunk

_____ Right lymphatic duct

_____ Subclavian trunk

_____ Thoracic duct

Figure 55.2 Label the diagram by placing the correct numbers in the spaces provided.

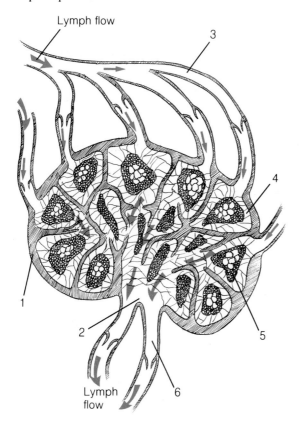

Lymph flow

Lymph flow

_____ Afferent lymphatic vessel

_____ Capsule

_____ Efferent lymphatic vessel

_____ Hilum

_____ Nodule

_____ Sinus

Procedure B—Lymph Nodes

1. Review the section entitled "Lymph Nodes" in chapter 19 of the textbook.
2. As a review activity, label figure 55.2.
3. Complete Part B of the laboratory report.
4. Observe the anatomical chart of the lymphatic system and the human manikin, and locate the clusters of lymph nodes in the following regions:

   cervical region

   axillary region

   inguinal region

   pelvic cavity

   abdominal cavity

   thoracic cavity

5. Palpate the lymph nodes in your cervical region. They are located along the lower border of the mandible and between the ramus of the mandible and the sternocleidomastoid muscle. They feel like small, firm lumps.
6. Obtain a prepared microscope slide of a lymph node and observe it, using low-power magnification. Identify the _capsule_ that surrounds the node and is composed mainly of collagenous fibers, the _nodules_ that appear as dense masses near the surface of the node, and the _sinus_ that appears as a narrow space between the nodules and the capsule.
7. Examine a nodule within the lymph node, using high-power magnification. Note that the nodule contains densely packed _lymphocytes_.

Name _____

Date _____

Section _____

Lymphatic System

Part A

Complete the following statements:

1. Lymphatic pathways begin as _____ .

2. The wall of a lymphatic capillary consists of a single layer of _____ cells.

3. Once tissue fluid is inside a lymph capillary, the fluid is called _____ .

4. The lymph capillaries in the intestinal villi are called _____ .

5. Lymphatic vessels have walls that are similar to those of _____ .

6. Some lymphatic vessels contain _____ that help prevent the backflow of fluid.

7. Lymphatic vessels usually lead to _____ that filter the fluid being transported.

8. The lymphatic trunk that drains the abdominal viscera is called the _____ trunk.

9. The lymphatic trunk that drains the head and neck is called the _____ trunk.

10. The _____ is the largest of the two lymphatic collecting ducts.

Part B

Complete the following statements:

1. Lymph nodes contain large numbers of white blood cells called _____ .

2. The indented region of a bean-shaped lymph node is called the _____ .

3. _____ are the structural units of a lymph node.

4. The spaces within a lymph node are called _____ .

5. Lymph enters a node through a(an) _____ lymphatic vessel.

6. The partially encapsulated lymph nodes in the pharynx are called _____ .

7. The aggregations of lymph nodules found within the inner lining of the small intestine are called

_____ .

8. The lymph nodes in the cervical region are associated with the lymphatic vessels that drain the

_____ .

9. The lymph nodes associated with the lymphatic vessels that drain the legs are located in the _____ region.

10. Lymph nodes contain _____ that engulf and destroy foreign particles, damaged cells, and cellular debris.

Laboratory Exercise 56

Structure of the Kidney

The two kidneys are the primary organs of the urinary system. They are located on each side, high on the posterior wall of the abdominal cavity and behind the parietal peritoneum. They perform a variety of complex activities that lead to the production of urine.

The other organs of the urinary system include the ureters, which transport urine away from the kidneys; the urinary bladder, which stores urine; and the urethra, which conveys urine to the outside of the body.

Purpose of the Exercise

To review the structure of the kidney, to dissect a kidney, and to observe the major parts of a nephron microscopically.

Learning Objectives

After completing this exercise, you should be able to:

1. Describe the location of the kidneys.
2. Locate and identify the major structural components of a kidney.
3. Identify the major features of a nephron.
4. Trace the path of filtrate through a renal tubule.
5. Trace the path of blood through the renal blood vessels.

Materials Needed:
    human manikin
    kidney model
    preserved pig (or sheep) kidney
    dissecting tray
    dissecting instruments
    long knife
    compound microscope
    prepared microscope slide of the following:
        kidney section

Procedure A—Kidney Structure

1. Review the section entitled "Structure of a Kidney" in chapter 20 of the textbook.
2. As a review activity, label figure 56.1.
3. Complete Part A of Laboratory Report 56 on page 385.
4. Observe the human manikin and the kidney model. Locate the following:

    kidneys

    ureters

    urinary bladder

    urethra

    renal sinus

    renal pelvis

    major calyces

    minor calyces

    renal papillae

    renal medulla

    renal pyramids

    renal cortex

    renal columns

    nephrons

5. To observe the structure of a kidney:
    a. Obtain a pig kidney, and rinse it with water to remove as much of the preserving fluid as possible.
    b. Carefully remove any fat from the surface of the specimen.

Figure 56.1 Label the major parts of the kidney.

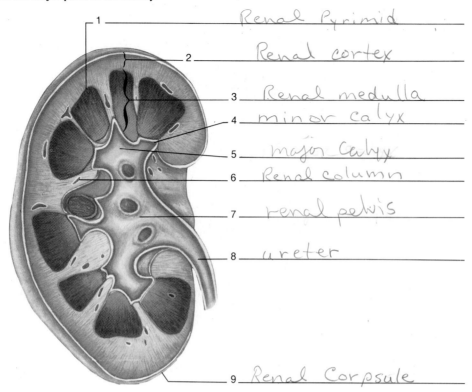

1. Renal Pyrimid
2. Renal cortex
3. Renal medulla
4. minor calyx
5. major Calyx
6. Renal column
7. renal pelvis
8. ureter
9. Renal Corpsule

c. Locate the following features:

renal capsule

hilum

renal artery

renal vein

ureter

d. Use a long knife to cut the kidney in half longitudinally along the frontal plane, beginning on the convex border.

e. Rinse the interior of the kidney with water, and locate the following:

renal pelvis

    major calyces

    minor calyces

renal cortex

    renal columns

renal medulla

    renal pyramids

Procedure B—The Renal Blood Vessels and Nephrons

1. Review the sections entitled "Renal Blood Vessels" and "Nephrons" in chapter 20 of the textbook.
2. As a review activity, label figures 56.2 and 56.3.
3. Complete Part B of the laboratory report.
4. Obtain a microscope slide of a kidney section and examine it, using low-power magnification. Locate the *renal capsule,* the *renal cortex* (which appears somewhat granular and may be more darkly stained than the other renal tissues), and the *renal medulla.*
5. Examine the renal cortex, using high-power magnification. Locate a *renal corpuscle.* These structures appear as isolated circular areas. Identify the *glomerulus,* which is the capillary cluster inside the corpuscle, and the *Bowman's capsule,* which appears as a clear area surrounding the glomerulus. Also note the numerous sections of renal tubules that occupy the spaces between renal corpuscles.
6. Prepare a labeled sketch of a representative section of renal cortex in Part C of the laboratory report.
7. Examine the renal medulla, using high-power magnification. Identify longitudinal and cross sections of various renal tubules. Note that these tubules are lined with simple epithelial cells, which vary in shape from squamous to cuboidal.
8. Prepare a labeled sketch of a representative section of renal medulla in Part D of the laboratory report.

Figure 56.2 Label the renal blood vessels.

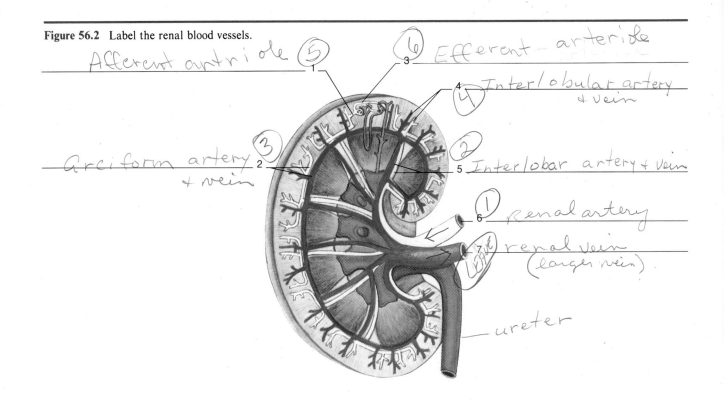

Afferent artriole (5)
Efferent arteriole 3
Interlobular artery + vein (4)
Arciform artery + vein (3) 2
Interlobar artery + vein (2) 5
renal artery (1) 6
renal vein (larger vein) 7
ureter

Figure 56.3 Label the major parts of the nephron and its blood vessels.

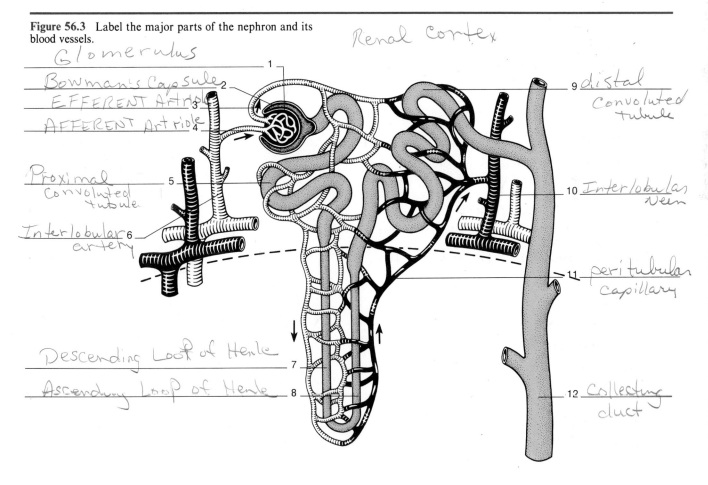

Renal Cortex

Glomerulus 1
Bowman's Capsule 2
EFFERENT Artriole 3
AFFERENT Artriole 4
Proximal Convoluted tubule 5
Interlobular artery 6
Descending Loop of Henle 7
Ascending Loop of Henle 8

distal convoluted tubule 9
Interlobular Vein 10
peritubular capillary 11
Collecting duct 12

Renal medulla

Name _____

Date _____

Section _____

Structure of the Kidney

Part A

Match the terms in column A with the descriptions in column B. Place the letter of your choice in the space provided.

Column A		Column B
A. calyx	E 1.	tissue forming a shell around the renal medulla
B. hilum	A 2.	portion of renal pelvis
C. nephron	H 3.	conical mass of tissue within renal medulla
D. renal column	F 4.	projection into kidney chamber
E. renal cortex	I 5.	hollow chamber within kidney
F. renal papilla	C 6.	functional unit of kidney
G. renal pelvis	D 7.	cortical tissue between renal pyramids
H. renal pyramid	G 8.	funnel-shaped end of ureter
I. renal sinus	B 9.	entrance to kidney chamber

Part B

Complete the following:

1. Distinguish between a renal corpuscle and a renal tubule. _renal corpuscle: glomerulus Brown's capsule_

    ren tubule: distal proximal convoluted tubules, ascending descending loop of henle, collecting duct

2. Number the following parts to indicate their respective positions in the renal tubule. Assign the number 1 to the part attached to Bowman's capsule.

    _3_ ascending loop of Henle

    _5_ collecting duct

    _2_ descending loop of Henle

    _4_ distal convoluted tubule

    _6_ papillary duct

    _1_ proximal convoluted tubule

    _7_ renal papilla

3. Number the following parts to indicate their respective positions in the blood pathway within the kidney. Assign the number 1 to the vessel attached to the renal artery.

    _1_ afferent arteriole

    _3_ efferent arteriole

    _2_ glomerulus

    _4_ peritubular capillary

    _5_ renal vein

4. Explain how the blood vessels associated with the renal corpuscle help to maintain relatively high blood pressure within the glomerulus. _____

_____

_____

_____

_____

_____

5. Define *juxtaglomerular apparatus.* _secretes renin, regulates p᷂ blood flow_

_____

_____

_____

_____

_____

Part C

In the space below, prepare a sketch of a representative section of renal cortex. Label the glomerulus, Bowman's capsule, and sections of renal tubules.

Part D

In the space below, prepare a sketch of a representative section of renal medulla. Label a longitudinal section and a cross section of a renal tubule.

Urinalysis

U rine is the end product of kidney functions, which include the removal of various waste substances from the blood and the maintenance of body fluid and electrolyte balances. Consequently, the composition of urine varies from time to time because of differences in dietary intake and physical activity. Also, the volume of urine produced by the kidneys varies with such factors as fluid intake, environmental temperature, relative humidity, respiratory rate, and body temperature.

An analysis of urine composition and volume often is used to evaluate the functions of the kidneys and other organs. This procedure, called *urinalysis,* involves observing the physical characteristics of a urine sample, testing for the presence of certain organic and inorganic substances, and examining the microscopic solids present in the sample.

Purpose of the Exercise

To perform the observations and tests commonly used to analyze the characteristics and composition of urine.

Learning Objectives

After completing this exercise, you should be able to:

1. Evaluate the color, transparency, and specific gravity of a urine sample.
2. Determine the pH of a urine sample.
3. Test a urine sample for the presence of glucose, protein, ketones, bilirubin, and hemoglobin.
4. Perform a microscopic study of urine sediment.
5. Evaluate the results of these observations and tests.

Materials Needed:
    disposable urine-collecting container
    paper towel
    urinometer cylinder
    urinometer hydrometer
    laboratory thermometer
    pH test paper

reagent strips to test for the presence of the following:
    glucose
    protein
    ketones
    bilirubin
    hemoglobin
compound microscope
microscope slide
coverslip
centrifuge
centrifuge tube
graduated cylinder, 10 ml
applicator stick
Sedi-stain

Warning
While performing the following tests, you should wear disposable plastic gloves so that skin contact with urine is avoided.

Procedure

1. Collect about 50 ml of urine in a clean, disposable container.
2. Place a sample of urine in a clean, transparent container. Describe the *color* of the urine. Normal urine varies from pale yellow to dark amber, depending on the presence of urochromes, which are pigments produced during the decomposition of hemoglobin. Dark urine indicates a high concentration of pigments.

    Abnormal urine colors include yellow-brown or green, due to elevated concentrations of bile pigments, and red to dark brown, due to the presence of blood. Certain foods, such as beets or carrots, and various drug substances also may cause color changes in urine, but in such cases the colors have no clinical significance. Enter the results of this and the following tests in Part A of Laboratory Report 57 on page 391.

3. Evaluate the *transparency* of the urine sample; that is, judge whether the urine is clear, slightly cloudy, or very cloudy. Normal urine is clear enough to see through. Cloudy urine indicates the presence of various substances that may include mucus, bacteria, epithelial cells, fat droplets, or inorganic salts.

4. Determine the *specific gravity* of the urine sample. Specific gravity is the ratio of the weight of something to the weight of an equal volume of water. For example, mercury (at 15°C) weighs 13.6 times as much as an equal volume of water; thus, it has a specific gravity of 13.6. Although urine is mostly water, it has substances dissolved in it and is slightly heavier than an equal volume of pure water. Thus, urine has a specific gravity of more than 1.000. Actually, the specific gravity of normal urine varies from 1.001 to 1.030.

   To determine the specific gravity of a urine sample:
   a. Pour enough urine into a clean urinometer cylinder to fill it about three-fourths full. Any foam that appears should be removed with a paper towel.
   b. Use a laboratory thermometer to measure the temperature of the urine.
   c. Gently place the urinometer hydrometer into the urine, and *make sure that the float is not touching the sides of the cylinder* (figure 57.1).
   d. Position your eye at the level of the urine surface, and determine which line on the stem of the hydrometer intersects the lowest level of the concave surface (meniscus) of the urine.
   e. Because liquids tend to contract and become denser as they are cooled, or to expand and become less dense as they are heated, it may be necessary to make a temperature correction to obtain an accurate specific gravity measurement. To do this, add 0.001 to the hydrometer reading for each 3° of urine temperature above 25°C or subtract 0.001 for each 3° below 25°C. Enter this calculated value in the chart of the laboratory report as the test result.

5. Reagent strips can be used to perform a variety of urine tests. In each case, directions for using the strips are found on the strip container. *Be sure to read them.*

   To perform each test:
   a. Obtain a urine sample and the proper reagent strip.
   b. Read the directions on the strip container.
   c. Dip the strip in the urine sample.
   d. Remove the strip and tap it to remove any excess liquid.

Figure 57.1 Float the hydrometer in the urine, making sure that it does not touch the sides of the cylinder.

   e. Then wait for the length of time indicated by the directions on the container before you compare the color of the test strip with the standard color scale on the side of the container. The value or amount represented by the matching color should be used as the test result and recorded in Part A of the laboratory report.

6. Perform the *pH test*. The pH of normal urine varies from 4.8 to 7.5, but most commonly it is near 6.0 (slightly acidic). The pH of urine may decrease as a result of a diet high in protein, or it may increase with a vegetarian diet.

7. Perform the *glucose test*. Normally, there is no glucose in urine. However, glucose may appear in the urine temporarily following a meal high in carbohydrates. Glucose also may appear in the urine in uncontrolled diabetes mellitus due to a deficiency of insulin.

8. Perform the *protein test*. Normally, proteins of large molecular size are not present in urine. However, those of small molecular sizes, such as albumins, may appear in trace amounts, particularly following strenuous exercise. Increased amounts of proteins also may appear as a result of kidney diseases in which the glomeruli are damaged, or as a result of high blood pressure.

9. Perform the *ketone test*. Ketones are products of fat metabolism. Usually they are present in urine in very small amounts. They may increase if the diet fails to provide adequate carbohydrate, as in the case of prolonged fasting or starvation or as a result of insulin deficiency (diabetes mellitus).

10. Perform the *bilirubin test*. Bilirubin, which results from hemoglobin decomposition, normally appears in very small amounts in urine. Its concentration may increase as a result of liver disorders which cause obstructions of the biliary tract.

Figure 57.2 Common types of urine sediment.

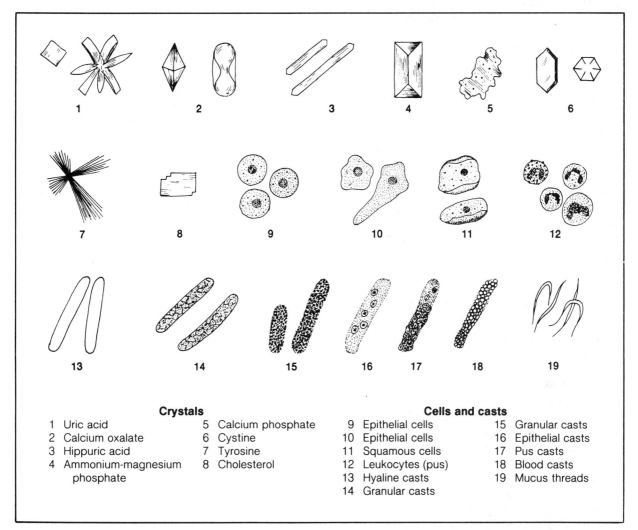

Crystals
1 Uric acid
2 Calcium oxalate
3 Hippuric acid
4 Ammonium-magnesium phosphate
5 Calcium phosphate
6 Cystine
7 Tyrosine
8 Cholesterol

Cells and casts
9 Epithelial cells
10 Epithelial cells
11 Squamous cells
12 Leukocytes (pus)
13 Hyaline casts
14 Granular casts
15 Granular casts
16 Epithelial casts
17 Pus casts
18 Blood casts
19 Mucus threads

11. Perform the *hemoglobin test*. Hemoglobin occurs in the red blood cells, and because such cells normally do not pass into the renal tubules, hemoglobin is not found in normal urine. Its presence in urine usually indicates a disease process, a transfusion reaction, or an injury to the urinary organs.

12. Complete Part A of the laboratory report.

13. A urinalysis usually includes a study of urine sediment—the microscopic solids present in a urine sample. This sediment normally includes mucus, certain crystals, and a variety of cells, such as the epithelial cells that line the urinary tubes and an occasional white blood cell. Other types of solids, such as casts or red blood cells, may indicate a disease or injury if they are present in excess. (Casts are cylindrical masses of cells or other substances that form in the renal tubules and are flushed out by the flow of urine.)

To observe urine sediment:

a. Stir or shake a urine sample thoroughly to suspend the sediment, which tends to settle to the bottom of the container.

b. Pour 5 ml of urine into a clean centrifuge tube, and centrifuge it for five minutes at slow speed.

c. Carefully decant the liquid from the sediment in the bottom of the centrifuge tube, as directed by your laboratory instructor.

d. Use an applicator stick or medicine dropper to remove some of the sediment, and place it on a clean microscope slide.

e. Add a drop of Sedi-stain to the sample, and add a coverslip.

f. Examine the sediment with low- and high-power magnifications.

g. Identify the kinds of solids present with the aid of figure 57.2.

h. In Part B of the laboratory report, make a sketch of each type of sediment.

14. Complete Part B of the laboratory report.

laboratory report 57

Name _____

Date _____

Section _____

Urinalysis

Part A

1. Enter your observations and test results in the following chart and complete the chart:

Urine Characteristics	Observations and Test Results	Normal Values	Evaluations
Color			
Transparency			
Specific gravity			
pH			
Glucose			
Protein			
Ketones			
Bilirubin			
Hemoglobin			

2. Write a statement to summarize the results of the urinalysis. _____

_____

_____

_____

_____

_____

3. Why do you think it is important to refrigerate a urine sample if an analysis cannot be performed immediately after collecting it? _____

_____

_____

_____

Part B

1. Make a sketch for each type of sediment you observed. Label those shown in figure 57.2.

2. Write a statement to summarize the results of the urine sediment study. _____

_____

_____

_____

_____

_____

_____

Cat Dissection: Urinary System

In this laboratory exercise, you will dissect the urinary organs of the embalmed cat. As you observe these structures, compare them with the corresponding human organs by observing the parts of the human manikin.

Purpose of the Exercise

To examine the urinary organs of the cat and to compare them with those of the human.

Learning Objectives

After completing this exercise, you should be able to:

1. Locate and identify the urinary organs of the cat.
2. Identify the corresponding organs in the human manikin.
3. Compare the urinary organs of the cat with those of the human.

Materials Needed:

human manikin	dissecting instruments
embalmed cat	disposable gloves
dissecting tray	

Procedure

1. Place the embalmed cat in a dissecting tray with its ventral side up.
2. Open the abdominal cavity, and remove the liver, stomach, and spleen.
3. Push the intestines to one side, and locate the *kidneys* in the dorsal abdominal wall on either side of the vertebral column. Note that the kidneys are located behind the *parietal peritoneum* (retroperitoneal).
4. Carefully remove the parietal peritoneum and the fat surrounding the kidneys. Locate the following, using figure 58.1 and reference plate 19 as guides:

   ureters

   renal arteries

   renal veins

5. Locate the *adrenal glands,* which lie medially and above the kidneys and are surrounded by connective tissues. Note that the adrenal glands are attached to blood vessels and are separated by the abdominal aorta and inferior vena cava.
6. Expose the ureters by cleaning away the connective tissues along their lengths. Note that they enter the *urinary bladder* on the dorsal surface.
7. Examine the urinary bladder, and note how it is attached to the abdominal wall by folds of peritoneum that form ligaments. Locate the *medial ligament* on the ventral surface of the bladder that connects it with the linea alba and the two *lateral ligaments* on the dorsal surface.
8. Remove the connective tissue from around the urinary bladder. Use a sharp scalpel to open the bladder, and examine its interior. Locate the openings of the ureters and the urethra on the inside.
9. Expose the *urethra* at the base of the urinary bladder.
10. Remove one kidney, and section it longitudinally along the frontal plane (figure 58.2 and reference plate 21). Identify the following features:

renal capsule	renal papilla
renal cortex	renal sinus
renal medulla	renal pelvis
renal pyramid	

11. Discard the organs and tissues that were removed from the cat, as directed by the laboratory instructor.
12. Complete Laboratory Report 58 on page 395.

Figure 58.1   The urinary organs of a female cat.

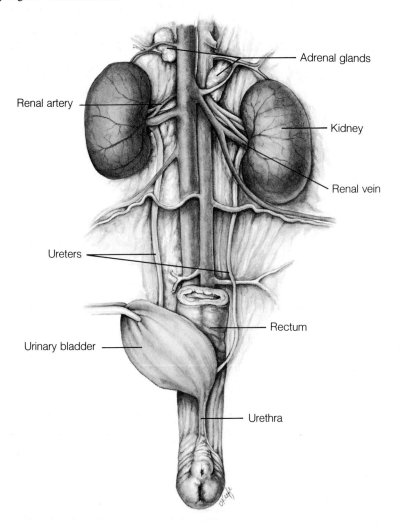

Adrenal glands

Renal artery

Kidney

Renal vein

Ureters

Rectum

Urinary bladder

Urethra

Figure 58.2   Frontal section of a cat's kidney.

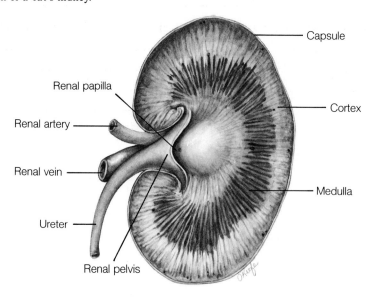

Capsule

Renal papilla

Renal artery

Cortex

Renal vein

Medulla

Ureter

Renal pelvis

Name _____

Date _____

Section _____

Cat Dissection: Urinary System

Complete the following:

1. Compare the positions of the kidneys in the cat with those in the human. _____
   _____
   _____
   _____

2. Compare the locations of the adrenal glands in the cat with those in the human. _____
   _____
   _____
   _____
   _____

3. What parts of the cat urinary system are retroperitoneal? _____
   _____
   _____
   _____

4. Describe the wall of the cat urinary bladder. _____
   _____
   _____
   _____

5. Compare the renal pyramids and renal papillae of the human kidney with those of the cat. _____
   _____
   _____
   _____
   _____

Laboratory Exercise 59

Male Reproductive System

The organs of the male reproductive system are specialized to produce and maintain the male sex cells, to transport these cells together with supporting fluids to the female reproductive tract, and to produce and secrete male sex hormones.

These organs include the testes, in which sperm cells and male sex hormones are produced, and sets of internal and external accessory organs. The internal organs include various tubes and glands, while the external organs are the scrotum and the penis.

Purpose of the Exercise

To review the structure and functions of the male reproductive organs and to examine some of these organs microscopically.

Learning Objectives

After completing this exercise, you should be able to:

1. Locate and identify the organs of the male reproductive system.
2. Describe the functions of these organs.
3. Recognize sections of the testis, epididymis, and penis microscopically.
4. Identify the major features of these microscopic sections.

Materials Needed:
  human manikin
  model of the male reproductive system
  anatomical chart of the male reproductive system
  compound microscope
  prepared microscope slides of the following:
    testis section
    epididymis, cross section
    penis, cross section

Procedure A—Male Reproductive Organs

1. Review the sections entitled "Testes," "Male Internal Accessory Organs," and "Male External Reproductive Organs" in chapter 22 of the textbook.
2. As a review activity, label figures 59.1, 59.2, and 59.3.
3. Observe the human manikin and the models and anatomical chart of the male reproductive system. Locate the following features:

testes

inguinal canal

spermatic cord

epididymis

vas deferens

ejaculatory duct

seminal vesicles

prostate gland

bulbourethral (Cowper's) glands

scrotum

penis

  corpora cavernosa

  corpus spongiosum

  tunica albuginea

  glans penis

  external urethral meatus

  prepuce

  crura

  bulb

4. Complete Part A of Laboratory Report 59 on page 402.

Figure 59.1 Label the major features of the male
reproductive system.

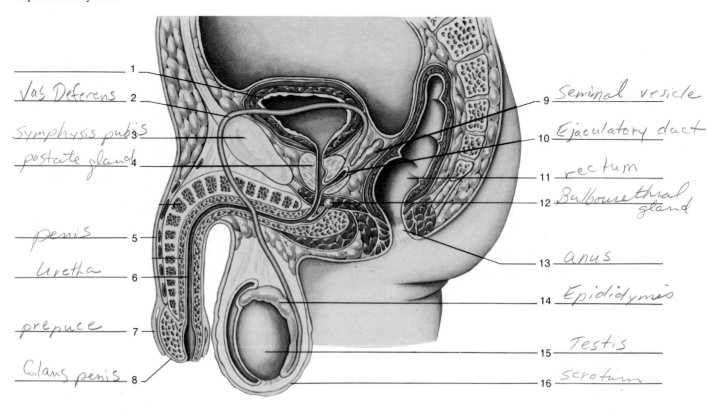

1 _____

Vas Deferens 2

Symphysis pubis 3

postate gland 4

penis 5

Uretha 6

prepuce 7

Glans penis 8

9 *Seminal vesicle*

10 *Ejaculatory duct*

11 *rectum*

12 *Bulbourethral gland*

13 *anus*

14 *Epididymis*

15 *Testis*

16 *scrotum*

Figure 59.2 Label this cross section of the penis.

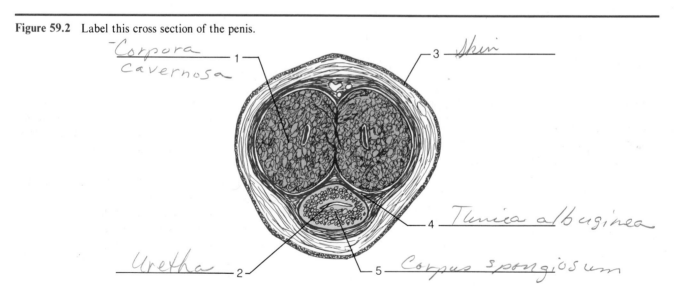

Corpora Cavernosa 1

Uretha 2

3 *Skin*

4 *Tunica albuginea*

5 *Corpus spongiosum*

Figure 59.3 Label the diagram by placing the correct numbers in the spaces provided.

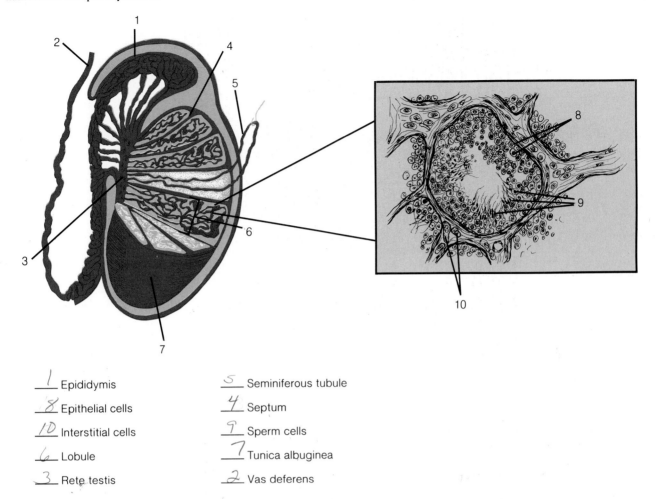

_1_ Epididymis

_8_ Epithelial cells

_10_ Interstitial cells

_6_ Lobule

_3_ Rete testis

_5_ Seminiferous tubule

_4_ Septum

_9_ Sperm cells

_7_ Tunica albuginea

_2_ Vas deferens

Procedure B—Microscopic Anatomy

1. Obtain a microscope slide of human testis section and examine it, using low-power magnification (figure 59.4). Locate the thick _fibrous capsule_ (tunica albuginea) on the surface and the numerous sections of _seminiferous tubules_ inside.

2. Focus on some of the seminiferous tubules, using high-power magnification (figure 59.5). Locate the _epithelium,_ which forms the inner lining of each tube. Within this epithelium, identify some _supporting cells_ (Sertoli's cells), which have pale, oval-shaped nuclei, and some _spermatogenic cells,_ which have smaller, round nuclei. Near the lumen of the tube, find some darkly stained, elongated heads of developing sperm cells. In the spaces between adjacent seminiferous tubules, locate some isolated _interstitial cells_ (cells of Leydig).

3. Prepare a labeled sketch of a representative section of the testis in Part B of the laboratory report.

4. Obtain a microscope slide of a cross section of _epididymis_ (figure 59.6). Examine its wall, using high-power magnification. Note the elongated,

pseudostratified columnar epithelial cells that comprise most of the inner lining. These cells have nonmotile cilia on their free surfaces. Also note the thin layer of smooth muscle and connective tissue surrounding the tube.

5. Prepare a labeled sketch of the epididymis wall in Part B of the laboratory report.

6. Obtain a microscope slide of a _penis_ cross section, and examine it with low-power magnification (figure 59.7). Identify the following features:

corpora cavernosa

corpus spongiosum

tunica albuginea

urethra

skin

7. Prepare a labeled sketch of a penis cross section in Part B of the laboratory report.

8. Complete Part B of the laboratory report.

Figure 59.4 Micrograph of a human testis.

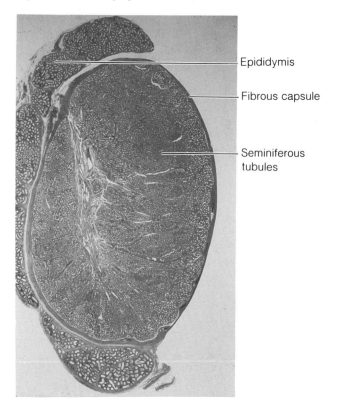

— Epididymis

— Fibrous capsule

— Seminiferous tubules

Figure 59.5 Micrograph of seminiferous tubules (×50).

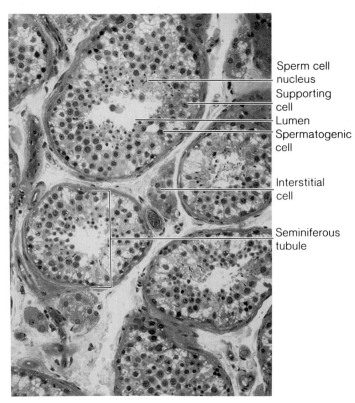

Sperm cell nucleus

Supporting cell

Lumen

Spermatogenic cell

Interstitial cell

Seminiferous tubule

Figure 59.6 Micrograph of the epididymis (×50).

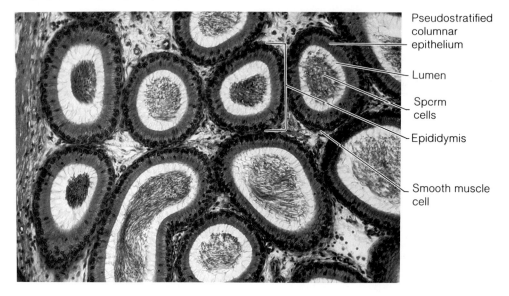

Pseudostratified columnar epithelium

Lumen

Sperm cells

Epididymis

Smooth muscle cell

Figure 59.7 Cross section of the penis (×5).

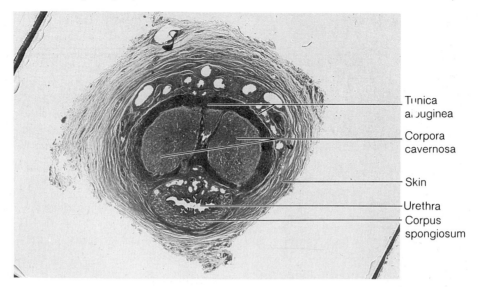

Tunica albuginea

Corpora cavernosa

Skin

Urethra

Corpus spongiosum

laboratory report 59

Name _____

Date _____

Section _____

Male Reproductive System

Part A

Complete the following statements:

1. The testes are suspended by _____ within the scrotum.

2. The testes originate from masses of tissues located near the developing _____ .

3. The descent of the testes is aided by a fibromuscular cord called the _____ .

4. As the testes descend, they pass through the _____ of the abdominal wall.

5. Connective tissue divides a testis into many _____ , which contain seminiferous tubules.

6. The _____ is a highly coiled tube on the surface of the testis.

7. The _____ cells of the epithelium that line the seminiferous tubule give rise to sperm cells.

8. Undifferentiated sperm cells are called _____ .

9. _____ is the process by which sperm cells are formed.

10. The number of chromosomes normally present in a human sperm cell is _____ .

11. The anterior end of a sperm head, called the _____ , aids the penetration of an egg cell membrane.

12. Sperm cells undergo maturation while they are stored in the _____ .

13. The secretion of the seminal vesicles is rich in the monosaccharide called _____ .

14. Prostate gland secretion helps neutralize seminal fluid because the secretion is _____ .

15. The secretion of the bulbourethral glands provides _____ , which aids sexual intercourse.

16. The pH of seminal fluid is slightly _____ .

17. The dartos muscle is in the wall of the _____ .

18. The sensitive, cone-shaped end of the penis is called the _____ .

19. _____ is the movement of sperm cells and various glandular secretions into the urethra.

20. _____ is the process by which seminal fluid is forced out through the urethra.

Part B

1. Prepare a labeled sketch of a representative section of the testis.

2. Prepare a labeled sketch of a region of the epididymis wall.

3. Prepare a labeled sketch of a penis cross section.

4. Briefly describe the function of each of the following:
   a. spermatogenic cell

   _____

   _____

   _____

   b. interstitial cell

   _____

   _____

   _____

   c. epididymis

   _____

   _____

   _____

   d. corpora cavernosa

   _____

   _____

   _____

Laboratory Exercise 60

Female Reproductive System

The organs of the female reproductive system are specialized to produce and maintain the female sex cells, to transport these cells to the site of fertilization, to provide a favorable environment for a developing offspring, to move the offspring to the outside, and to produce female sex hormones.

These organs include the ovaries, which produce the egg cells and female sex hormones, and sets of internal and external accessory organs. The internal accessory organs include the uterine tubes, uterus, and vagina. The external organs are the labia majora, labia minora, clitoris, and vestibular glands.

Purpose of the Exercise

To review the structure and functions of the female reproductive organs and to examine some of their features microscopically.

Learning Objectives

After completing this exercise, you should be able to:

1. Locate and identify the organs of the female reproductive system.
2. Describe the functions of these organs.
3. Recognize sections of the ovary, uterine tube, and uterus wall microscopically.
4. Identify the major features of these microscopic sections.

Materials Needed:
human manikin
model of the female reproductive system
anatomical chart of the female reproductive system
compound microscope

prepared microscope slides of the following:
ovary section with maturing follicles
uterine tube, cross section
uterus wall section

For Demonstration:

prepared microscope slides of the following:
uterus wall, early proliferative phase
uterus wall, secretory phase
uterus wall, early menstrual phase

Procedure A—Female Reproductive Organs

1. Review the sections entitled "Ovaries," "Female Internal Accessory Organs," and "Female External Reproductive Organs" in chapter 22 of the textbook.
2. As a review activity, label figures 60.1, 60.2, 60.3, and 60.4.
3. Observe the human manikin and the model and anatomical chart of the female reproductive system. Locate the following features:

ovaries

   medulla

   cortex

broad ligament

suspensory ligament

ovarian ligament

uterine tubes

   infundibulum

   fimbriae

Figure 60.1 Label the ligaments associated with the female internal reproductive organs.

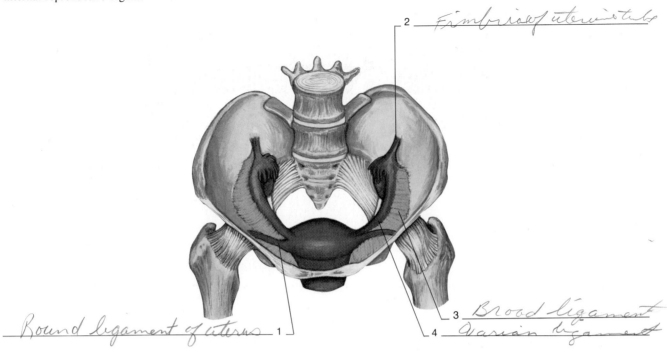

2 _Fimbriae of uterine tube_

3 _Broad ligament_

4 _Ovarian ligament_

Round ligament of uterus 1

Figure 60.2 Label the major features of the female reproductive system.

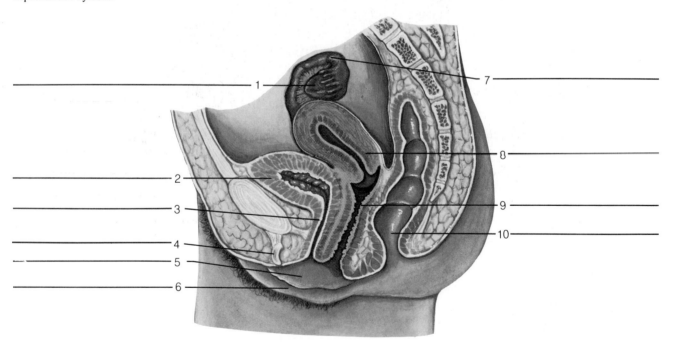

Figure 60.3 Label these parts of the female reproductive system.

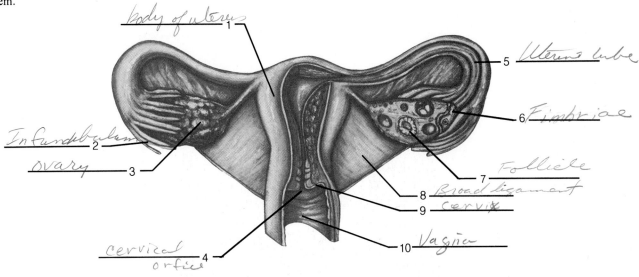

Body of uterus — 1

Infundibulum — 2

ovary — 3

cervical orfice — 4

5 — Uterine tube

6 — Fimbriae

7 — Follicle

8 — Broad ligament

9 — cervix

10 — Vagina

Figure 60.4 Label this diagram by placing the correct numbers in the spaces provided.

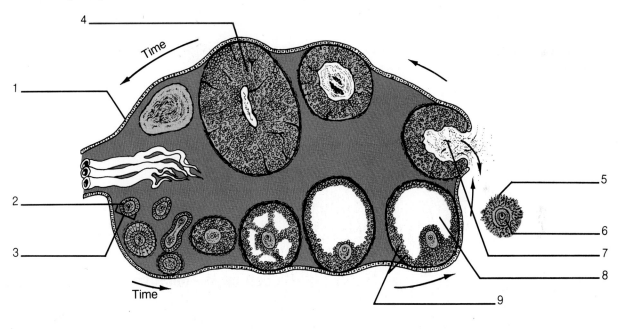

_____ Corona radiata

_____ Corpus luteum

_____ Follicular cells

_____ Follicular fluid

_____ Germinal epithelium

_____ Ovulation

_____ Primary oocyte

_____ Primary follicle

_____ Secondary oocyte

Figure 60.5 Micrograph of the ovary (×30).

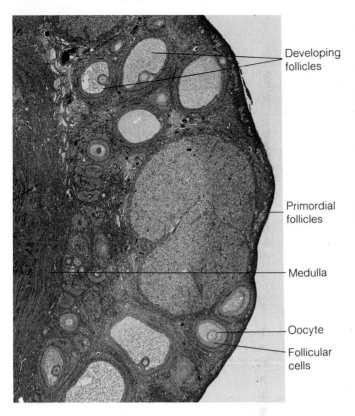

- Developing follicles
- Primordial follicles
- Medulla
- Oocyte
- Follicular cells

Figure 60.6 Micrograph of the ovarian cortex (×100).

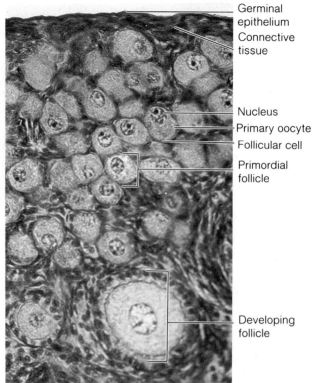

- Germinal epithelium
- Connective tissue
- Nucleus
- Primary oocyte
- Follicular cell
- Primordial follicle
- Developing follicle

uterus

    cardinal ligament

    round ligament

    body

    cervix

    cervical orifice

    endometrium

    myometrium

    perimetrium

vagina

    fornices

    vaginal orifice

    hymen

    mucosal layer

    muscular layer

    fibrous layer

mons pubis

labia majora

labia minora

clitoris

    glans

vestibule

vestibular glands

vestibular bulbs

4. Complete Part A of Laboratory Report 60 on page 411.

Procedure B—Microscopic Anatomy

1. Obtain a microscope slide of an ovary section with maturing follicles, and examine it with low-power magnification (figure 60.5). Locate the outer layer, or *cortex,* which is composed of densely packed cells, and the inner layer, or *medulla,* which consists largely of loose connective tissue.

2. Focus on the cortex of the ovary, using high-power magnification (figure 60.6). Note the thin layer of small cuboidal cells on the free surface. These cells comprise the *germinal epithelium.* Also locate some *primordial follicles* just beneath the germinal epithelium. Note that each follicle consists of a single, relatively large *primary oocyte* with a prominent nucleus and a covering of *follicular cells.*

3. Prepare a labeled sketch of the ovarian cortex in Part B of the laboratory report.

Figure 60.7 Cross section of the uterine tube (×8).

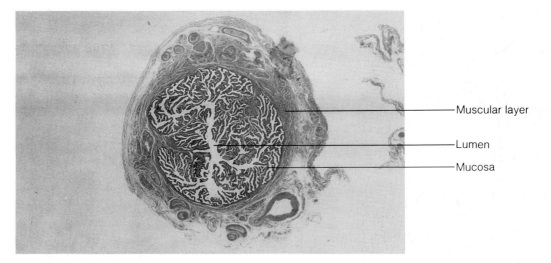

Muscular layer

Lumen

Mucosa

Demonstration

Observe the slides in the demonstration microscopes. Each slide contains a section of uterine mucosa taken during a different phase in the menstrual cycle. In the *early proliferative phase*, note the simple columnar epithelium on the free surface of the mucosa and the numerous sections of tubular uterine glands in the tissues beneath the epithelium. In the *secretory phase*, note that the endometrium is thicker and that the uterine glands appear more extensive and that they are coiled. In the *early menstrual phase*, note that the endometrium is thinner because its surface layer has been lost. Also note that the uterine glands are less apparent and that the spaces between the glands contain many leukocytes. What is the significance of these changes? _____

_____

_____

_____

4. Use low-power magnification to search the ovarian cortex for maturing follicles in various stages of development. Prepare three labeled sketches in Part B of the laboratory report to illustrate the changes that occur in a follicle as it matures.
5. Obtain a microscope slide of a cross section of a uterine tube. Examine it, using low-power magnification (figure 60.7). Note that the shape of the lumen is very irregular.
6. Focus on the inner lining of the uterine tube with high power. Note that the lining is composed of *simple columnar epithelium* and that some of the epithelial cells are ciliated on their free surfaces.
7. Prepare a labeled sketch of a representative region of the wall of the uterine tube in Part B of the laboratory report.
8. Obtain a microscope slide of the uterus wall section. Examine it, using low-power magnification (figure 60.8), and locate the following:

Figure 60.8 A section of the uterine wall (×10).

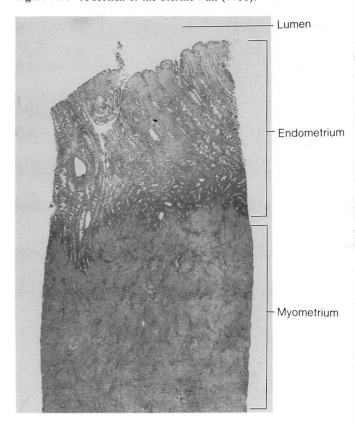

Lumen

Endometrium

Myometrium

endometrium
myometrium
perimetrium

9. Prepare a labeled sketch of a representative section of the uterine wall in Part B of the laboratory report.
10. Complete Part B of the laboratory report.

409

laboratory report 60

Name _____

Date _____

Section _____

Female Reproductive System

Part A

Complete the following statements:

1. The ovaries are located in the _____ cavity.

2. The largest of the ovarian attachments is called the _____ ligament.

3. The ovarian cortex appears granular because of the presence of_____ .

4. The cuboidal cells on the free surface of the ovary are called the_____ .

5. A primary oocyte is surrounded by _____ cells.

6. When a primary oocyte divides, a secondary oocyte and a(an) _____ are produced.

7. Primordial follicles are stimulated to develop into primary follicles by the hormone called _____ .

8. _____ is the process by which a secondary oocyte is released from the ovary.

9. Uterine tubes also are called _____ .

10. The _____ is the funnel-shaped expansion at the end of a uterine tube.

11. The _____ ligament connects the upper end of the uterus to the pelvic wall.

12. A portion of the uterus called the _____ extends downward into the upper portion of the vagina.

13. The inner lining of the vagina is called the _____ .

14. The myometrium is composed largely of _____ tissue.

15. The hymen is composed of connective tissue and _____ epithelium.

16. The structures that surround the openings of the urethra and vagina comprise the_____ .

17. The rounded mass of fatty tissue overlying the symphysis pubis of the female is called the_____ .

18. The female organ that corresponds to the male penis is the _____ .

19. The _____ of the female corresponds to the bulbourethral glands of the male.

20. Orgasm is accompanied by a series of reflexes involving the _____ regions of the spinal cord.

Part B

1. Prepare a labeled sketch of a representative region of the ovarian cortex.

2. Prepare a series of three labeled sketches to illustrate follicular maturation.

3. Prepare a labeled sketch of a representative section of the wall of a uterine tube.

4. Prepare a labeled sketch of a representative section of uterine wall.

5. Complete the following:
   a. Explain how the germinal epithelium gives rise to primordial follicles. _____

   _____

   _____

   _____

   _____

   _____

   b. Describe the fate of a mature follicle. _____

   _____

   _____

   _____

   _____

   _____

   c. Describe the function of the cilia in the lining of the uterine tube. _____

   _____

   _____

   _____

   _____

   _____

   d. Briefly describe the changes that occur in the uterine lining during a menstrual cycle. _____

   _____

   _____

   _____

   _____

   _____

Cat Dissection: Reproductive Systems

In this laboratory exercise, you will dissect the reproductive system of the embalmed cat. If you have a female cat, begin with Procedure A. If you have a male cat, begin with Procedure B. After completing the dissection, exchange cats with someone who has dissected one of the opposite sex and examine its reproductive organs.

As you observe the cat reproductive organs, compare them with the corresponding human organs by examining the models of the human reproductive systems.

Purpose of the Exercise

To examine the reproductive organs of the embalmed cat and to compare them with the corresponding organs of the human.

Learning Objectives

After completing this exercise, you should be able to:

1. Locate and identify the reproductive organs of a cat.
2. Identify the corresponding organs in models of the human reproductive systems.
3. Compare the reproductive organs of the cat with those of the human.

Materials Needed:
   embalmed cat
   dissecting tray
   dissecting instruments
   bone cutters
   disposable gloves
   models of human reproductive systems

Procedure A—Female Reproductive System

1. Place the embalmed female cat in a dissecting tray with its ventral side up, and open its abdominal cavity.
2. Locate the small oval ovaries just posterior to the kidneys (figure 61.1 and reference plate 21).
3. Examine an ovary, and note that it is suspended from the dorsal body wall by a fold of peritoneum called the *mesovarium.* Another attachment, the *ovarian ligament,* connects the ovary to the tubular *uterine horn* (figure 61.1 and reference plate 21).
4. Near the anterior end of the ovary, locate the funnel-shaped *infundibulum,* which is at the end of the *fallopian tube,* or *oviduct.* Note the tiny projections, or *fimbriae,* that create a fringe around the edge of the infundibulum. Trace the fallopian tube around the ovary to its connection with the uterine horn near the attachment of the ovarian ligament.
5. Examine the uterine horn. Note that it is suspended from the body wall by a fold of peritoneum, the *broad ligament,* which is continuous with the mesovarium of the ovary. Also note the fibrous *round ligament,* which extends from the uterine horn laterally and posteriorly to the body wall (figure 61.1 and reference plate 21).
6. To observe the remaining organs of the reproductive system more easily, remove the left leg by severing it near the hip with bone cutters. Then, use the bone cutters to remove the left anterior portion of the pelvic girdle. Also remove the necessary muscles and connective tissue to expose the structures shown in figure 61.2.

Figure 61.1 Ventral view of the female cat's reproductive system.

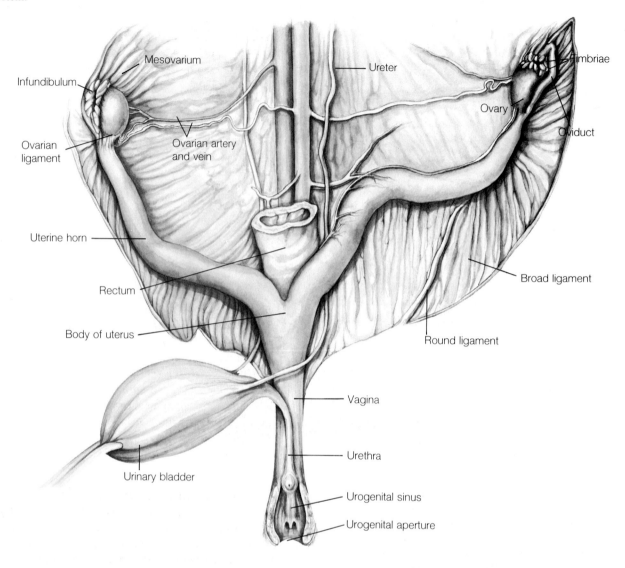

7. Trace the uterine horns posteriorly, and note that they unite to form the *uterine body,* which is located between the urethra and rectum. The uterine body is continuous with the *vagina,* which leads to the outside.

8. Trace the *urethra* from the urinary bladder posteriorly, and note that it and the vagina open into a common chamber, called the *urogenital sinus.* The opening of this chamber, which is ventral to the anus, is called the *urogenital aperture.* Locate the *labia majora,* which are folds of skin on either side of the urogenital aperture.

9. Use scissors to open the vagina along its lateral wall, beginning at the urogenital aperture and continuing to the body of the uterus. Note the *urethral orifice* in the ventral wall of the urogenital sinus, and locate the small rounded *cervix* of the uterus, which projects into the vagina at its deep end. The *clitoris* is located in the ventral wall of the urogenital sinus near its opening to the outside.

10. Complete Part A of Laboratory Report 61 on page 421.

Figure 61.2 Lateral view of the female cat's reproductive system.

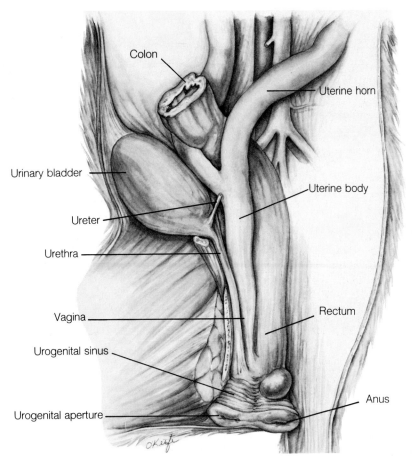

Colon

Uterine horn

Urinary bladder

Uterine body

Ureter

Urethra

Vagina

Rectum

Urogenital sinus

Anus

Urogenital aperture

Procedure B—Male Reproductive System

1. Place the embalmed male cat in a dissecting tray with its ventral side up.
2. Locate the *scrotum* between the legs. Make an incision along the midline of the scrotum, and expose the *testes* inside. (See reference plate 22.) Note that the testes are separated by a septum and that each testis is enclosed in a sheath of connective tissue.
3. Remove the sheath surrounding the testes, and locate the convoluted *epididymis* on the dorsal surface of each testis (figure 61.3).
4. Locate the *spermatic cord* on the right side, leading away from the testis. This cord contains the *vas deferens*, which is continuous with the epididymis as well as with the nerves and blood vessels that supply the testis on that side. Trace the spermatic cord to the body wall where its contents pass through the *inguinal canal* and enter the pelvic cavity (figure 61.3).
5. Locate the *penis* and identify the *prepuce*, which forms a sheath around the end of the penis. Make an incision through the skin of the prepuce, and expose the *glans* of the penis. Note that the glans has minute spines on its surface.

6. To observe the remaining structures of the reproductive system more easily, remove the left leg by severing it near the hip with bone cutters. Then use the bone cutters to remove the left anterior portion of the pelvic girdle. Also remove the necessary muscles and connective tissues to expose the organs shown in figure 61.4.
7. Trace the vas deferens from the inguinal canal to the penis. Note that the vas deferens loops over the ureter within the pelvic cavity and passes downward behind the urinary bladder to join the urethra. Locate the *prostate gland*, which appears as an enlargement at the junction of the vas deferens and urethra.
8. Trace the urethra to the penis. Locate the *bulbourethral glands*, which form small swellings on either side at the proximal end of the penis.
9. Use a sharp scalpel to cut across the penis, and examine the resulting cross section. Identify the *corpora cavernosa*, each of which contains many blood spaces surrounded by a sheath of connective tissue.
10. Complete Part B of the laboratory report.

Figure 61.3 Ventral view of the male cat's reproductive system.

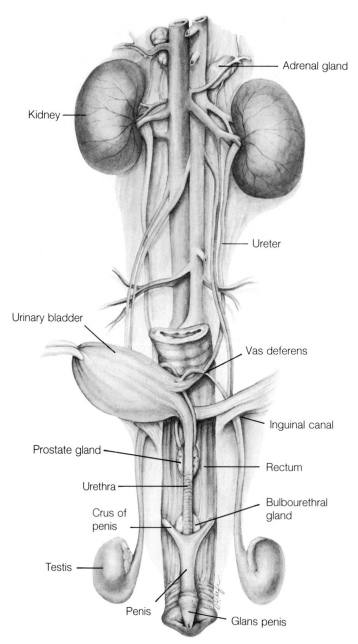

Figure 61.4 Lateral view of the male cat's reproductive system.

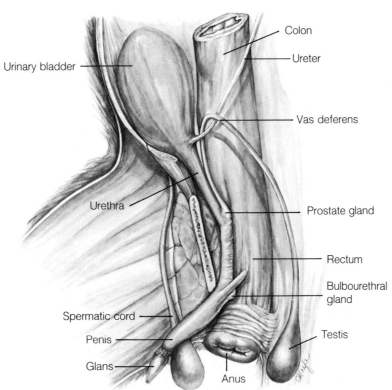

Urinary bladder

Colon

Ureter

Vas deferens

Urethra

Prostate gland

Rectum

Bulbourethral gland

Spermatic cord

Penis

Glans

Anus

Testis

Name _____

Date _____

Section _____

Cat Dissection: Reproductive Systems

Part A

Complete the following:

1. Compare the relative lengths and paths of the uterine tubes (fallopian tubes) of the cat and the human. _____

_____

_____

_____

_____

_____

2. How do the shape and structure of the uterus of the cat compare with that of the human? _____

_____

_____

_____

_____

_____

3. What do you think is the function of the uterine horns of the cat? _____

_____

_____

_____

_____

_____

4. Compare the relationship of the urethra and the vagina in the cat and in the human. _____

_____

_____

_____

_____

_____

Part B

Complete the following:

1. Compare the glans of the penis in the cat and in the human. _____
_____
_____
_____
_____

2. How do the location and the relative size of the prostate gland of the cat compare with that of the human? _____
_____
_____
_____
_____

3. What glands associated with the human vas deferens are missing in the cat? _____
_____
_____

Fertilization and Early Development

Fertilization is the process by which the nuclei of an egg cell and a sperm cell come together and combine their chromosomes, forming a single cell called a zygote.

Ordinarily, before fertilization can occur in a human female, an egg cell must be released from the ovary and be carried into a uterine tube. Also, seminal fluid containing sperm cells must be deposited in the vagina, and some of these sperm cells must travel through the uterus and into the uterine tubes. Although many sperm cells may reach an egg cell, only one will participate in the fertilization of the egg.

Shortly after fertilization, the zygote undergoes division (mitosis) to form two cells. These two cells become four, they in turn divide into eight, and so forth. The resulting mass of cells continues to grow and undergoes developmental changes that give rise to an offspring.

Purpose of the Exercise

To review the process of fertilization, to observe sea urchin eggs being fertilized, and to examine embryos in early stages of development.

Learning Objectives

After completing this exercise, you should be able to:

1. Describe the process of fertilization.
2. Describe the early developmental stages of a sea urchin.
3. Describe the early developmental stages of a human.
4. Identify the major features of human embryo models.

Materials Needed:

sea urchin egg suspension
sea urchin sperm suspension
compound microscope
depression microscope slide
coverslip
medicine droppers
prepared microscope slide of the following:
    sea urchin embryos (early and late cleavage)
models of human embryos

For Optional Activity:

Vaseline
toothpick

For Demonstration:

preserved mammalian embryos

Procedure A—Fertilization

1. Review the sections entitled "Fertilization" and "Early Embryonic Development" in chapter 22 of the textbook.
2. As a review activity, label figure 62.1.
3. Complete Part A of Laboratory Report 62 on page 427.
4. Although it is difficult to observe fertilization in animals in which the process occurs internally, it is possible to view forms of external fertilization. For example, egg and sperm cells can be collected from sea urchins, and the process of fertilization can be observed microscopically. To make this observation:
   a. Place a drop of sea urchin egg-cell suspension in the chamber of a depression slide, and add a coverslip.

Figure 62.1 Label the diagram by placing the correct numbers in the spaces provided.

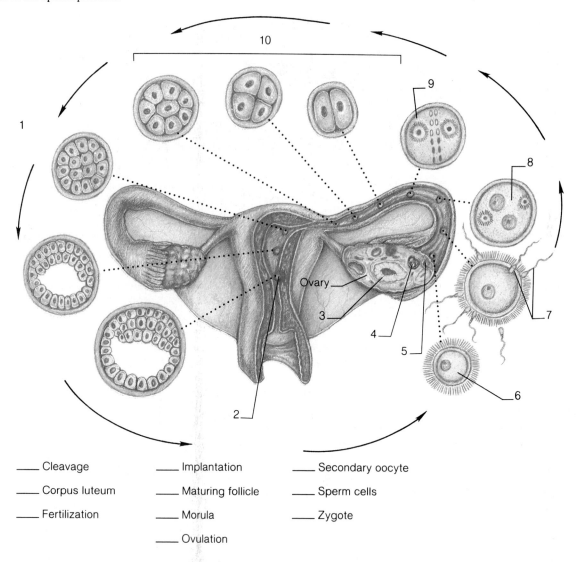

_____ Cleavage	_____ Implantation	_____ Secondary oocyte
_____ Corpus luteum	_____ Maturing follicle	_____ Sperm cells
_____ Fertilization	_____ Morula	_____ Zygote
	_____ Ovulation	

b. Examine the egg cells, using low-power magnification.

c. Focus on a single egg cell with high-power magnification, and sketch the cell in Part B of the laboratory report.

d. Remove the coverslip, and add a drop of sea urchin sperm-cell suspension to the depression slide. Replace the coverslip, and observe the sperm cells with high-power magnification as they cluster around the egg cells.

e. Observe the egg cells with low-power magnification once again, and watch for the appearance of *fertilization membranes*. Such a

membrane forms as soon as an egg cell is penetrated by a sperm cell; it looks like a clear halo surrounding the egg.

f. Focus on a single fertilized egg cell, and sketch it in Part B of the laboratory report.

Optional Activity

Use a toothpick to draw a thin line of Vaseline around the chamber of the depression slide containing the fertilized sea urchin egg cells. Place a coverslip over the chamber, and gently press it into the Vaseline to seal the chamber and prevent the liquid inside from evaporating. Keep the slide in a cool place so that the temperature never exceeds 22°C. Using low-power magnification, examine the slide every thirty minutes, and look for the appearance of 2-, 4-, and 8-cell stages of developing sea urchin embryos.

Figure 62.2 Label the major features of this early embryo and the structures associated with it.

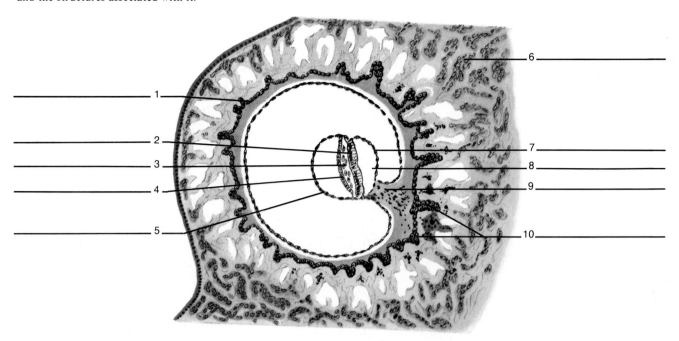

Procedure B—Sea Urchin Early Development

1. Obtain a prepared microscope slide of developing sea urchin embryos. This slide contains embryos in various stages of cleavage. Search the slide using low-power magnification, and locate embryos in 2-, 4-, 8-, and 16-cell stages.
2. Prepare a sketch of each stage in Part C of the laboratory report.

Procedure C—Human Early Development

1. Review the sections entitled "Period of Cleavage" and "Embryonic Stage" in chapter 23 of the textbook.
2. As a review activity, label figures 62.2 and 62.3.
3. Complete Parts D and E of the laboratory report.
4. Observe the models of human embryos, and identify the following features:

blastomeres

morula

blastocyst

inner cell mass

trophoblast

embryonic disk

primary germ layers

   ectoderm

   endoderm

   mesoderm

connecting stalk

chorion

chorionic villi

lacunae

amnion

amniotic fluid

umbilical cord

   umbilical arteries

   umbilical vein

yolk sac

allantois

placenta

Demonstration

Observe the preserved mammalian embryos that are on display. In addition to observing the developing external body structures, identify such features as the chorion, chorionic villi, amnion, yolk sac, umbilical cord, and placenta. What special features provide clues as to the type of mammal these embryos represent? _____

_____

_____

_____

Figure 62.3 Label the structures associated with this seven-week-old embryo.

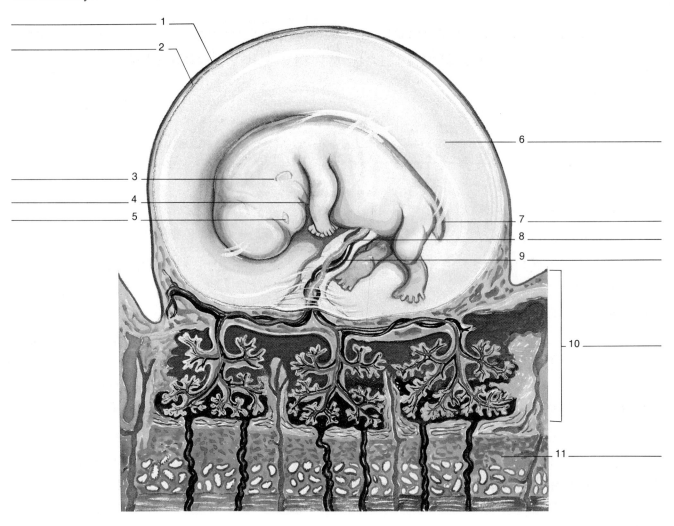

Fertilization and Early Development

Part A

Complete the following statements:

1.   The zona pellucida surrounds a(an) _____ .

2.   Enzymes secreted by the _____ of a sperm cell help it to penetrate an egg cell.

3.   The cell resulting from fertilization is called a(an) _____ .

4.   The cell resulting from fertilization divides by the process of _____ .

5.   _____ is the phase of development during which cellular divisions result in smaller and smaller daughter cells.

6.   _____ on the inner lining of the uterine tube aid in moving a developing embryo.

7.   The trip from the site of fertilization to the uterine cavity takes about _____ days.

8.   The hollow ball of cells formed early in development is called a(an) _____ .

9.   A human offspring is called a(an) _____ until the end of the seventh week of development.

10. After the seventh week, a developing human is called a(an) _____ .

Part B

Prepare sketches of the following:

Single sea urchin egg (____X)

Fertilized sea urchin egg

Part C

Prepare sketches of the following:

2-cell sea urchin embryo

4-cell sea urchin embryo

8-cell sea urchin embryo

16-cell sea urchin embryo

Part D

Match the terms in column A with the descriptions in column B. Place the letter of your choice in the space provided.

Column A	Column B
A. blastocyst	____ 1. germ layer that gives rise to muscle and bone
B. chorionic villi	____ 2. blastocyst cells that give rise to embryo proper
C. connecting stalk	____ 3. spaces that become filled with blood
D. ectoderm	____ 4. hollow ball of cells
E. endoderm	____ 5. cells forming wall of blastocyst
F. inner cell mass	____ 6. solid ball of about sixteen cells
G. lacunae	____ 7. inner layer of embryonic disk
H. mesoderm	____ 8. extensions of trophoblast
I. morula	____ 9. tissue connecting embryo proper to developing placenta
J. trophoblast	____ 10. outer layer of embryonic disk

Part E

Complete the following statements:

1. The membrane that separates the embryonic blood from the maternal blood is called the _____ .

2. The embryonic membrane that is attached to the edge of the embryonic disk and surrounds the developing body is called the _____ .

3. The umbilical cord contains three blood vessels, two of which are _____ .

4. Eventually the amniotic cavity is surrounded by a double-layered membrane called the _____ .

5. The _____ functions to form blood cells during the early stages of development.

6. The _____ gives rise to the cells that later become sex cells.

7. The _____ gives rise to the umbilical blood vessels.

8. The embryonic stage is completed by the end of the _____ week of development.

9. All essential external and internal body parts are formed during the _____ stage of development.

10. After an embryo is clearly recognizable as a human being, it is called a(an) _____ .

Appendix

Preparation of Solutions

Amylase solution
Place 0.5 gm of bacterial amylase in a graduated cylinder or volumetric flask. Add distilled water to the 100 ml level. Stir until dissolved.

Benedict's solution
Prepared solution is available.

Caffeine, 0.2%
Place 0.2 gm of caffeine in a graduated cylinder or volumetric flask. Add distilled water to the 100 ml level. Stir until dissolved.

Calcium chloride, 2.0%
Place 2.0 gm of calcium chloride in a graduated cylinder or volumetric flask. Add distilled water to the 100 ml level. Stir until dissolved.

Glucose, 10%
Place 10 gm of glucose in a graduated cylinder or volumetric flask. Add distilled water to the 100 ml level. Stir until dissolved.

IKI solution
Add 20 gm of potassium iodide to 1 liter of distilled water and stir until dissolved. Then add 4.0 gm of iodine and stir again until dissolved. Solution should be stored in a dark stoppered bottle.

Limewater
Add an excess of calcium hydroxide to 1 liter of distilled water. Stopper the bottle and shake thoroughly. Allow the solution to stand for twenty-four hours. Pour the supernatant fluid through a filter. Store the clear filtrate in a stoppered container.

Methylene blue
Dissolve 0.3 gm of methylene blue powder in 30 ml of 95% ethyl alcohol. In a separate container, dissolve 0.01 gm of potassium hydroxide in 100 ml of distilled water. Mix the two solutions.

Physiological saline solution
Place 0.9 gm of sodium chloride in a graduated cylinder or volumetric flask. Add distilled water to the 100 ml level. Stir until dissolved.

Potassium chloride, 5%
Place 5.0 gm of potassium chloride in a graduated cylinder or volumetric flask. Add distilled water to the 100 ml level. Stir until dissolved.

Quinine sulfate, 0.5%
Place 0.5 gm of quinine sulfate in a graduated cylinder or volumetric flask. Add distilled water to the 100 ml level. Stir until dissolved.

Red blood cell diluting fluid
Dissolve the following salts in 400 ml of distilled water:

    1.0 gm mercuric chloride
    4.4 gm sodium sulfate (anhydrous)
    2.0 gm sodium chloride

Ringer's solution (frog)
Dissolve the following salts in 1 liter of distilled water:

    6.50 gm sodium chloride
    0.20 gm sodium bicarbonate
    0.14 gm potassium chloride
    0.12 gm calcium chloride

Sodium chloride solutions

a. *0.85% solution.* Place 0.85 gm of sodium chloride in a graduated cylinder or volumetric flask. Add distilled water to the 100 ml level. Stir until dissolved.

b. *1.0% solution.* Place 1.0 gm of sodium chloride in a graduated cylinder or volumetric flask. Add distilled water to the 100 ml level. Stir until dissolved.

c. *2.0% solution.* Place 2.0 gm of sodium chloride in a graduated cylinder or volumetric flask. Add distilled water to the 100 ml level. Stir until dissolved.

d. *5.0% solution.* Place 5.0 gm of sodium chloride in a graduated cylinder or volumetric flask. Add distilled water to the 100 ml level. Stir until dissolved.

Starch, 1% solution

Add 10 gm of cornstarch to 1 liter of distilled water. Heat until the mixture boils. Cool the liquid and pour it through a filter. Store the filtrate in a refrigerator.

Sucrose, 5% solution

Place 5.0 gm of sucrose in a graduated cylinder or volumetric flask. Add distilled water to the 100 ml level. Stir until dissolved.

White blood cell diluting fluid

Mix 5 ml of concentrated hydrochloric acid in 495 ml of distilled water.

Wright's stain

Prepared solution is available.

Credits

Index